SOLAR SYSTEM MAGNETIC FIELDS

GEOPHYSICS AND ASTROPHYSICS MONOGRAPHS

Editor

B. M. McCORMAC, *Lockheed Palo Alto Research Laboratory, Palo Alto, Calif., U.S.A.*

Editorial Board

R. GRANT ATHAY, *High Altitude Observatory, Boulder, Colo., U.S.A.*
W. S. BROECKER, *Lamont-Doherty Geological Observatory, Palisades, New York, U.S.A.*
P. J. COLEMAN, JR., *University of California, Los Angeles, Calif., U.S.A.*
G. T. CSANADY, *Woods Hole Oceanographic Institution, Woods Hole, Mass., U.S.A.*
D. M. HUNTEN, *University of Arizona, Tucson, Ariz., U.S.A.*
C. DE JAGER, *The Astronomical Institute, Utrecht, The Netherlands*
J. KLECZEK, *Czechoslovak Academy of Science, Ondřejov, Czechoslovakia*
R. LÜST, *President Max-Planck Gesellschaft für Förderung der Wissenschaften, München, F.R.G.*
R. E. MUNN, *University of Toronto, Toronto, Ont., Canada*
Z. ŠVESTKA, *The Astronomical Institute, Utrecht, The Netherlands*
G. WEILL, *Service d'Aéronomie, Verrières-le-Buisson, France*

SOLAR SYSTEM MAGNETIC FIELDS

Edited by

E. R. PRIEST

St. Andrews University, Scotland

D. REIDEL PUBLISHING COMPANY

A MEMBER OF THE KLUWER ACADEMIC PUBLISHERS GROUP

DORDRECHT / BOSTON / LANCASTER / TOKYO

Library of Congress Cataloging in Publication Data

CIP

Main entry under title:

Solar system magnetic fields.

(Geophysics and astrophysics monographs)
Based on lectures presented at Summer School on Solar System Plasmas, held in Sept. 1984 at Imperial College with the support of the Science and Engineering Research Council; aimed at Ph. D. students.
Includes bibliographies and index.
1. Space plasmas. 2. Magnetosphere. 3. Magnetohydrodynamics. 4. Magnetic fields (Cosmic physics) I. Priest, E. R. (Eric Ronald), 1943–
II. Summer School on Solar System Plasmas (1984 : Imperial College)
III. Series.
QC809.P5S63 1986 523.2 85-24397
ISBN 90-277-2137-8
ISBN 90-277-2138-6 (pbk.)

Published by D. Reidel Publishing Company,
P.O. Box 17, 3300 AA Dordrecht, Holland.

Sold and distributed in the U.S.A. and Canada
by Kluwer Academic Publishers,
190 Old Derby Street, Hingham, MA 02043, U.S.A.

In all other countries, sold and distributed
by Kluwer Academic Publishers Group,
P.O. Box 322, 3300 AH Dordrecht, Holland.

Printed in The Netherlands

Participants at the Summer School on Solar System Plasmas,
Imperial College, September 1984

CONTENTS

PREFACE

In September 1984 a Summer School on Solar System Plasmas was held at Imperial College with the support of the Science and Engineering Research Council. An excellent group of lecturers was assembled to give a series of basic talks on the various aspects of the subject, aimed at Ph.D. students or researchers from related areas wanting to learn about the plasma physics of the solar system. The students were so appreciative of the lectures that it was decided to write them up as the present book.

Traditionally, different areas of solar system science, such as solar and magnetospheric physics, have been studied by separate communities with little contact. However, it has become clear that many common themes cut right across these distinct topics, such as magnetohydrodynamic instabilities and waves, magnetic reconnection, convection, dynamo activity and particle acceleration. The plasma parameters may well be quite different in the Sun's atmosphere, a cometary tail or Jupiter's magnetosphere, but many of the basic processes are similar and it is by studying them in different environments that we come to understand them more deeply. Furthermore, direct in situ measurements of plasma properties at one point in the solar wind or the magnetosphere complement the more global view by remote sensing of a similar phenomenon at the Sun. Clearly, much can be gained in the future from cross-fertilization between the different branches of solar system science, and it is hoped that this glimpse at a few of them can help in some small way.

E R Priest
St Andrews, May 1985

ACKNOWLEDGEMENTS

The authors gratefully acknowledge permission to reproduce the following copyright figures: Fig. 1.2 (S Martin), Fig. 1.3 (G Newkirk), Fig. 1.4 (J Harvey, National Solar Observatory), Fig. 1.5 (D Webb, American Science and Engineering), Fig. 1.6 (H Zirin, Big Bear Solar Observatory), Fig. 1.9 (R Tousey, Naval Research Laboratory), Fig. 1.10 (D Rust, American Science and Engineering), Figs. 8.3 and 8.4 (E Marsch), Fig. 8.6 (J Belcher), Fig. 8.12 (M Thomsen), Fig. 9.7 (J Phillips, C Russell), Fig. 9.9 (J Spreiter), Fig. 9.16 (V Vasyliunas), Fig. 10.1 (Hale Observatories), Fig. 10.2 (W Ip), Fig. 10.3 (H Alfvén), Fig. 10.4 (A Delsemme), Fig. 10.5 (J Brandt), Fig. 10.6 (H Keller), Figs. 10.8-10.11 (H Schmidt).

The editor is also most grateful to Shiela Wilson for her efficiency and good humour.

CHAPTER 1

INTRODUCTION TO SOLAR ACTIVITY

E R Priest
Applied Mathematics Department
The University
St Andrews KY16 9SS

This first chapter is meant as an introduction and preparation for the more specialised chapters to follow. Its aim is to summarise some of the basic properties of the Sun, to remind the reader of the magnetohydrodynamic equations and to outline some of the major problems in solar activity. (For more details see e.g. Parker (1979), Priest (1982).)

Scientists are interested in the Sun for several reasons. Firstly, it affects in many complex ways the Earth and its magnetic environment. Secondly, the Sun is our closest star, which makes solar physics by far the most highly developed branch of astrophysics. The imagination of theorists like myself is certainly inspired by the observations, but it is also constrained by them much more than in more exotic branches of astronomy. Thirdly, the solar atmosphere represents a cosmic environment where we can observe the basic properties of a magnetised plasma at high magnetic Reynolds number (see equation 1.7). However, the main reason for my own interest is that I find the Sun intrinsically an object of great fascination, with a rich variety of phenomena and mathematical problems that we are only just beginning to tackle.

1.1 SOME BASIC PROPERTIES OF THE SUN

The Sun is a sphere of radius 700,000km, whose core contains a successful fusion reactor operating at a density and temperature of about $10^{32}\mathrm{m}^{-3}$ and $1.5 \times 10^{7}\mathrm{K}$ (Figure 1.1). The top 200,000km of the interior is a turbulent convective zone, with turbulence on several discrete scales, which at the surface show up as granulation, mesogranulation, supergranulation and giant cells on scales of 1000 km, 10,000 km, 30,000 km and 300,000 km, respectively. The atmosphere consists of three parts, the *photosphere, chromosphere* and *corona*. The photosphere is a thin surface layer, only 500 km deep, from which most of the observable light comes. It reveals global oscillations with amplitudes of only a few cm s^{-1} or less (e.g. Deubner, 1975). Until recently, we had virtually no knowledge of the solar interior, but now these oscillations are being used to probe the interior in the same way that seismology enables us to deduce the earth's internal structure. So far a thousand normal modes of oscillation have been discovered and the radial variation of sound speed and angular velocity inside the Sun have been inferred (e.g. Gough, 1983; Duvall *et al.*, 1984).

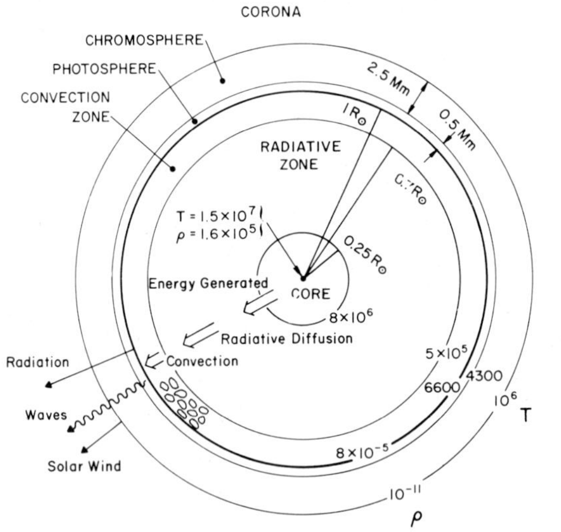

Figure 1.1 The solar interior

Above the photosphere and into the chromosphere (of depth 2000 km) and corona, the density falls rapidly from $10^{23}\mathrm{m}^{-3}$ through $10^{17}\mathrm{m}^{-3}$ to $10^{14}-10^{15}\mathrm{m}^{-3}$, while the temperature increases slightly from 6000 K to 10^4K and then suddenly jumps through a narrow transition layer to 2×10^6K. Correspondingly, the sound speed increases from 10 $\mathrm{km\,s}^{-1}$ through 20 $\mathrm{km\,s}^{-1}$ to 200 $\mathrm{km\,s}^{-1}$. The magnetic field strength varies between one and a few thousand gauss in the photosphere and typically one and a few hundred gauss in the chromosphere and corona. The corresponding Alfvén speed (see equation 1.9) in the three regions is .01 - 10 $\mathrm{km\,s}^{-1}$, $10-10^3$ $\mathrm{km\,s}^{-1}$ and 10^2-10^4 $\mathrm{km\,s}^{-1}$, while the plasma beta (see equation (1.12)) is 10^6-1, $10^{-4}-1$, $10^{-4}-1$. Also, the ion Larmor radius is $10^{-3}-1$m, $5\times10^{-3}-5\times10^{-1}$m and $10^{-1}-10$m, in the photosphere, chromosphere and corona, respectively, while the particle mean-free path is 10^{-7}m, 10^{-1}m and 3×10^5m. The plasma eventually becomes collisionless in the outer corona beyond a few solar radii.

A picture of the photosphere reveals dark *sunspots*, often in pairs where a large magnetic flux tube breaks through the surface in one spot and goes back down through another. An Hα photograph of the chromosphere (Figure 1.2) shows a bright *active region* surrounding a sunspot group, in which a *flare* occasionally takes place. Also, thin dark structures called *filaments* or *prominences* meander across the disc. They are of two types, namely long and thick *quiescent filaments*, usually located far from active regions, and much smaller *plage filaments* within active regions. At times of solar eclipse the corona is revealed with plasma structures outlining the magnetic field (Figure 1.3). There are closed regions, where the magnetic field is

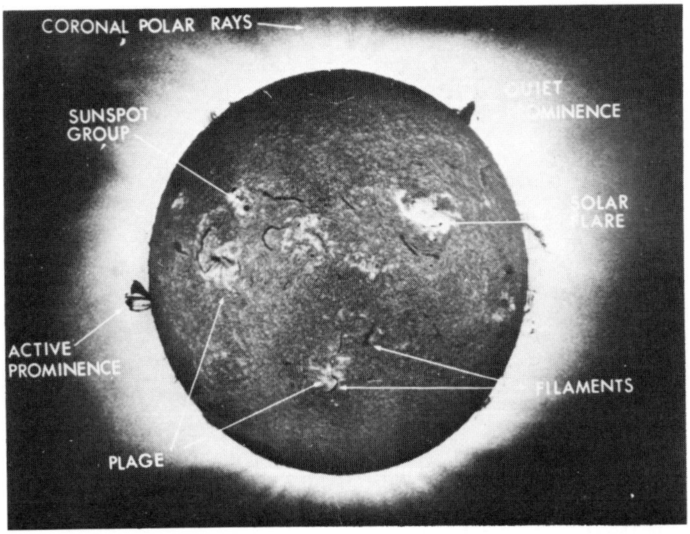

Figure 1.2 A composite including the chromosphere (courtesy S Martin)

strong enough to contain the plasma, and also open magnetic regions, where plasma streams out from the Sun as the fast *solar wind*. However, the mechanism for accelerating it is unknown and the slow solar wind may well originate in some manner from predominantly closed regions.

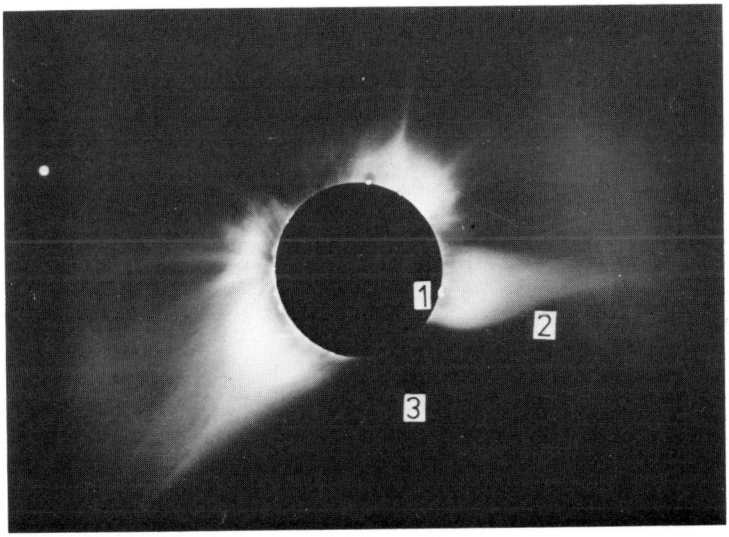

Figure 1.3 The corona, showing (1) prominence, (2) streamer (3) coronal hole (courtesy G Newkirk)

Our understanding of the Sun's atmosphere has changed dramatically over the past ten years, because of high-resolution observations from the ground and also because of space observations from the Skylab and Solar Maximum Mission satellites (see Zirker (1977), Sturrock (1980), Orrall (1981), Kundu and Woodgate (1985)). The old view of the Sun was of a spherically symmetric, homogeneous atmosphere, with the magnetic field only important in sunspots and the solar wind heated and driven by acoustic waves propagating up from the photosphere.

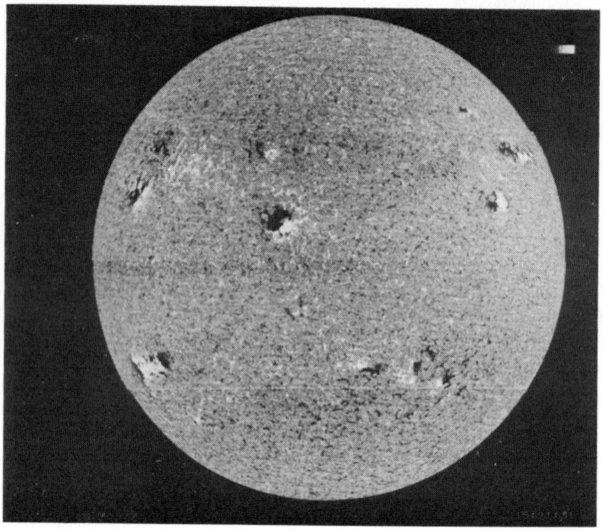

Figure 1.4 Photospheric magnetic field (courtesy J Harvey, National Solar Observatory)

We now realise that the Sun is far from uniform and contains much complicated structure. In the photosphere the magnetic field is concentrated to form tiny intense flux tubes with kilogauss field strengths at the boundaries of supergranule cells, as shown in Figure 1.4 where light and dark give opposite magnetic polarities. The corona too is highly structured and is believed to be heated magnetically. Soft X-ray images from Skylab (Figure 1.5) reveal dark regions of open magnetic field, called *coronal holes*, along which the fast solar wind is escaping. *X-ray bright points* are small regions of intense heating due to magnetic field interaction. Also, such X-ray images show up myriads of loop structures, both within and between active regions.

All these discoveries are dominated by the magnetic field and its subtle nonlinear interaction with the plasma atmosphere, and they are well modelled by the equations of magnetohydrodynamics. Important topics in solar MHD include the following:

dynamo theory - how the solar magnetic field is generated in a cyclic manner that oscillates with an eleven-year period (see Moffatt (1978) and Chapter 7);

magnetoconvection - the way turbulent motions concentrate magnetic flux (see Proctor and Weiss (1982) and Chapter 6);

sunspots - how they are cooled and how waves and flows are driven

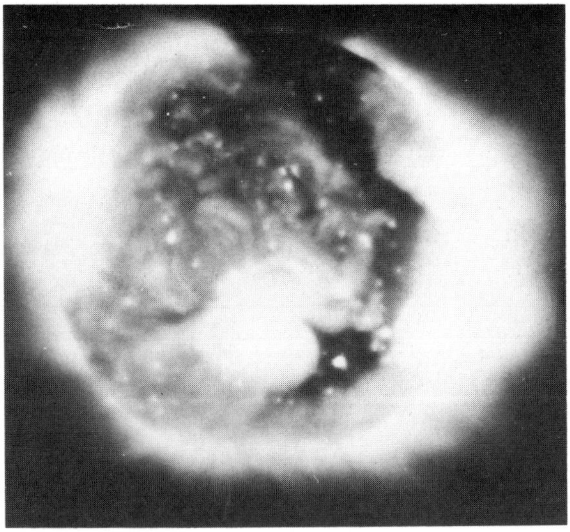

Figure 1.5 The corona in soft X-rays (courtesy D Webb)

(see Cram and Thomas (1982));

magnetic loops - their thermal structure and the flows and waves that are ducted along them (see Priest (1981b) and Chapter 3);

magnetic reconnection - the way that magnetic field lines break and rapidly reconnect (see Hones (1984), Priest (1985) and Chapter 5);

prominences - how they are formed and supported (see Tandberg-Hanssen (1974) and Jensen *et al.* (1979));

coronal heating - the details of the as-yet unknown, but presumably magnetic, mechanism (see e.g. Chiuderi (1981), Heyvaerts and Schatzman (1980) and Hollweg (1981));

solar flares - the way in which magnetic energy is slowly built up and suddenly released (see Svestka (1976), Sturrock (1980) and Priest (1981a)).

1.2 BASIC EQUATIONS OF MAGNETOHYDRODYNAMICS

The fundamental equations for the plasma velocity ($\mathbf{v}$), magnetic field ($\mathbf{B}$), plasma density (ρ), electric current ($\mathbf{j}$), plasma pressure (p) and temperature (T) are the induction equation

$$\frac{\partial \mathbf{B}}{\partial t} = \nabla \times (\mathbf{v} \times \mathbf{B}) + \eta \nabla^2 \mathbf{B}, \qquad (1.1)$$

the equation of motion

$$\rho \frac{d\underset{\sim}{v}}{dt} = -\underset{\sim}{\nabla} p + \underset{\sim}{j} \times \underset{\sim}{B} + \rho \underset{\sim}{g}, \tag{1.2}$$

the mass continuity equation

$$\frac{d\rho}{dt} = -\rho(\underset{\sim}{\nabla}.\underset{\sim}{v}), \tag{1.3}$$

the energy equation

$$\frac{\rho^\gamma}{\gamma-1}\frac{d}{dt}\left(\frac{p}{\rho^\gamma}\right) = -\nabla.(\kappa\nabla T) - \rho^2 Q(T) + \frac{j^2}{\sigma} + h\rho, \tag{1.4}$$

where $\underset{\sim}{\nabla}.\underset{\sim}{B} = 0$

$$\underset{\sim}{j} = \underset{\sim}{\nabla} \times \underset{\sim}{B}/\mu \tag{1.5}$$

is Ampère's law, μ is the magnetic permeability and

$$p = \frac{k_B}{m}\rho T = n k_B T = \frac{\mathcal{R}}{\tilde{\mu}}\rho T \tag{1.6}$$

is the perfect gas law in terms of the mean particle mass (m), the Botlzmann constant (k_B) and the particle number density (n).

The significance of equation (1.1), in which $\eta = (\mu\sigma)^{-1}$ is the *magnetic diffusivity* and σ is the electrical conductivity, is that changes in magnetic field strength are caused by transport of the magnetic field with the plasma (as represented by the first term on the right), together with diffusion of the magnetic field through the plasma (the second term on the right). In order of magnitude, the ratio of the first to the second term on the right is the *magnetic Reynolds number*

$$R_m = \frac{v_0 \ell_0}{\eta} \tag{1.7}$$

in terms of a typical plasma speed (v_0) and length-scale (ℓ_0) for magnetic variations. For length-scales comparable with the size of typical coronal structures, the magnetic Reynolds number is enormous (say, $10^6 - 10^8$) and so diffusion is negligible and the magnetic field is effectively frozen to the plasma. It is only in intense current concentrations such as filaments or sheets, where ℓ_0 is extremely small (kilometres or even less), that diffusion and therefore reconnection can take place. A related dimensionless parameter is the *Lundquist number*

$$S = \frac{v_A \ell_0}{\eta}, \tag{1.8}$$

where

$$v_A = \frac{B}{(\mu\rho)^{\frac{1}{2}}} \tag{1.9}$$

is the *Alfvén speed.* It may be written as the ratio

$$S = \frac{\tau_d}{\tau_A} \tag{1.10}$$

of the *magnetic diffusion time* ($\tau_d = \ell_0^2/\eta$) to the *Alfvén travel time* ($\tau_A = \ell_0/v_A$).

Equation (1.2) shows that the plasma is acted on by forces due to a plasma pressure gradient, the magnetic field and gravity. The magnetic force may be rewritten using (1.5) as

$$\underset{\sim}{j} \times \underset{\sim}{B} = -\underset{\sim}{\nabla}(B^2/(2\mu)) + (\underset{\sim}{B}.\underset{\sim}{\nabla})\underset{\sim}{B}/\mu, \tag{1.11}$$

which represents the sum of a magnetic pressure force acting from regions of high to low magnetic pressure ($B^2/(2\mu)$) and a magnetic tension force acting towards the centre of curvature of curved magnetic field lines. A useful parameter is the *plasma beta*

$$\beta = \frac{p}{B^2/(2\mu)}, \tag{1.12}$$

which measures the ratio of the pressure gradient to the magnetic force. It may also be written as

$$\beta = \frac{2}{\gamma}\frac{c_s^2}{v_A^2}, \tag{1.13}$$

where γ is the ratio of specific heats and

$$c_s = \left(\frac{\gamma p}{\rho}\right)^{\frac{1}{2}} \tag{1.14}$$

is the *sound speed.*

In equation (1.4), which is appropriate for the corona, entropy changes are caused by the terms on the right, namely thermal conduction, optically thin radiation, ohmic heating and a small-scale heating term, which is assumed to be proportional to density with a constant of proportionality h. κ is the (tensor) coefficient of thermal conduction. The optically thin radiation is proportional to density squared and has a temperature dependence ($Q(T)$) which possesses a maximum at about 10^5K and a minimum at about 10^7K. Lower in the atmosphere the radiation becomes optically thick, conduction becomes radiative rather than Coulomb and the small-scale heating term disappears. Also, in some applications viscous terms are added to the momentum and energy equations.

A few features of the solutions to the MHD equations are mentioned below. More details can be found in the books by Cowling (1976), Roberts (1967) and Priest (1982).

1.2.1 Magnetohydrostatics

Slowly changing structures, such as sunspots, prominences or coronal loops, are in approximate equilibrium under a balance between various forces. In particular, if flow speeds are much smaller than both the sound speed, the Alfvén speed and the free-fall speed $(g\ell)^{\frac{1}{2}}$, equation (1.2) reduces to

$$\underset{\sim}{0} = -\underset{\sim}{\nabla} p + \underset{\sim}{j} \times \underset{\sim}{B} + \rho \underset{\sim}{g}. \tag{1.15}$$

Since the $\underset{\sim}{j} \times \underset{\sim}{B}$ force is perpendicular to $\underset{\sim}{B}$, the component of (1.15) along the magnetic field may be written

$$0 = -\frac{dp}{dz} - \rho g \tag{1.16}$$

when gravity acts in the negative z-direction. From the perfect gas law (1.6)

$$\rho = \frac{mp}{k_B T}, \tag{1.17}$$

and so ρ may be eliminated between (1.16) and (1.17) and the resulting equation integrated to give

$$p = p_0 \exp - \int_0^z \frac{dz}{\Lambda}, \tag{1.18}$$

where

$$\Lambda = \frac{k_B T}{mg} \tag{1.19}$$

is the *pressure scale-height*. If, in particular, the temperature is uniform, (1.18) becomes

$$p = p_0 \exp(-z/\Lambda), \tag{1.20}$$

and so the pressure decreases exponentially with height along each magnetic field line. The scale-height is typically 100 km in the photosphere, 500 km in the chromosphere and 100,000 km in the corona, so that as one rises up above the solar surface the pressure falls off extremely rapidly at first and later much more slowly.

In the particular case when $\beta << 1$ and the vertical distance under consideration is much less than Λ/β, (1.15) reduces to

$$\underset{\sim}{0} = \underset{\sim}{j} \times \underset{\sim}{B} \qquad (1.21)$$

and the magnetic field is said to be *force-free*, where

$$\underset{\sim}{j} = \underset{\sim}{\nabla} \times \underset{\sim}{B}/\mu \quad \text{and} \quad \underset{\sim}{\nabla}.\underset{\sim}{B} = 0.$$

Very little is known about general solutions to the nonlinear equation (1.21) in spite of its apparent simplicity (Low, 1982). Even less is known about (1.15), where the pressure and density terms provide a coupling to the energy equation. Only rather simple solutions have been investigated for prominences, sunspots and coronal loops.

Since (1.21) implies that the current (and therefore $\underset{\sim}{\nabla} \times \underset{\sim}{B}$) is parallel to the magnetic field, it may be rewritten

$$\underset{\sim}{\nabla} \times \underset{\sim}{B} = \alpha \underset{\sim}{B}, \qquad (1.22)$$

where α is some function of position, which is constant along each magnetic field line since the divergence of (1.22) gives

$$0 = \underset{\sim}{B}.\underset{\sim}{\nabla}\alpha.$$

If in particular α is uniform, taking the same constant value on every field line, the curl of (1.22) gives

$$(\nabla^2 + \alpha^2)\underset{\sim}{B} = \underset{\sim}{0}, \qquad (1.23)$$

and we have a *linear* force-free field. In this case, many solutions are known, such as, for instance,

$$(B_x, B_y, B_z) = B_0 e^{-\ell z}\left(\frac{\ell}{k}\cos kx,\ \frac{\alpha}{k}\cos kx,\ -\sin kx\right), \qquad (1.24)$$

where

$$\ell^2 = k^2 + \alpha^2.$$

This may model a coronal arcade with the y-axis being the axis of the arcade and the x-axis representing the solar surface. Seen from above the magnetic field lines are straight, and inclined at an angle $\tan^{-1}(\alpha/\ell)$ to the x-axis.

1.2.2 Waves

For a uniform plasma in a uniform magnetic field ($\underset{\sim}{B}_0$), equations (1.1)-(1.4) may be linearised by writing variables in the form

$$\underset{\sim}{B} = \underset{\sim}{B}_0 + \underset{\sim}{B}_1 \exp i(\omega t - \underset{\sim}{k}.\underset{\sim}{r}) \qquad (1.25)$$

and retaining only terms linear in B_1/B_0. Here ω is the wave frequency and $\underset{\sim}{k}$ is the wavenumber vector in a direction θ, say, to the magnetic field $\underset{\sim}{B}_0$. When the frequency is so high that terms on the right of (1.4) are negligible, the variations are adiabatic and the modes become particularly simple.

For example, in the absence of gravitational and pressure effects ($g = p_0 = 0$), there are two distinct magnetic modes, namely the *shear Alfvén wave*, with a dispersion relation

$$\omega = k\, v_A \cos\theta, \qquad (1.26)$$

and the *compressional Alfvén wave*, for which

$$\omega = k\, v_A . \qquad (1.27)$$

When the plasma is not cold ($p_0 \neq 0$) and the magnetic field is negligible we have a sound wave

$$\omega = k\, c_s . \qquad (1.28)$$

In the presence of a magnetic field this mode couples to the compressional Alfvén wave to give slow and fast *magnetoacoustic waves*.

Gravity gives rise to a gravity wave with dispersion relation

$$\omega = N \sin\theta_g, \qquad (1.29)$$

in terms of the Brunt frequency

$$N = (\gamma - 1)^{\frac{1}{2}} \frac{g}{c_s} \qquad (1.30)$$

and the inclination of the direction of propagation ($\underset{\sim}{k}$) to the force of gravity. In general, this too couples to the magnetoacoustic modes.

The above simple picture is made much more complex by considering a non-uniform basic state (see Roberts (1982) and Chapter 3). A hierarchy of different modes appears when the medium is structured in the form of, for example, an interface, a slab or a flux tube. A further complication arises when the disturbances are no longer linear. For example, the magnetoacoustic modes can steepen to form shock waves, with a *slow shock* causing the magnetic field to rotate towards the shock normal as it passes and a *fast shock* making it rotate the other way and so increasing the field strength rather than decreasing it. Also, intermediate waves (or finite-amplitude Alfvén waves) can cause the tangential component of the magnetic field to reverse its sign while the magnitude of the field is unchanged.

1.2.3 Instabilities

Instabilities may be discovered by seeking perturbation solutions to (1.1) - (1.4) about an equilibrium $\underset{\sim}{B}_0$ (in general non-uniform) in the form

$$\underset{\sim}{B}_0 + \underset{\sim}{B}_1 e^{\omega t}.$$

The magnetic field is found to modify some of the basic plasma (or fluid) instabilities (see Chapter 4) such as:

Rayleigh-Taylor instability, which has a growth-rate $\omega = (gk)^{\frac{1}{2}}$ and occurs when a dense plasma is supported against gravity on top of a rarer plasma;

Kelvin-Helmholtz instability, which takes place when one plasma streams over another one;

convective instability when a layer of plasma is heated from below;

radiative instability, which is driven by the optically thin radiative loss term $\rho^2 Q(T)$ in the energy equation. For example, if we write $Q(T) = Q_0 T^\alpha$, where Q_0 and α are piecewise constant functions of T, the energy equation (1.4) at constant pressure (and neglecting thermal conduction and joule heating) becomes

$$\rho\, c_p \frac{\partial T}{\partial t} = h\rho - Q_0 \rho^2 T^\alpha. \tag{1.31}$$

Thus, variations of temperature from a uniform state (T_0) in which radiation balances coronal heating ($h\rho_0$) are determined by

$$c_p \frac{\partial T}{\partial t} = h\left(1 - \frac{T^{\alpha-1}}{T_0^{\alpha-1}}\right).$$

This equation implies that a cooling from the equilibrium (i.e. $T < T_0$) will continue (i.e. $\partial T/\partial t < 0$) if $\alpha < 1$. In other words, the equilibrium is unstable if T_0 is bigger than about 10^5K.

Several instabilities are peculiar to the magnetic field (see Chapter 4). For instance, the surface of a magnetic flux tube tends to form ripples due to the *flute instability*. Also a local bending of the tube tends to be unstable to the *kink mode*. In addition, a sheared magnetic field can be unstable to the breaking and reconnection of field lines. One such reconnection instability is the *tearing mode* (e.g. Van Hoven, 1981), whose growth-rate is

$$\omega \sim (\tau_d \, \tau_A)^{-\frac{1}{2}}, \tag{1.32}$$

namely the geometric mean of the diffusion and Alfvén frequencies. Since the Alfvén travel time (τ_A) is typically 1 sec in the corona and the diffusion time is typically 300 years, this instability has a typical growth-time of about a day.

1.3 SUNSPOTS

Sunspots have been observed since at least the fourth century BC, and the fact that they are the sites of strong magnetic fields was discovered at the beginning of this century, but the complexity of their structure and behaviour has been appreciated only recently (Cram and Thomas, 1982). A large spot is perhaps 20,000 km across, with a field of 3000 G, and it consists of a central umbra with a temperature of 4,100 K, surrounded by a filamentary penumbra where the field is fanning out. Magnetostatic models for spots have been constructed, although the Evershed outflow of 6 km s^{-1} is not well-understood, nor is the way that sunspots decay away and disappear.

Spots may be unipolar, bipolar or complex and may last for up to a hundred days. They occur in two zones north and south of the equator and the number varies with an eleven-year period, although very few seem to have been present during the Maunder minimum (1645-1715). At the beginning of a sunspot cycle they appear at high latitudes, and as time proceeds their location moves towards the equator. Also, the polarity of bipolar spot groups tends to have a definite sense, which reverses at the start of a new cycle. The dynamo mechanism, by which the magnetic field of sunspots is thought to be generated, is described in Chapter 7 and Moffatt (1978).

Sunspots are thought to be cooled because the strong magnetic field inhibits convection and so stops sunspot plasma mixing efficiently with the hotter layers below (see Chapter 6). Also, they are thought to form when a large flux tube rises through the convective zone by magnetic buoyancy. Such an upwards force acts because a tube, of field strength B and in lateral equilibrium with an environment with plasma p_0, will contain plasma of pressure (p) given by

$$p + \frac{B^2}{2\mu} = p_0. \tag{1.33}$$

Thus

$$p < p_0$$

and, if the tube is not too much cooler than the environment,

$$\rho < \rho_0,$$

which means that the tube is less dense than its surroundings and so will tend to rise. However, the speed of rise is not well-established when account is taken of diffusive and drag effects. Also, other uncertainties are the way a sunspot disappears as well as the properties of other flux tubes, since sunspots are only the largest member of a whole hierarchy of tubes present in the photosphere.

1.4 PROMINENCES

In an Hα picture of the solar chromosphere (Figure 1.2a), the thin dark ribbons known as filaments or prominences comprise two different types, namely *plage* (or active-region) *filaments* inside active regions and the much larger *quiescent filaments* (Tandberg-Hanssen, 1974). However, such filaments are in reality vertical sheets of plasma up in the corona, typically 100-500 times cooler and 100-500 times denser than the surrounding medium. They can remain apparently stable for months, probably supported against gravity by a magnetic field, even though it is difficult for us to keep a similar laboratory plasma stable for a second.

Figure 1.6 Prominence at the limb (courtesy H Zirin)

When seen from the side (Figure 1.6), a prominence shows much internal structure in the form of thin *threads* of diameter 300 km containing a slow downflow at 1 km s^{-1}. Such threads have not been explained, although they may be caused by a resistive or interchange instability modified by gravity and thermal effects. Another strange feature of prominences is the *feet* which project down to the chromosphere and are not understood at all.

A large quiescent prominence may be 200,000 km long, 50,000 km high and 6,000 km thick with a density 10^{16}-10^{17}m^{-3}, a temperature 5000 - 8000 K and a magnetic field of 5 - 10 G. Plage filaments are much smaller, lower and denser ($\gtrsim 10^{17} m^{-3}$) with stronger magnetic fields (20-200 G). Prominences always lie above a reversal in the line-of-sight component of the photospheric magnetic field, and this has inspired two basic models, namely the Kippenhahn-Schlüter (1957) model with the prominence supported by a dip in the field lines at the summit of an arcade and the Kuperus-Raadu (1974) model, for which the horizontal magnetic field in the prominence has the opposite direction due to the presence of an X-type neutral point below the prominence.

Recent observations by Leroy *et al.* (1983) of the magnetic fields in prominences have shown that two-thirds of the cases (namely tall ones with weak fields) are of Kuperus-Raadu type, while one-third (the low-lying cases with stronger fields) are of Kippenhahn-Schlüter type. Furthermore, observations of prominences on the disc reveal steady upflows, both in Hα within the prominence at $3\ km\,s^{-1}$ and also in the transition region both sides of the prominence at $6-10\ km\,s^{-1}$.

1.4.1 Prominence Formation

Following Parker (1953) and Field (1965), consider a uniform hot equilibrium with a balance

$$0 = h\rho_0 - Q_0\rho_0^2 \tag{1.34}$$

between coronal heating ($h\rho$) and radiation ($Q_0\rho^2$). Perturb this at constant p, so that

$$\rho\, c_p \frac{\partial T}{\partial t} = h\rho - Q_0\rho^2 + \kappa_{\parallel} \frac{\partial^2 T}{\partial s^2}, \tag{1.35}$$

where s is the distance along a magnetic field line of length L, say. Suppose the temperature has the form

$$T = T_0 + T_1 \exp \omega t \sin \pi s/L,$$

so that the perturbation vanishes at the ends of the field line (s = 0, L). Then (1.35) gives the growth-rate as

$$\omega = \frac{Q_0\rho_0}{c_p T_0} - \frac{\kappa_{\parallel}\pi^2}{\rho_0 L^2} \tag{1.36}$$

and for short field lines thermal conduction keeps the plasma stable ($\omega < 0$), while for field lines longer than a critical value the radiative instability takes place ($\omega > 0$) and the plasma can cool down from coronal temperatures.

The inclusion of thermal conduction in the equilibrium (Hood and Priest, 1979) leads to a hot equilibrium (at a few million degrees) when the loop length or pressure are small, but when they exceed critical values the plasma loses thermal equilibrium and it rapidly cools down by a thermal catastrophe to temperatures of typically 10^4K. The effect of a magnetic field (Heyvaerts, 1974; Steinolfson and Van Hoven, 1984) and of the nonlinear development of thermal instability (Hildner, 1974; Malherbe *et al.*, 1984) have also been considered briefly, although a fuller treatment is required.

The process of prominence formation in a force-free magnetic arcade has been investigated (Smith and Priest, 1979) by solving along each field line the equations of hydrostatic equilibrium

$$\frac{dp}{dz} = -\rho g \tag{1.37}$$

and energy balance

$$\underset{\sim}{\nabla}\cdot(\kappa_{\shortparallel}\underset{\sim}{\nabla}T) = \rho^2 Q(T) - h\rho . \tag{1.38}$$

The result is that when the arcade is sheared too much no hot equilibrium is possible between two heights which depend on the shear. This suggests a dynamic model in which hot plasma is sucked up along the field lines into the region of thermal nonequilibrium, where it cools and slowly dribbles through the magnetic field.

1.4.2 Magnetostatic Support

The original model for support of plasma in a curved magnetic field against gravity (Kippenhahn and Schlüter, 1957) assumed the temperature (T) and horizontal components (B_x,B_y) of the magnetic field to be uniform and the pressure (p), density (ρ) and vertical magnetic field (B_z) to be functions of the horizontal distance (x) from the prominence axis. Then the horizontal and vertical components of force balance are

$$p + \frac{B^2}{2\mu} = \text{constant}, \tag{1.39}$$

$$\rho g = \frac{dB_z}{dx}\ \frac{B_x}{\mu} , \tag{1.40}$$

so that the magnetic field supports the plasma against gravity (equation (1.40)) and also compresses it (equation (1.39)). After substituting for p from (1.6) and eliminating ρ, these equations may be solved to give

$$B_z = B_0 \tanh \frac{x}{w}, \quad p = \frac{B_0^2}{2\mu} \operatorname{sech}^2 \frac{x}{w} , \tag{1.41}$$

where B_0 is the constant external vertical field component and the half-width of the prominence is

$$w = \frac{B_x \Lambda}{B_0} . \tag{1.42}$$

The temperature may be allowed to be non-uniform in this model by coupling (1.39) and (1.40) to an energy balance equation (Milne *et al.*, 1979). Also, models for the external field may be set up by assuming that the electric current vanishes everywhere except in the prominence. In other words, from the vanishing of both curl $\underset{\sim}{B}$ and div$\underset{\sim}{B}$, the magnetic field satisfies Laplace's equation

$$\nabla^2 \underset{\sim}{B} = \underset{\sim}{0}. \tag{1.43}$$

In particular, two-dimensional models may be constructed (Anzer, 1972; Malherbe and Priest, 1983) by writing

$$B_y + i\,B_x = f(Z), \tag{1.44}$$

an analytic function of the complex variable $Z = x + iy$ except in the prominence, which is regarded as a cut in the complex plane from $Z = ip$ to $Z = iq$, say. For example

$$B_y + iB_x = -\frac{B_0[(p^2+Z^2)(q^2+Z^2)]^{\frac{1}{2}}}{Z(Z+ih)^2} - \frac{B_1}{Z} \tag{1.45}$$

and

$$B_y + iB_x = \frac{B_0[(p^2+Z^2)(q^2+Z^2)]^{\frac{1}{2}}}{Z} + B_1(Z - ip) \tag{1.46}$$

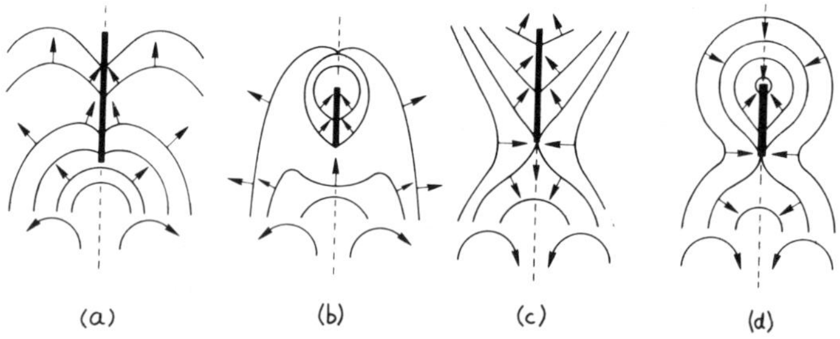

Figure 1.7 Prominence models (Malherbe & Priest, 1983)

give reasonable models of Kippenhahn-Schlüter and Kuperus-Raadu type, respectively (Figure 1.7). Also, as shown in Figure 1.7b and 1.7d, models may be constructed that are locally of the same types but contain magnetic islands, so that, when a strong magnetic component is added out of the plane of the Figure and such a prominence erupts, it may give the helical structure that is often observed. Furthermore, the steady upflows may be understood as the response to slow footpoint motion towards the axis of a Kuperus-Raadu prominence. Such a motion would be expected if the prominence lies along the boundary of a giant cell.

1.5 THE CORONA

As was mentioned in §1.1 coronal images such as Figure 1.5 have confirmed the previous impression from coronagraph pictures that the corona is highly structured with a variety of plasma features. X-ray bright points appear at a rate of 1500 per day and have a lifetime of typically 8 hours and a dimension of 20,000 km. They are no longer thought to be equivalent to small emerging flux regions (ephemeral active regions on magnetograms) but are regions of small-scale magnetic field interaction (Martin, 1983; Martin *et al.*, 1984; Harvey, 1984). The density and temperature in a coronal hole are about $5 \times 10^{11} m^{-3}$ and $1.6 \times 10^6 K$ at an altitude of one solar radius.

Coronal loops are of several different types. Interconnecting loops join different active regions and have lengths of 20,000 - 70,000 km, temperatues of about $2 \times 10^6 K$ and densities of about $0.7 \times 10^{15} m^{-3}$. Quiet-region loops have the same range of lengths, but tend to be cooler ($1.5 \times 10^6 - 2.1 \times 10^6 K$) and rarer ($0.2 \times 10^{15} - 1.0 \times 10^{15} m^{-3}$), while active-region loops are shorter (10,000 - 100,000 km), hotter ($2.2 \times 10^6 - 2.8 \times 10^6 K$) and denser ($0.5 \times 10^{15} - 5 \times 10^{15} m^{-3}$) on average.

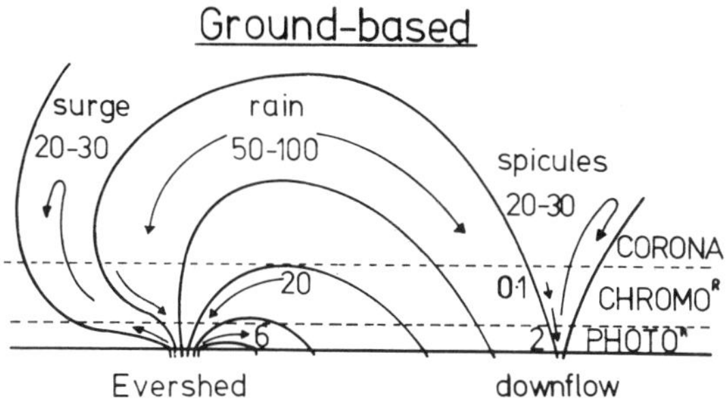

Figure 1.8 Schematic of active region flows

Figure 1.8 shows a diagram of a typical active region magnetic structure, with the preceding flux concentrated in a single sunspot and the following flux spread out to a hot plage region. A dominant feature is that active-region plasma is dynamic, showing continual activity and a wide range of flows, such as Evershed flow, surges, coronal rain and spicules. Recent space observations have revealed an even greater variety, both of long-lived large-scale and transient short-lived flows (see e.g. Priest, 1981b; Athay, 1981).

1.5.1 Models of the Corona

Many attempts have been made to model the coronal magnetic field and plasma properties. First of all, the magnetic field was assumed to be potential, so that (1.43) was solved between two spheres of radius $R_\odot$ and R_S, say. On the inner boundary ($r = R_\odot$) the normal component of the magnetic field was prescribed from observations at the photosphere. At the outer boundary ($r = R_S$) the solar wind was assumed to be so strong that it drags the magnetic field lines into a radial direction. The resulting magnetic field lines agree reasonably with observed X-ray and eclipse structures.

The next step was to construct some fully magnetohydrodynamic models (Pneuman and Kopp, 1971) for an outflowing solar wind with a dipole field prescribed at the solar surface. The solutions resemble a coronal streamer with closed field lines near the equator and close to the solar surface surmounted by a current sheet extending outwards from a cusp-type neutral point. Stimulated by the Skylab observations Kopp and Holzer (1976) then modelled a coronal hole by solving the hydrodynamic equations along a hole of given shape. More details of the properties of the solar wind can be found in Chapter 8.

Several models for the thermal structure of coronal loops have been set up. First of all, the energy balance equation

$$\frac{d}{ds}\left(\kappa_{\shortparallel}\frac{dT}{ds}\right) = \rho^2 Q(T) - h\rho \qquad (1.47)$$

was solved to give the temperature as a function of distance (s) along a single field line, where the coefficient of thermal conduction parallel to the magnetic field is written

$$\kappa_{\shortparallel} = \kappa_0 T^{5/2}, \qquad (1.48)$$

with κ_0 a constant. In order of magnitude, equating the conductive and radiative terms in (1.47) and assuming Q(T) is proportional to $T^{-1/2}$ gives

$$\frac{T^{7/2}}{L^2} \sim \frac{p^2}{T^2} T^{-1/2}$$

or $$T \sim (pL)^{1/3}. \qquad (1.49)$$

This scaling law (Rosner *et al.*, 1978) shows how the loop temperature increases roughly with its pressure and length (L). The first and simplest solutions to (1.47) assumed a uniform pressure, and subsequently the equation of hydrostatic equilibrium (1.16) was adopted instead (Wragg and Priest, 1981). As an alternative, the equation of motion has been coupled with (1.47) to model siphon flow along a loop driven by a pressure difference at the footpoints (Cargill and Priest, 1980; Noci, 1981).

1.5.2 Coronal Heating

In the photosphere at the temperature minimum a wave flux of 10^5 - $10^6 Wm^{-2}$ has been observed, but in the chromosphere only 4000 Wm^{-2} is required to provide the observed radiation in a coronal hole or a quiet region. In active regions it is somewhat more (20,000 Wm^{-2}). The heating requirement for the corona above a coronal hole is 600 Wm^{-2} to provide the enthalpy flux, whereas above a quiet region or an active region 300 Wm^{-2} or 5000 Wm^{-2}, respectively, is needed to provide the conductive flux and radiation.

Acoustic waves are generated at the photosphere and are thought to steepen into shocks at an altitude of 500 - 1000 km. They may heat the low chromosphere but are inappropriate for the upper chromosphere and corona, where a magnetic mechanism of some kind is thought to be operating. The reasons for this belief are that: the observed acoustic flux at 10^5K is only 10 Wm^{-2} (Athay and White, 1978; Mein *et al.*, 1980); strong magnetic field regions above sunspot groups, for instance, are found to be hotter than normal (Figure 1.5); stars without the strong convective zones that would generate acoustic waves are nevertheless found to possess hot coronae.

As far as magnetic waves are concerned, fast and slow magnetoacoustic modes are not generally favoured because they tend to damp low down in the atmosphere and may be reflected at the transition region (Hollweg, 1981). Alfvén waves may be channelled up through the photosphere along intense flux tubes and at certain resonant frequencies they are not reflected off the transition region but can enter the coronal part of a loop like a resonant cavity (Hollweg, 1979). However, the big problem is to know how they give up their energy, since in a uniform medium linear Alfvén waves tend to dissipate much too slowly. Three possible dissipation mechanisms have been proposed, as follows.

First of all, Alfvén waves may couple nonlinearly in weak magnetic fields and create slow magnetoacoustic waves, which can then dissipate rapidly (Wentzel, 1974,1977). Secondly, in a non-uniform medium with magnetic field $B_y(x)$, waves with velocities in the x-direction and wavenumber k may be resonantly absorbed at locations where

$$\omega = k_z v_A(x), \tag{1.50}$$

so that the imposed frequency (ω) matches the local Alfvén frequency (Hasegawa and Chen, 1976; Sedlacek, 1971; Rae and Roberts, 1981). A third possibility is that waves of the other polarization, with velocities in the z-direction, may develop indefinitely strong gradients due to phase mixing (Heyvaerts and Priest, 1983).

In response to footpoint motions slower than the Alfvén travel time, the coronal magnetic field tries to evolve through a series of equilibria. Parker (1972) and Syrovatsky (1978) have suggested that the result in a complicated magnetic field is the formation of many small current sheets at which rapid dissipation can take place (see also Ionson (1978)). The result of this *direct magnetic dissipation* is that the coronal magnetic field is continuously unstable and in a state of *tearing turbulence*. Recently, the technical means for evaluating the evolution of the large-scale field has been set up (Heyvaerts and Priest, 1984; Browning and Priest, 1984) and the mechanism appears to be efficient enough to provide the required heating. The large-scale magnetic field evolves through a series of linear force-free fields (1.23), where α is determined by the amount of *magnetic helicity* that is injected into the corona by the footpoint motions.

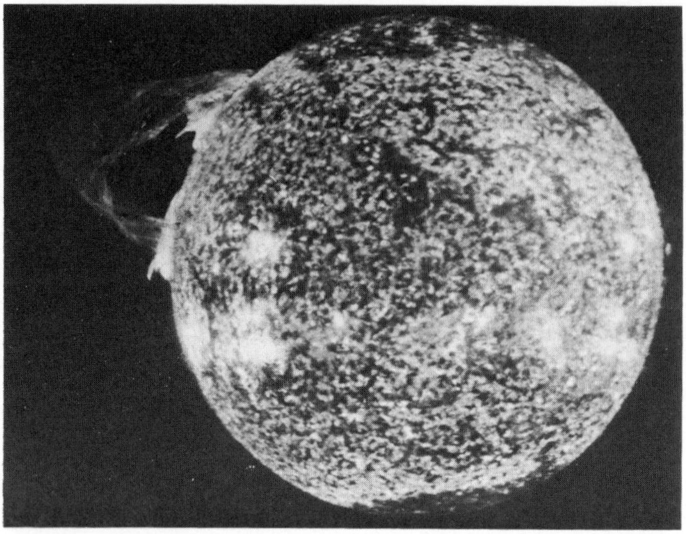

Figure 1.9 Erupting prominence (courtesy R Tousey)

1.6 SOLAR FLARES

A large flare is a dramatic event involving the release of up to 3×10^{25} J of energy from the magnetic field. It consists of three phases. The *preflare phase* lasts about half an hour, during which a large magnetic flux tube (a filament) rises very slowly and the surrounding plasma brightens slightly in soft X-rays as it is heated. At the *rise phase* (for between five minutes and an hour) the flux tube suddenly erupts much more rapidly and disappears from view (Figure 1.9). Simultaneously, there is a steep rise in the intensity of Hα and X-ray emission and two bright ribbons form in the chromosphere. The *main phase* lasts for between an hour and a day or more. The intensity declines slowly and the ribbons move apart, joined by an arcade of rising loops which reach an altitude of 100,000 km or more (Figure 1.10). A system of hot loops lies above a set of cold loops which contain downflowing plasma at 10^4K. The loops rise very rapidly at first (20 - 50 $\mathrm{km\,s^{-1}}$), later much more slowly (0.5 $\mathrm{km\,s^{-1}}$). The density and temperature in the hot loops may be $10^{17}\mathrm{m^{-3}}$ and 2×10^7K at first, declining to $10^{16}\mathrm{m^{-3}}$ and 5×10^6K after a few hours in a large event.

Some of the major problems are: why does the filament erupt? How is the plasma heated to such high temperatures? Where does the enormous amount of downflowing plasma come from? Figure 1.9 shows a large quiescent filament erupting, with apparently a substantial amount of twist. Figure 1.10 shows soft X-ray (above) and Hα (below) images before (left) and during (right) a large two-ribbon event. These have suggested that the overall picture of what is happening in a large flare is as follows (Figure 1.11).

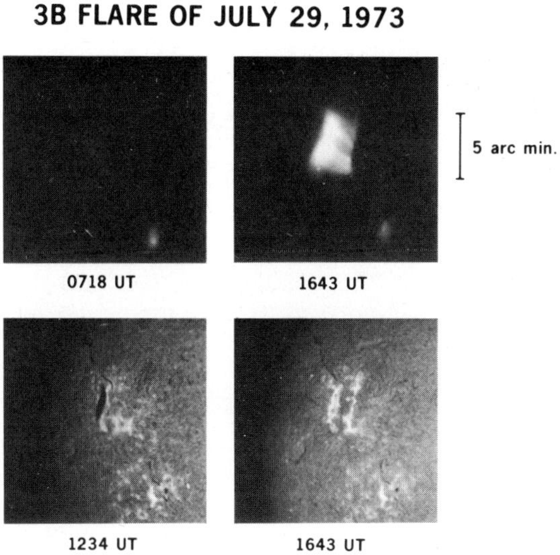

Figure 1.10 Flare in Hα and soft X-rays (courtesy D Rust)

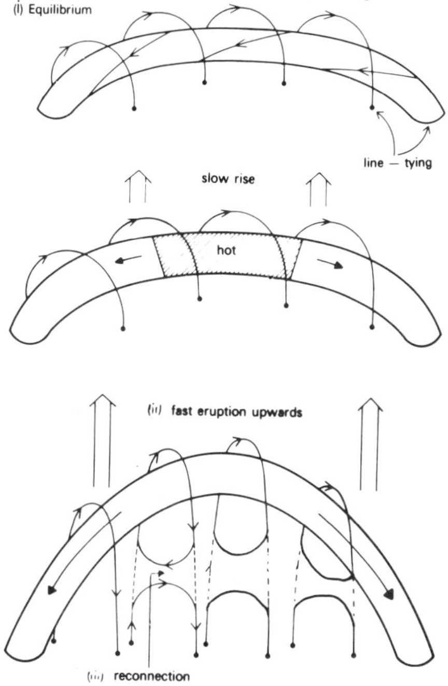

Figure 1.11 Overall behaviour in a large flare

During the preflare phase the flux tube with an overlying arcade of magnetic field lines rises slowly, possibly because of the onset of an *eruptive MHD instability* (see Van Hoven (1981) and Chapter 4). The overlying arcade of field lines is therefore stretched out, until at flare onset they begin to reconnect below the flux tube. The flux tube is no longer held down by the reconnected field lines and so erupts rapidly. During the main phase the reconnection continues and creates hot loops and Hα ribbons as the magnetic field closes down. Kopp and Pneuman (1976) explained the source of mass for the cold loops by suggesting that plasma first rises and is heated by shock waves generated by the reconnection, and then it cools and falls. Also, the shocks, being slow magnetoacoustic in nature, can indeed heat the plasma to the observed temperatures (Cargill and Priest, 1982), while numerical experiments have shown that fast mode shocks may be present too and the energy release process can be in an impulsive bursty manner (Forbes and Priest, 1982, 1983).

1.7 CONCLUSION

Many interesting magnetic processes are taking place in the solar atmosphere which are similar to those in other parts of the solar system, although often they are operating in different parameter regimes and different physical effects are important. Nevertheless, much can be learnt about these basic processes by studying them in different environments. The details of wave motions, instabilities and magnetic reconnection as they operate in the solar atmosphere, the solar wind, comets and magnetospheres are described in subsequent chapters together with dynamo and magnetoconvection processes. However, it is important to remember that many of these are hot topics of current research and are by no means completely understood. There is plenty of room to allow our imaginations to make new suggestions for research - provided they are not too distant from the observations.

REFERENCES

Anzer U (1972) *Solar Phys.* **24**, 324.

Athay R G (1981) Ch 4 of F Orrall (ed) *Skylab Active Region Workshop*, Colo. Ass. Univ. Press.

Athay R G and White O R (1978) *Astrophys. J.* **226**, 1135.

Browning P K and Priest E R (1984) submitted.

Cargill P and Priest E R (1980) *Solar Phys.* **65**, 251.

Cargill P and Priest E R (1982) *Solar Phys.* **76**, 357.

Chiuderi C (1981) in *Solar Phenomena in Stars and Stellar Systems* (ed R Bonnet and A Dupree) D Reidel, p 269.

Cowling T G (1976) *Magnetohydrodynamics*, Adam Hilger, Bristol, UK.

Cram L and Thomas J H (1982) *Physics of Sunspots*.

Deubner F L (1975) *Astron. Astrophys.* **44**, 371.

Duvall T L, Dziembowski W A, Goode P R, Gough D O, Harvey J W and Leibacher J W (1984) *Nature* **310**, 22.

Field G B (1965) *Astrophys. J.* **142**, 531.

Forbes T G and Priest E R (1982) *Solar Phys.* **81**, 303.

Forbes T G and Priest E R (1983) *Solar Phys.* **84**, 169.

Gough D O (1983) ed. *Problems of Solar and Stellar Oscillations*, D Reidel.

Harvey K L (1984) *Proc. 4th European Solar Physics Meeting*, Utrecht.

Hasegawa A and Chen L (1976) *Phys. Fluids* **19**, 1924.

Heyvaerts J (1974) *Astron. Astrophys.* **37**, 65.
Heyvaerts J and Priest E R (1983) *Astron. Astrophys.* **117**, 220.
Heyvaerts J and Priest E R (1984) *Astron. Astrophys.* **137**, 63.
Heyvaerts J and Schatzman E (1980) in *Proc. Japan-France Seminar on Solar Phys.* (ed F Moriyama and J Henoux) p 77.
Hildner E (1974) *Solar Phys.* **35**, 123.
Hollweg J V (1979) *Solar Phys.* **62**, 227.
Hollweg J V (1981) in *Solar Active Regions,* ed F Q Orrall, Colo. Ass. Univ. Press.
Hones E (1984) *Magnetic Reconnection in Space and Laboratory Plasmas,* American Geophysical Union.
Hood A W and Priest E R (1979) *Astron. Astrophys.* **77**, 233.
Ionson J A (1978) *Astrophys. J.* **226**, 650.
Jensen E, Maltby P and Orrall F Q (1979) *Physics of Solar Prominences, IAU Colloq.* No 44.
Kippenhahn R and Schlüter A (1957) *Zs. Ap.* **43**, 36.
Kopp R A and Holzer T E (1976) *Solar Phys.* **49**, 43.
Kopp R A and Pneuman G W (1976) *Solar Phys.* **50**, 85.
Kundu M and Woodgate B (1985) 'Energetic Phenomena on the Sun', Proc. SMM Workshop, NASA.
Kuperus M and Raadu M A (1974) *Astron. Astrophys.* **31**, 189.
Leroy J L, Bommier V and Sahal-Brechot S (1983) *Solar Phys.* **83**, 135.
Low B C (1982) *Rev. Geophys. Space Phys.* **20**, 145.
Malherbe J M and Priest E R (1983) *Astron. Astrophys.* **123**, 80.
Malherbe J M, Forbes T G and Priest E R (1984) *Proc. 4th European Meeting on Solar Physics,* Utrecht.
Martin S F (1983) *Proc. Symp. on Small-Scale Dynamics in Stellar Atmospheres,* Sacramento Peak.
Martin S F, Livi S H B, Wang J and Shi Z (1984) *Proc. Workshop of Solar Vector Magnetic Fields,* Huntsville.
Mein P, Mein N, Schmieder B (1980) in *Proc. Japan-France Seminar on Solar Phys.* (ed F Moriyama and J C Henoux) p 70.
Milne A, Priest E R and Roberts B (1979) *Astrophys. J.* **232**, 304.
Moffatt H K (1978) *Magnetic Field Generation in Electrically Conducting Fluids,* Cambridge University Press, England.
Noci G (1981) *Solar Phys.* **69**, 63.
Orrall F Q (1981) *Proc. Skylab Active Region Workshop,* Colo. Ass. Univ. Press.
Parker E N (1953) *Astrophys. J.* **117**, 431.
Parker E N (1972) *Astrophys. J.* **174**, 499.
Parker E N (1979) *Cosmical Magnetic Fields,* Oxford University Press, England.
Priest E R (1981a) *Solar flare MHD,* Gordon and Breach, London.
Priest E R (1981b) Ch 9 of Orrall F (ed), *Solar Active Regions,* Colo. Ass. Univ. Press.
Priest E R (1982) *Solar MHD,* D Reidel, Dordrecht.
Priest E R (1985) *Rep. Prog. Phys.* **48**, No 7.
Pneuman G W and Kopp R A (1971) *Solar Phys.* **18**, 258.
Proctor M R E and Weiss N O (1982) *Rep. Prog. Phys.* **45**, 1317.
Rae I C and Roberts B (1981) *Geophys. Astroph. Fluid Dynamics* **18**, 197.
Roberts B (1982) *Physics of Sunspots* (ed L Cram & J H Thomas) p 369.

Roberts P H (1967) *An Introduction to Magnetohydrodynamics*, Longmans, London.
Rosner R, Tucker H W and Vaiana G S (1978) *Astrophys. J.* **222**, 317.
Sedlacek Z (1971) *J. Plasma Phys.* **5**, 239.
Smith E A and Priest E R (1979) *Solar Phys.* **53**, 25.
Steinolfson R S and Van Hoven G (1984) *Astrophys. J.* **276**, 291.
Sturrock P A (1980) *Solar Flares*, Colo. Ass. Univ. Press.
Svestka Z (1976) *Solar Flares*, D Reidel.
Syrovatsky S J (1978) *Solar Phys.* **58**, 89.
Tandberg-Hanssen E H (1974) *Solar Prominences*, D Reidel.
Van Hoven G (1981) Ch 4 of *Solar Flare MHD* (ed E R Priest) Gordon and Breach.
Wentzel D G (1974) *Solar Phys.* **39**, 129.
Wentzel D G (1977) *Solar Phys.* **52**, 163.
Wragg M A and Priest E R (1981) *Solar Phys.* **70**, 293.
Zirker J B (1977) *Coronal Holes and High Speed Wind Streams*, Colo. Ass. Univ. Press.

CHAPTER 2

AN INTRODUCTION TO MAGNETOSPHERIC MHD

D. J. Southwood
The Blackett Laboratory
Imperial College of Science and Technology
London SW7 2BZ

2.1 INTRODUCTION

The idea behind this paper is to describe the structure of the magnetosphere using as far as possible notions drawn from magnetohydrodynamics. The formation of a magnetic cavity in the solar wind, the internal circulation pattern, overall observed particle morphology, and much of the spectrum of low frequency magnetic pulsations recorded on Earth and in space can be explained by such an approach.

The terrestrial magnetosphere is in many respects a good example of a magnetohydrodynamic (MHD) system. However much of the ill-understood physics occurs where the MHD approach breaks down and the holes in our knowledge can dominate one's attention. Here, however, we shall emphasise how much we know and then it is natural to start from MHD. We use the notion of field line motion to describe the effect of electric fields, $\underset{\sim}{E}$, perpendicular to $\underset{\sim}{B}$. To a first approximation most plasma constituents throughout most of the magnetosphere and indeed the ionosphere down to $\sim$ 100 km altitude move with a velocity $\underset{\sim}{u}$ such that

$$\underset{\sim}{E} = - \underset{\sim}{u} \wedge \underset{\sim}{B} \tag{1}$$

When (1) holds the field is frozen into the flow; any two plasma particles moving with velocity $\underset{\sim}{u}$ on the same field line remain on the same field line as the flow progresses. Equation (1) is the basis of our use of MHD. Equation (1) breaks down when the magnetic field is very weak. It may also not apply where the plasma is very hot and where strong currents are required to flow along the field, for instance above the auroral zones.

Next we should define terms. The magnetosphere we shall take to be that part of the Earth's ionised atmosphere where collisions are negligible. Being collision-free it is made up of a variety of populations of different origin and subsequent history which can coexist often without efficiently exchanging energy or mixing. However the Larmor gyration of individual particles about the field combined with the frozen in field behaviour allow the use of fluid descriptions. Next note that the field is effectively frozen in down to altitudes as low as $\sim$ 100 km well inside the collisional domain, which we shall call the ionosphere. Ion-neutral collisions are dominant and one may think of the ionosphere as a frictional boundary layer ($\lesssim$ 1000 km thick) at the feet of magnetospheric flux tubes. However there are circumstances in which the "boundary" can drive the system; the neutral atmosphere inertia is large and ionospheric motion such as tidal oscillations in which ions are moved with the neutrals are imposed on the magnetosphere by virtue of the MHD constraint. Here we shall concentrate on high latitude (solar wind) driven plasma motions whose time scales are short compared with the ion-neutral inertia time (many hours) and for which these ionospheric ion-neutral collisions constitute a simple sink of momentum (friction).

Equation (1) justifies the use of an MHD approach. However almost as important conceptually is the momentum equation. The dominant body forces are

$$- \underset{\sim}{\nabla} \cdot \underset{\approx}{p} + \underset{\sim}{j} \wedge \underset{\sim}{B} \tag{2}$$

The pressure commonly needs to be treated as anisotropic. However unless the field varies significantly on the scale of a particle Larmor radius one can treat it as gyrotropic. Thus

$$\underset{\approx}{p} = p_\perp \underset{\approx}{1} + (p_\parallel - p_\perp)\, \hat{\underset{\sim}{e}}\hat{\underset{\sim}{e}}$$

where $\hat{\underset{\sim}{e}}$ is the unit vector along $\underset{\sim}{B}$ and $\underset{\approx}{1}$ is the unit dyadic. The magnetic force is often usefully regarded in terms of Maxwell stresses. The total stress tensor then takes the form

$$\underset{\approx}{T} = \left(p_\perp + \frac{B^2}{2\mu_0}\right) \underset{\approx}{1} + \left(p_\parallel - p_\perp + \frac{B^2}{\mu_0}\right) \hat{\underset{\sim}{e}}\hat{\underset{\sim}{e}}$$

Whether magnetic or plasma pressures dominate is a function of position within the magnetosphere. (The field pressure varies by ten orders of magnitude between magnetotail and ionosphere.) The dominant forces are also a function of time scale of the process considered.

2.2 WHY IS THERE A MAGNETOSPHERE?

Interplanetary space is filled by the solar wind, ionised material streaming outwards from the solar corona. The Earth's magnetic field is compressed but its pressure is sufficient to hold off and divert the flow. Thus a cavity forms in the solar wind. The cavity is the Earth's magnetosphere. To a first approximation, the Earth's magnetic field is entirely contained within the cavity. However there are clear features of the cavity which suggest that simple containment of the field by pressure balance is not the full explanation.

On the sunward side the magnetosphere boundary, the magnetopause, is found some 10 R_E (R_E = Earth radius = 6.4 x 10^3 km) above the equator, its position varying considerably in response to changes in solar wind pressure but also with the external field direction. On the nightside a detectable magnetic tail or wake is found as far as 1000 R_E

downstream (Intriligator et al., 1969). For the field to be stretched out behind Earth in an extended tail of such a magnitude there must be some form of viscous momentum transfer at the magnetopause. The dominant process is now generally believed to be through direct connection taking place between exterior and interior fields. The same process can also explain the tendency for the dayside magnetopause to be closer to the Earth when the external field is southward (Fairfield, 1971).

Dungey (1961) first proposed that the Earth's fields could thread the magnetopause and that reconnection took place between the interior and exterior fields. A consequence of Dungey's model is that the imposition of solar wind flow on a portion of the terrestrial flux tubes gives rise to a large scale plasma circulation system much as viscosity produces a circulation within a falling rain drop. The basis of Dungey's proposed circulation system is shown in Figure 1 reproduced from the original paper. The field topology is schematised. Magnetospheric field lines divide into two classes, closed field lines at lower latitudes which have both feet in the ionosphere and open field lines over the polar caps where the field extends into interplanetary space. There is an electric field pointing at right angles to the plane of the sketch. This field (see equation (1)) corresponds to antisolar motion on the open field lines which connect to interplanetary space and to a sunward flow at the equator of the closed field lines. Let us look at this more closely. The antisolar flow over the polar cap leads to the field being dragged out to form the magnetotail. In the tail there is a motion towards the centre plane across which the field reverses. In the centre plane MHD breaks down and reconnection breaks the field. A newly created field line rejoins the solar wind behind the Earth whilst the newly formed closed field line moves back towards the sun as part of the circulation return flow. On the dayside reconnection also takes place as closed tubes open to complete the cyclic motion. The general disturbance circulation pattern with day to night motion over the polar caps and a return flow on lower latitude flux tubes is known to exist (e.g. Heppner, 1972).

Dungey's model of solar wind-magnetosphere interaction is not the only one put forward to explain the circulation (or convection) system known to exist from day to night over the polar caps during geomagnetic disturbances with a return flow at lower latitudes. Axford and Hines (1961) proposed a simple viscous interaction (e.g. due to diffusion of particles through the magnetopause). There is evidence (Cowley, 1982) of both types of process occurring. However the correlation recognised by Fairfield and Cahill (1966) between southward interplanetary field components and geomagnetic activity militates strongly in favour of the Dungey mechanism being dominant when it operates (Cowley, 1982). The topology of the magnetic fields is rather different when the external field is northward (Dungey, 1963) in the reconnection model. Polar cap convection cells can be set up but the observational situation is not clear (Rieff and Burch, 1985).

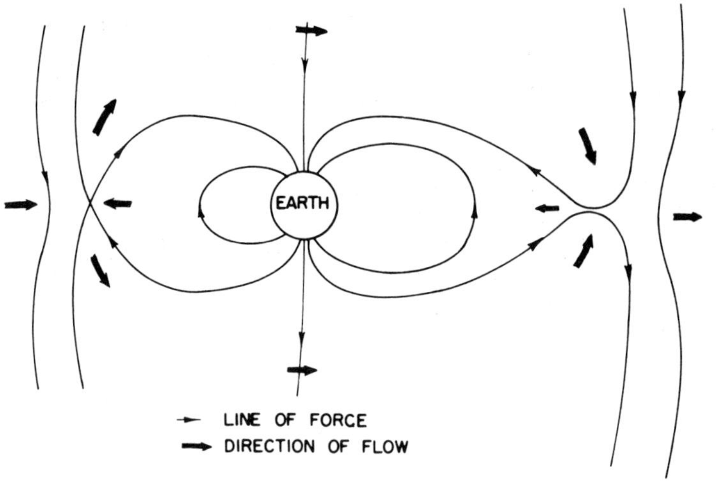

Figure 2.1 Dungey's open magnetosphere model. The figure is topologically correct but not to scale.

2.3 THE OPEN MAGNETOSPHERE MORPHOLOGY

Figure 1 is only topologically correct; it does not purport to show the expected magnetospheric morphology. Even if there is a field component crossing the magnetopause a boundary of sorts may well be detectable. A more realistic picture of the open magnetosphere which outlines the particle populations known to exist in the magnetosphere is given in Figure 2 (Southwood, 1985).

Solar wind plasma enters fairly directly on flux tubes leading to the dayside auroral zone. Magnetic mirroring carries most of this plasma back up the field to provide a plasma mantle on the polar cap field lines as they move tailwards. In the deepest closed field region is a cold dense plasma of terrestrial origin, the plasmasphere, on flux tubes which effectively corotate about the Earth and never move out far enough in the convection system to reach the magnetopause and open. On the closed flux tubes that do open periodically during convection one finds the hot ring current plasma (1-10 keV mean energy). On flux tubes outside the plasmasphere there is a general upflux of cold ionospheric plasma to return the flux tube to an equilibrium like that within the plasmasphere. On the polar cap field lines which are open this effect leads to the polar wind (broad arrow). On the nightside the tail lobe is populated by polar wind plasma on the Earthward end of flux tubes and mantle plasma which drifts towards the centre of the tail. The plasma which actually drifts into the X-type magnetic neutral point where reconnection is taking place may be predominantly of solar or terrestrial origin according to the location of the neutral line. Neutral line acceleration and further compression on flux tubes moving

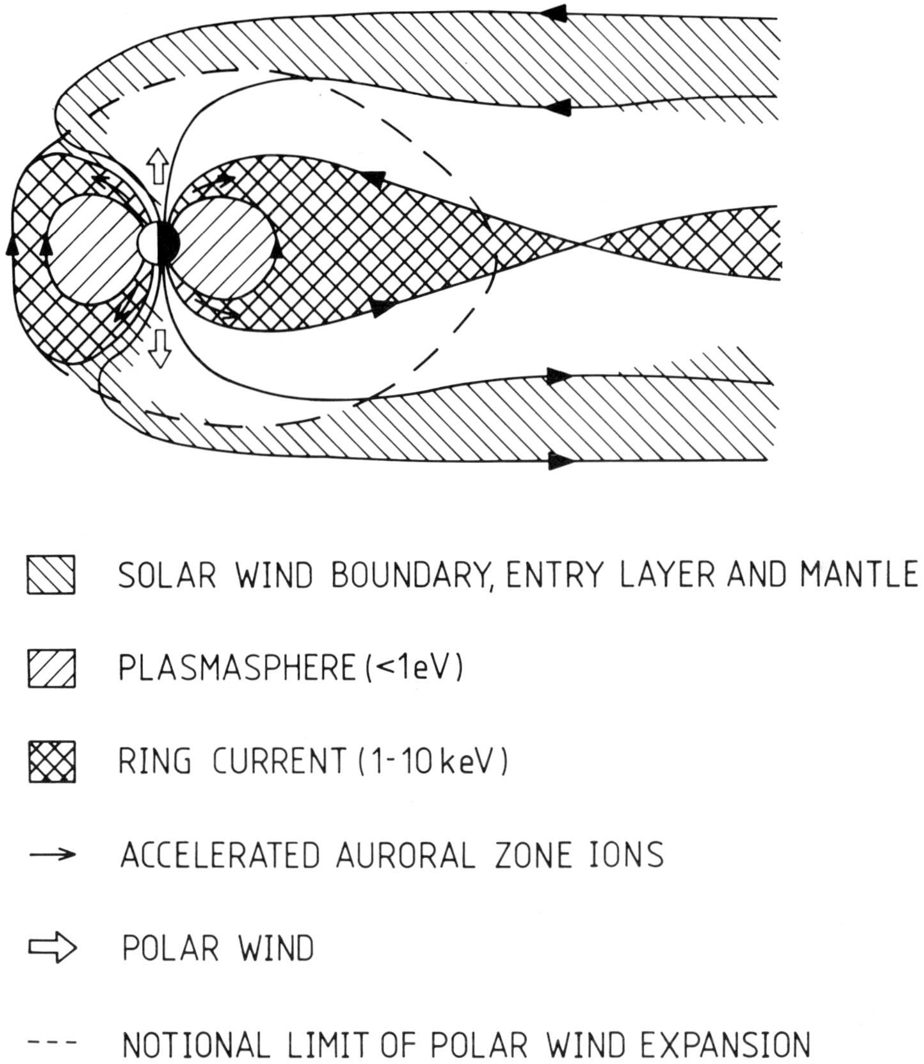

Figure 2.2 The open magnetosphere particle morphology.

Earthwards converts polar wind or solar wind plasma to a hotter ring current plasma. Ultimately ring current is lost out of the system by atmospheric collision (e.g. charge exchange) at the feet of the flux tubes or when the flux tube opens on the dayside. (Note how the magnetosphere acts as a source of accelerated particles for the heliosphere.) The last major source of plasma is represented by the arrows over the auroral zones. Substantial anomalous heating in the two dimensions perpendicular to $\underset{\sim}{B}$, in all likelihood driven by the current systems we discuss below, produces substantial upward fluxes of heavy (O^+) ions from the topside auroral ionosphere. The plasma has energy $\sim$ 100 eV and is eventually assimilated into the ring current after scattering at high altitude. Cowley (1980) describes much of the overall morphological consequences of the open model in more detail.

2.4 MOMENTUM TRANSFER

Unless reconnection is occurring on a section of a flux tube one expects equation (1) to hold at all points on a magnetospheric flux tube. Consider now the circulation system we have described above. The source of momentum on any flux tube in the polar cap must be the solar wind. At the feet of the flux tubes ion-neutral collisions extract momentum from the flow. Equally well on the closed field lines the ionosphere must oppose the flow and there must be a driving force (pressure) at the equator. In steady state there must be a general momentum transfer down the flux tube in order that the circulation be maintained.

The stress is easily pictured in MHD. For instance, in the polar cap the solar wind end of the flux tubes is pulling the ionospheric feet through the neutrals. The field is bent in order that there is a component of field tension in the direction of the flow. Commonly in magnetospheric physics it is useful to use a description involving electrical currents rather than just Maxwell stress. In the current description momentum transfer along the magnetic field is accomplished by current flow parallel to the field. Figure 3 shows an idealised configuration designed to make the point clear (Southwood and Hughes, 1983).

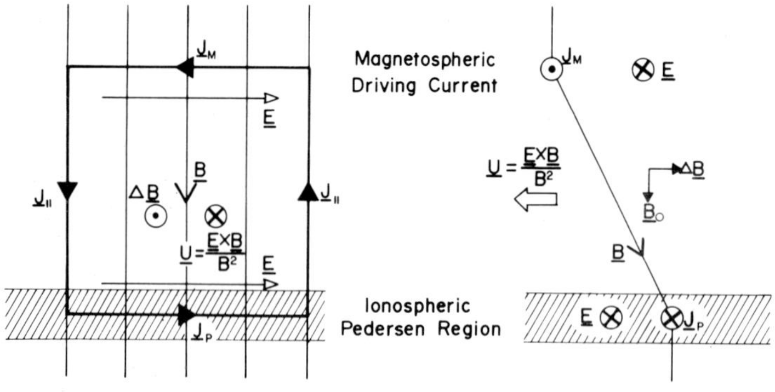

Figure 2.3 An illustration of the transfer of stress by means of field aligned current flow.

MHD makes no attempt to describe the mechanics of parallel current flow. The MHD limit corresponds to zero electron inertia. In fact the electron mobility along the field may not be infinite. Where substantial current flow is required from the ionosphere large potential differences may be set up along the field. This is another circumstance where MHD may break down (see, e.g., Knight, 1973). Some component of auroral acceleration is believed to originate thus (Lyons and Williams, 1984).

A moment's consideration reveals that the circulation system of the magnetosphere described above has an associated gross field aligned current system. Momentum is transferred to the polar cap ionosphere and indeed also to lower latitudes to drive the return flow on closed flux tubes from the night side by downward currents entering the ionosphere on the dawn side and flowing up out of the ionosphere at dusk. These are the region I currents detected by polar orbiting spacecraft (Iijima and Potemra, 1976). A schematic dawn-dusk cross-section of the magnetosphere displaying the electric field and current configuration is shown in Figure 4.

Magnetic field lines are electric equipotentials in steady state (cf. equation (1)). Bearing this in mind, note how Figure 4 illustrates how the open model magnetosphere leads to the notion of an interplanetary electric potential being imposed over the polar cap ionosphere which serves to drive the dissipative ionospheric currents shown. There is another important current present which is not marked in Figure 4, that at the magnetopause. At least on the dayside and over the pole the associated current opposes the antisolar flow ($\underset{\sim}{j} \wedge \underset{\sim}{B}$ is towards the sun). This current is associated with the sharp bend in the field and in the MHD picture corresponds to the field tension force associated with the slingshot (catapult) configuration. Even where MHD is not valid the cross field reconnection current must be fed from somewhere else on the flux tubes involved, thus witnessing the fact that the energy and momentum transfer effected in the magnetopause reconnection process is fed at the expense of the solar wind. Field aligned currents within the solar wind must be present both to drive ionospheric and magnetopause/reconnection currents.

It was noted above that the region I currents drive horizontal motions in the ionosphere at mid-latitude equatorward of the auroral zone. On these flux tubes the flow is returning to the dayside. The flow is driven by back pressure larger at the equator of the flux tube. A curious feature is the shielding of the return flow from very low latitudes (Vasyliunas, 1972; Jaggi and Wolf, 1973). The shielding is evident in the existence of a region II parallel current system at mid-latitudes which switches off the flow at lower latitudes. An MHD explanation of the phenomenon is possible (Southwood, 1977) but it is not surprising that the effect was predicted using an approach based on current conservation. Quasi-steady flow in the closed field region is very much controlled by the distribution of ionospheric conductivities and MHD considerations such as momentum balance do not place strong constraints.

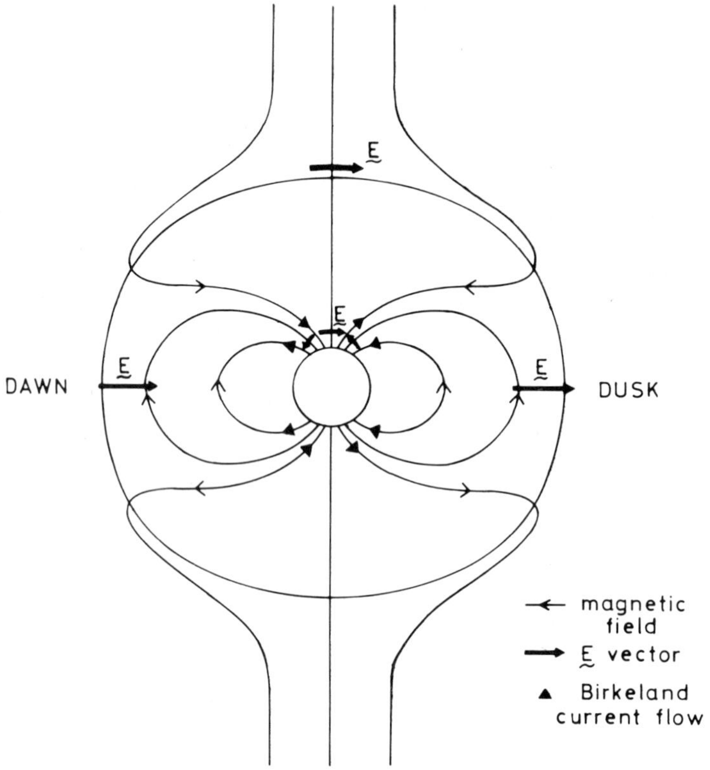

Figure 2.4 A schematic of electric field and current flow in the dawn-dusk meridian in the open magnetosphere model.

2.5 MAGNETOSPHERIC SUBSTORMS

The circulation pattern and associated pattern of stresses described thus far rarely if ever occur in a steady manner for any extended period. Rather the flow proceeds in bursts which occur in association with a variety of other phenomena on a global scale, ground magnetic perturbations, injection and enhancement of the ring current energetic particle population, and the breakup and dramatic poleward motion of the aurora.

The concept of a global magnetospheric perturbation that came to be called a substorm arose in the late sixties from the work of several groups (see, e.g., Jelley and Brice, 1967, McPherron et al., 1968). The topic was actively pursued in the seventies; a good collection of papers is found in the book edited by Akasofu (1979).

There is little doubt that a southward turning of the interplanetary magnetic field heralds increased geomagnetic activity, and magnetospheric substorms in particular. Also it is agreed that a series of substorms will lead to the enhanced energetic particle ring current population whose drift about the Earth depresses the equatorial field at the surface. Over a period of days the field may become depressed by a hundred or more nanotesla and then recover. Such a cumulative disturbance which occurs over a matter of two or three days is called a geomagnetic storm.

Details of the substorm process, and indeed all interactions which take place as the field imposed on the magnetosphere turns southward, still contain areas of scientific controversy. One concern is how rapidly enhanced circulation follows southward turnings of the exterior field, the key question being whether dayside reconnection may often proceed for some time without corresponding reconnection in the magnetotail. Without tail reconnection the magnetic flux in the tail lobe increases. Many workers (e.g. McPherron et al., 1973) would argue that the substorm is initiated by changed conditions in the neutral sheet in the tail centre. The substorm manifests itself as a dramatic reorientation of the field on the nightside of Earth; a spacecraft in the vicinity of synchronous orbit ($\sim 5.6\ R_E$ altitude) sees the field return to a dipolar configuration from a more tail-like form which usually has developed during the hour or more beforehand. Note however that on a spacecraft time and space structure may be confused. The author's sympathy is with the view represented by McPherron and numerous others but the reader is urged to note the controversy that currently remains.

The manner in which reconnection drives the system is as yet not clear, there being clear evidence of patchy reconnection regularly occurring (Rijnbeek et al., 1984) and furthermore of a residual viscous (non-reconnection) driven momentum transfer at the magnetosphere (see e.g. Cowley, 1982). However there are those who still feel the significance of reconnection has been overemphasised.

2.6 MAGNETOHYDRODYNAMIC WAVES

There are three distinct MHD wave modes, fast, transverse and slow. They occur in many different magnetospheric contexts. It is worthwhile to start by outlining the dynamical properties of each mode.

The MHD dispersion relation is

$$(\omega^2 - k_{\shortparallel}^2 A^2)(\omega^4 - \omega^2 k^2 (c_s^2 + A^2) + k_{\shortparallel}^2 k^2 A^2 c_s^2) = 0 \qquad (3)$$

where $k_{\shortparallel} = \underset{\sim}{k}.\underset{\sim}{B}/B$, c_s is sound speed and A is the Alfvén speed, $B/(\mu_0\rho)^{\frac{1}{2}}$. The first factor represents the transverse or intermediate or shear Alfven mode, whose dispersion relation

$$\omega^2 = k_{\shortparallel}^2 A^2$$

implies its group velocity is always parallel to B, i.e. it is field guided. The quartic factor represents the fast and slow compressional i.e. magnetoacoustic modes. The fast mode propagates fairly isotropically. The slow mode phase velocity is zero for perpendicular propagation. The limiting form of the phase velocity for $|\underset{\sim}{k}| >> k_{\shortparallel}$ is

$$\omega^2 = \frac{c^2 A^2 k_{\shortparallel}^2}{c_s^2 + A^2}$$

As ω is independent of the perpendicular component of $\underset{\sim}{k}$, it too is field guided in this limit.

The dynamical roles of the three modes are each different. The fast mode is rather like a sound wave in a gas. The motion in the wave is compressional, both magnetic and plasma pressure changing in phase and the wave can carry energy in any direction efficiently. In contrast the slow mode is not able to carry energy across the field at all. Its motion is also compressional but of a subtler form as particle and field pressure vary in antiphase. Both field and particle pressure forces come into play in motion across field; this is inefficiently driven in the slow mode. Motion is predominantly along the field. In a strong field the motion resembles a sound wave in a pipe, the rigidity of the pipe being proportional to the field strength.

Whereas the fast mode is excited when there is an imbalance of pressure across the field the slow mode serves primarily to balance up pressure imbalance along the field. The Alfvén mode's role is also subtle. It is not compressional and is strictly field guided. It does carry parallel current. As pointed out earlier parallel currents serve to distribute perpendicular stress along the field and thus the Alfvén wave is excited if stress balance across the field varies at one point on the field; an Alfvén surge reestablishes equilibrium by bending the field adjusting the direction of the field tension force.

The wave modes are likely to be excited when information needs to be transmitted concerning changed flow or forces acting from one point of the system to another. Not unexpectedly there are characteristic magnetic oscillations recorded both in space and on the ground when substorms are initiated (see, e.g., Southwood and Stuart, 1979). The signals, known as pi2 pulsations, are transients associated with rapid reconfiguration of the magnetic field, enhanced flow and the consequent rearrangement of field aligned currents between ionosphere and equatorial magnetosphere.

The sharp initiation of the substorm also provides an interesting instance of the breakdown of MHD wave theory. The effect is directly due to the absence of collisions. The reconfiguration of the magnetic field on the nightside takes place in a matter of a minute or so. In the outer magnetosphere the initial surge in flow can take place on a time scale short compared with the bounce time back and forth along the magnetic field of the ring current energetic plasma. There is large field motion near the equator matched by little near the Earth as the originally distended tail field lines collapse to a more dipolar configuration. The plasma near the equator of the field is accelerated by the collapse but that far off the equator is not. After the collapse there is a hotter population near the equator and a higher pressure there. Fluid theory would predict that a slow mode perturbation would propagate back and forth along the flux tubes to eliminate the pressure imbalance. As Quinn and Southwood (1982) describe, what is actually seen is an entirely ballistic (collisionless) response; the pressure perturbation in particles of a given energy is found to bounce back and forth along the field at the bounce frequency appropriate to that energy thus providing a dispersive energy dependent signature in a particle spectrometer. The slow mode is ill-described by the MHD approximation because of this very effect. However in other (high pressure) regions clear fluid like slow mode behaviour is detected (e.g. Southwood and Saunders, 1984).

As well as being excited as transients, in a bounded system waves can give rise to normal mode disturbances of the system as a whole. A major success early on in magnetospheric theory was the discovery that the guided Alfvén wave mode did give rise to oscillatory signals with a standing structure along the field. Dungey's (1954) prediction was borne out by studies of signals at magnetically conjugate sites on Earth (e.g. Sugiura and Wilson, 1964).

Subsequently it has been recognised that the MHD wave modes are coupled by the field inhomogeneity and special resonant behaviour occurs on magnetic shells where the Alfvén mode dispersion relation is satisfied. The theory once again has been very successful in explaining observations. A review is given by Southwood and Hughes (1983). Recently attempts have been initiated to discuss coupling of fast mode global scale eigenmodes to the standing resonant Alfvén mode oscillations which are localised to particular magnetic shells (e.g. Kivelson and Southwood, 1985). The subject is too large to review in this introductory paper and still looks to be a continuing active research area.

REFERENCES

Akasofu S I (1979) (Ed.) *Dynamics of the Magnetosphere*, D. Reidel, Dordrecht.

Axford W I & C O Hines (1961) *Can. J. Phys.* 39, 1433.

Cowley S W H (1980) *Space Sci. Rev.* 26, 217.

Cowley S W H (1982) *Rev. Geophys. Space Phys.* 20, 531.

Dungey J W (1954) *Report 69, Ionos Res. Lab.*, Pa. State Univ.

Dungey J W (1961) *Phys. Rev. Lett.* 6, 47.

Dungey J W (1963) in *Geophysics, the Earth's Environment*, p.505, ed. C. Dewitt, J. Hieblot & A. Lebeau, Gordon and Breach, New York.

Fairfield D H (1971) *J. Geophys. Res.* 76, 6700.

Fairfield D H & L J Cahill, Jr. (1966) *J. Geophys. Res.* 71, 155.

Heppner J P (1972) *Planet. Space Sci.* 20, 1475.

Iijima T & T A Potemra (1976) *J. Geophys. Res.* 81, 2165.

Intriligator D S, J H Wolfe, D D McKibbin & H R Collard (1969) *Planet. Space Sci.* 17, 321.

Jaggi R K & R A Wolf (1973) *J. Geophys. Res.* 78, 2852.

Jelley D H & N M Brice (1967) *J. Geophys. Res.* 72, 5919.

Kivelson M G & D J Southwood (1985) *Geophys. Res. Lett.* 12.

Knight S (1973) *Planet. Space Sci.* 21, 741.

Lyons L R & D Williams (1984) *Quantitative aspects of magnetospheric physics*, D. Reidel, Dordrecht.
McPherron R L, C T Russell & M P Aubry (1973) *J. Geophys. Res.* 78, 3131.
McPherron R L, G K Parks, F V Coroniti & S H Ward (1968) *J. Geophys. Res.* 73, 1697.
Quinn H & D J Southwood (1982) *J. Geophys. Res.* 87, 536.
Rieff P H & J L Burch (1985) *J. Geophys. Res.* 90, 1595.
Rijnbeek R P, S W H Cowley, D J Southwood & C T Russell (1984) *J. Geophys. Res.* 89, 786.
Southwood D J (1977) *J. Geophys. Res.* 82, 5512.
Southwood D J (1985) in *Physics of Ionosphere-Magnetosphere Connection*, ed. E R Schmerling, Space Research Series, Pergamon Press, London.
Southwood D J & W J Hughes (1983) *Space Sci. Rev.* 35, 301.
Southwood D J & M A Saunders (1984) in *Proc. IMS Symposium*, ed. J G Roederer, ESA Special Publication.
Southwood D J & W F Stuart (1979) in *Dynamics of the Magnetosphere*, p.385, ed. S-I Akasofu, Reidel, Dordrecht.
Sugiura M & C R Wilson (1964) *J. Geophys. Res.* 69, 1211.
Vasyliunas V M (1972) in *Earth's Magnetospheric Processes*, p.29, ed. B M McCormac, D. Reidel, Dordrecht, Netherlands.

CHAPTER 3

MAGNETOHYDRODYNAMIC WAVES

B Roberts
Department of Applied Mathematics
University of St Andrews
St Andrews
Fife KY16 9SS
Scotland

3.1 STRUCTURING AND STRATIFICATION

The solar atmosphere, from the photosphere to the corona, is strongly structured by magnetic field and stratified by gravity. In such a medium the propagation of magnetohydrodynamic (mhd) waves is extremely complicated, with the well-known results of a uniform medium providing only a limited guidance as to the behaviour of the modes in the inhomogeneous atmosphere of the Sun. As an illustration of solar stratification it is amusing to observe that in a region extending upwards from the photosphere to the low chromosphere, over which the temperature decreases from a photospheric value of 6400 °K to the 4200 °K at the temperature minimum and finally increases again to reach 6400 °K in the chromosphere, the plasma density declines by five orders of magnitude!

Magnetic fields introduce structuring into the atmosphere. At the photospheric level, the field is found to occur in magnetic clumps or flux tubes, which are isolated from their neighbours. The field strengths are high, generating a hierarchy ranging from 1.5 kG in intense flux tubes to 3 kG in sunspots. Intense flux tubes are the 'building blocks' of the photospheric field; they are found to reside preferentially in the down-draughts of supergranules and have a radius of about 10^2km. Possessing such a small scale, they are below Earth-bound telescopic resolution and are consequently to be the subject of several future space missions (e.g., the Solar Optical Telescope). The physics of intense tubes is reviewed at length in Parker (1979) and Spruit (1981a), and more briefly in Spruit and Roberts (1983).

Sunspots represent the largest spatial scale of photospheric structuring, possessing diameters typically in the $4 \times 10^3 - 10^4$ km range. Between the extremes of the intense tube and the sunspot, knots and pores occur; field strengths are in the range 1.5 - 2 kG and diameters are of order 10^3 km. The physics of sunspots has recently been discussed in Cram and Thomas (1981).

Above the photospheric layers, in the low chromosphere, the isolated flux tubes rapidly expand to merge with their neighbours, completely filling the chromosphere and corona with magnetic field

(Figure 3.1). The photosphere, then, may be viewed as comprising two media: a magnetic region where the plasma beta (β) (see Chapter 1) is of order unity, and a weak-field region where β is very high. In the chromosphere, β is everywhere of order unity or lower; in the corona, the magnetic field is generally dominant with $\beta \ll 1$.

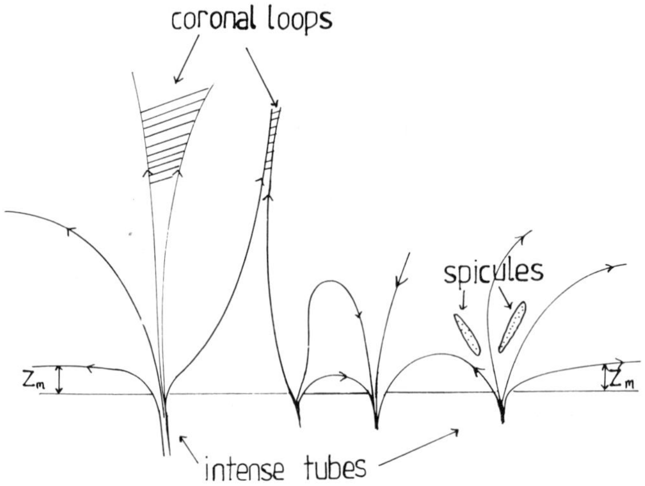

Figure 3.1 Magnetic field at the solar surface (after Spruit, 1981a)

The merging level z_m of intense flux tubes is difficult to determine observationally (Giovanelli, 1980; Giovanelli and Jones, 1982), and only simple theoretical models (e.g. Anzer and Galloway, 1983) have so far been developed to predict it. A level of about 1000 km gives a rough guide, with merging being at a lower level in *active regions* and at a higher level in *quiet regions* (see Chapter 1), depending upon the spacing between isolated tubes (see Spruit, 1981a; Roberts, 1984b).

Magnetic structuring is also present in the corona. Whereas both structuring and stratification are significant in the photosphere, stratification is less important in the corona because many coronal structures (such as loops) have scales less than or at most comparable with the local pressure scale-height. Thus, the coronal plasma is in some ways easier to treat analytically than the flux tubes of the photosphere. Nonetheless, each region has its own set of problems. In the corona density and temperature inhomogeneities are common.

To investigate the propagation of hydromagnetic waves in the solar atmosphere we proceed as follows. We begin by examining waves in magnetically structured atmospheres, ignoring the complications introduced by gravity. Those complications are considered in Section 3.7, once we have obtained some insight into the effects of structuring. The governing equations are given in Chapter 1, namely the equations of momentum and continuity for a gas (Eqs (1.2) with $g = 0$, (1.3) and (1.6)). In the momentum equation, the current $\underset{\sim}{j}$ is linked to the magnetic field $\underset{\sim}{B}$ by Ampere's law (Eq (1.5)) and the gas is taken to be a perfect conductor (so $\eta = 0$ in the induction equation (1.1)). Adiabatic energy exchange is assumed (so that the right-hand side of (1.4) is ignored).

3.2 WAVES IN A MAGNETICALLY STRUCTURED ATMOSPHERE

Consider the *equilibrium state*

$$\underset{\sim}{B} = B_o(x)\hat{\underset{\sim}{z}}, \quad p = p_o(x), \quad \rho = \rho_o(x), \quad T = T_o(x), \tag{3.1}$$

representing a plasma with pressure p_o, density ρ_o and temperature T_o embedded in a unidirectional magnetic field B_o. All equilibrium qualities are functions of x in a cartesian coordinate system xyz, with the magnetic field aligned along the z-axis. The momentum equation demands that

$$\frac{d}{dx}\left[p_o + \frac{B_o^2}{2\mu}\right] = 0, \tag{3.2}$$

expressing the fact that the *total* (gas p_o plus magnetic $B_o^2/2\mu$) pressure is constant.

Linear *perturbations* about the equilibrium (3.1) and (3.2) are readily sought, and lead to the system of equations (e.g. Roberts, 1981, 1984)

$$\left.\begin{aligned}
&\frac{\partial\rho}{\partial t} + \operatorname{div}\rho_o\underset{\sim}{v} = 0,\\
&\rho_o\frac{\partial\underset{\sim}{v}}{\partial t} = -\nabla\left[p + \frac{1}{\mu}\underset{\sim}{B}_o\cdot\underset{\sim}{B}\right] + \frac{1}{\mu}(\underset{\sim}{B}_o\cdot\nabla)\underset{\sim}{B} + \frac{1}{\mu}(\underset{\sim}{B}\cdot\nabla)\underset{\sim}{B}_o,\\
&\frac{\partial\underset{\sim}{B}}{\partial t} = \operatorname{curl}(\underset{\sim}{v}\times\underset{\sim}{B}_o),\ \operatorname{div}\underset{\sim}{B} = 0,\\
&\frac{\partial p}{\partial t} + \underset{\sim}{v}\cdot\nabla p_o = c_s^2\left[\frac{\partial\rho}{\partial t} + \underset{\sim}{v}\cdot\nabla\rho_o\right],
\end{aligned}\right\} \tag{3.3}$$

describing small-amplitude disturbances with velocity $\underset{\sim}{v} = (v_x, v_y, v_z)$, magnetic field $\underset{\sim}{B} = (B_x, B_y, B_z)$, pressure p and density ρ.

It is convenient to introduce the magnetic pressure perturbation $p_m \left[\equiv \frac{1}{\mu} \underset{\sim}{B}_0 \cdot \underset{\sim}{B}\right]$ and the perturbation p_T in total pressure:

$$p_T = p + p_m. \tag{3.4}$$

Then Eqs (3.3) yield

$$\rho_0 \left[\frac{\partial^2}{\partial t^2} - v_A^2 \frac{\partial^2}{\partial z^2}\right] \underset{\sim}{v}_\perp + \nabla_\perp \left[\frac{\partial p_T}{\partial t}\right] = 0 \tag{3.5}$$

and

$$\left[\frac{\partial^2}{\partial t^2} - c_s^2 \frac{\partial^2}{\partial z^2}\right] v_\parallel - c_s^2 \,\mathrm{div}\left[\frac{\partial \underset{\sim}{v}_\perp}{\partial z}\right] = 0, \tag{3.6}$$

where $\underset{\sim}{v}_\perp = (v_x, v_y, 0)$ and $\underset{\sim}{v}_\parallel = v_\parallel \hat{\underset{\sim}{z}} = (0,0,v_z)$ are the velocities perpendicular and parallel to the applied field $\underset{\sim}{B}_0$, and $\nabla_\perp = \left[\frac{\partial}{\partial x}, \frac{\partial}{\partial y}, 0\right]$. The velocity components are related to the pressure perturbations p_T and p_m by

$$\frac{\partial p_T}{\partial t} = -\rho_0 \left[c_s^2 \frac{\partial v_\parallel}{\partial z} + (c_s^2 + v_A^2)\mathrm{div}\, \underset{\sim}{v}_\perp\right],$$

$$\frac{\partial p_m}{\partial t} = \left[\frac{dp_0}{dx}\right] v_x - \rho_0 v_A^2 \,\mathrm{div}\, \underset{\sim}{v}_\perp. \tag{3.7}$$

In the above $c_s(x)$ $(= (\gamma p_0(x)/\rho_0(x))^{\frac{1}{2}})$ and $v_A(x)$ $(= B_0(x)/(\mu\rho_0(x))^{\frac{1}{2}})$ are the usual sound and Alfvén speeds.

At this stage it is convenient to introduce Fourier components,

$$v_x = v_x(x)e^{i(\omega t - k_y y - k_z z)}, \text{ etc.} \tag{3.8}$$

We may then eliminate v_y and v_z in favour of $v_x(x)$ and $p_T(x)$ to yield the pair of ordinary differential equations

$$\frac{dp_T}{dx} + \frac{i\rho_0}{\omega}(\omega^2 - k_z^2 v_A^2)v_x = 0, \tag{3.9}$$

$$(\omega^2 - k_z^2 v_A^2)(\omega^2 - k_z^2 c_T^2)\frac{dv_x}{dx}$$

$$+ \frac{i\omega}{\rho_o(c_s^2 + v_A^2)}(\omega^2 - \omega_s^2(x))(\omega^2 - \omega_f^2(x))p_T = 0, \tag{3.10}$$

where $\omega_s(x)$ and $\omega_f(x)$ satisfy

$$\omega_s^2 + \omega_f^2 = (k_y^2 + k_z^2)(c_s^2 + v_A^2), \quad \omega_s^2\omega_f^2 = k_z^2(k_y^2 + k_z^2)c_s^2 v_A^2. \tag{3.11}$$

The significance of the frequencies ω_s and ω_f will emerge later (Section 3.3.2). The cusp speed c_T, defined by

$$c_T(x) = c_s v_A/(c_s^2 + v_A^2)^{\frac{1}{2}},$$

is both sub-sonic and sub-Alfvénic it plays an important role in the subsequent discussion.

The pair of equations (3.9) and (3.10) may be reduced to a single differential equation by eliminating p_T or v_x. Eliminating p_T gives

$$\frac{d}{dx}\left\{\frac{\rho_o(c_s^2 + v_A^2)(\omega^2 - k_z^2 v_A^2)(\omega^2 - k_z^2 c_T^2)}{(\omega^2 - \omega_s^2)(\omega^2 - \omega_f^2)}\frac{dv_x}{dx}\right\}$$

$$+ \rho_o(\omega^2 - k_z^2 v_A^2)v_x = 0. \tag{3.12}$$

Equation (3.12) governs the behaviour of hydromagnetic waves in an inhomogeneous medium. It is a complicated equation, not yet fully explored. It has two singularities (Appert *et al*, 1974): an *Alfvén singularity* ($\omega^2 = k_z^2 v_A^2(x)$), and a *cusp singularity* ($\omega^2 = k_z^2 c_T^2(x)$). These singularities are associated with the occurrence of *continuous spectra* in mhd (see the extensive discussion in Goedbloed (1983) and in Adam (1981,1982)). Their occurrence is a reflection of the highly anisotropic nature of the Alfvén wave and the slow magnetoacoustic wave - see Section 3.3. There are two other factors of particular interest in Eq (3.12), namely the locations where $\omega^2 = \omega_s^2(x)$ or where $\omega^2 = \omega_f^2(x)$. These *cut- off points* are not singularities, but are associated with the change from oscillatory to evanescent behaviour in a wave, corresponding to wave reflection or wave trapping.

3.3 WAVES IN A UNIFORM MEDIUM

In a uniform medium the coefficients in Eqs (3.9), (3.10) and (3.12) are all constants. It follows, then, that Eq (3.12) may be written

$$\rho_o(\omega^2 - k_z^2 v_A^2)\left\{\frac{(c_s^2 + v_A^2)(\omega^2 - k_z^2 c_T^2)}{(\omega^2 - \omega_s^2)(\omega^2 - \omega_f^2)}\frac{d^2 v_x}{dx^2} + v_x\right\} = 0. \tag{3.13}$$

Furthermore, we may now Fourier analysis in the x-direction by writing $v_x \propto e^{-ik_x x}$. Substituting in Eq (3.13) then yields the general dispersion relation for hydromagnetic modes:

$$(\omega^2 - k_z^2 v_A^2)(\omega^4 - \omega^2(c_s^2 + v_A^2)k^2 + k^2 k_z^2 c_s^2 v_A^2) = 0, \tag{3.14}$$

where $k^2 = k_x^2 + k_y^2 + k_z^2$. It is evident that there are three roots (in ω^2) to consider.

The dispersion relation (3.14) characterises the behaviour of the three hydromagnetic waves but for further insight into the individual behaviour of these waves it is necessary to investigate the behaviour of the fluid and magnetic perturbations. Of particular interest are the inter-relationships between magnetic pressure, gas pressure and field-line tension $\mathfrak{T}$ $\left[\equiv \frac{1}{\mu}(\mathbf{B}_0.\nabla)\mathbf{B}\right]$, for it is these forces which give rise to wave propagation and it is the interaction of these forces which determines the individual characters of the three modes. From Eq (3.3) we may show that

$$p_m = \left[\frac{\omega^2 - k_z^2 c_s^2}{\omega^2}\right] \left[\frac{v_A^2}{c_s^2}\right] p, \quad p_T = \left[\frac{\omega^2 - k_z^2 c_T^2}{\omega^2}\right] \left[\frac{c_s^2 + v_A^2}{c_s^2}\right] p, \tag{3.15}$$

and

$$p_T = \frac{\rho_0}{\omega k_x}(\omega^2 - k_z^2 v_A^2) v_x, \quad \rho c_s^2 = p = \rho_0 \left[\frac{\omega}{k_z}\right] v_z. \tag{3.16}$$

The tension force $\mathfrak{T}$ is given by

$$\mathfrak{T} = ik_z \left[\frac{k_z}{\omega}\rho_0 v_A^2 v_x, \frac{k_z}{\omega}\rho_0 v_A^2 v_y, -p_m\right]. \tag{3.17}$$

We use these relations shortly.

3.3.1 The Alfvén Wave

One solution of the dispersion relation (3.14) is

$$\omega^2 = k_z^2 v_A^2 = v_A^2 k^2 \cos^2\theta, \tag{3.18}$$

where we have introduced the angle θ that the wave vector $\mathbf{k} = (k_x, k_y, k_z)$ makes with the applied magnetic field $B_0\hat{\mathbf{z}}$. This is the Alfvén wave (Alfvén, 1942). It has a phase-speed c ($\equiv \omega/k$) $= \pm v_A$, and a group velocity $\mathbf{c}_g$ ($\equiv (\partial\omega/\partial k_x, \partial\omega/\partial k_y, \partial\omega/\partial k_z)) = \pm v_A\hat{\mathbf{z}}$. Inspection of Eqs (3.15) and (3.16) reveals that p, p_m and p_T are all zero as are ρ and v_z. Thus, motions in an Alfvén wave are perpendicular to $\mathbf{B}_0$, involving no compression of the plasma. Explicitly, we may write the velocity in the Alfvén mode as proportional to

$$\underset{\sim}{v}^{(\text{Alfvén})} = \frac{v_y}{k_x}(-k_y, k_x, 0), \tag{3.19}$$

showing an incompressible (div $\underset{\sim}{v} = 0$), *transverse* (i.e. perpendicular to $\underset{\sim}{k}$) motion akin to the transverse waves on an elastic string. The Alfvén wave is driven by the tension force. It is an anisotropic mode, unable to propagate across the magnetic field (since $\omega = 0$ for $\theta = \pi/2$, $k_z = 0$). Note, from Eq (3.17), that the tension force is parallel to the velocity $\underset{\sim}{v}^{(\text{Alfvén})}$ but perpendicular to the wave vector $\underset{\sim}{k}$.

3.3.2 Magnetoacoustic Waves

Aside from the root $\omega^2 = k_z^2 v_A^2$, the dispersion relation (3.14) yields

$$\omega^4 - \omega^2(c_s^2 + v_A^2)k^2 + k^4 c_s^2 v_A^2 \cos^2\theta = 0, \tag{3.20}$$

which may be solved thus:

$$2\frac{\omega^2}{k^2} = (c_s^2 + v_A^2) \pm [(c_s^2 + v_A^2)^2 - 4c_s^2 v_A^2 \cos^2\theta]^{\frac{1}{2}}. \tag{3.21}$$

The two solutions correspond to the *fast* and *slow magnetoacoustic* modes. A comparison of Eqs (3.11) and (3.20) makes clear that the frequencies ω_f and ω_s arising in Eq (3.10) are related to the fast and slow waves with $k_y^2 + k_z^2$ replaced by k^2.

It is of interest to note that Eq (3.20) may be rewritten in the form

$$k_x^2 + k_y^2 = n_o^2, \tag{3.22}$$

where we have introduced

$$n_o^2 = \frac{(\omega^2 - k_z^2 c_s^2)(\omega^2 - k_z^2 v_A^2)}{(c_s^2 + v_A^2)(\omega^2 - k_z^2 c_T^2)}. \tag{3.23}$$

Equation (3.20) reveals that the slow wave has longitudinal phase-speed, ω/k_z, intermediate between c_T and the minimum of c_s and v_A; the fast wave has longitudinal phase-speed above the maximum of c_s and v_A. For propagation along the field, $\theta = 0$ and $\omega^2 = k^2 c_s^2$, $k^2 v_A^2$; propagation perpendicular ($\theta = \pi/2$) to the field yields $\omega^2 = k^2(c_s^2 + v_A^2)$, and $\omega^2 = 0$ (more precisely, $\omega^2 \sim k^2 c_T^2 \cos^2\theta = k_z^2 c_T^2$ as $\theta \to \pi/2$).

Thus, the slow wave - like the Alfvén wave - cannot propagate across field-lines. The fast wave, on the other hand, acquires its greatest phase-speed when propagating across the field. This behaviour may be readily understood when we consider the relationship between the magnetic pressure perturbation p_m and the gas pressure perturbation p. From Eq (3.15), we see that p_m and p are in phase (i.e. $p_m/p > 0$) when $\omega^2 > k_z^2 c_s^2$, and out of phase (i.e. $p_m/p < 0$) when

$\omega^2 < k_z^2 c_s^2$. Thus, in the *slow* wave the magnetic pressure p_m and gas pressure p are always *out of phase*; furthermore, as the angle of propagation θ approaches $\pi/2$, $p_m \sim -p$ and the total pressure perturbation p_T falls to zero. Also, the tension force $\underset{\sim}{T}$ (given in Eq (3.17)) falls to zero as $\theta \to \pi/2$.

By contrast, for the *fast* wave p_m and p are always *in phase*, and p_T rises to its maximum value (relative to p) when θ approaches $\pi/2$. The tension force is again zero when $\theta = \pi/2$. Thus, at an angle of propagation of $\pi/2$ the fast wave acquires its maximum propagation speed, driven entirely by gas and magnetic pressure variations acting in unison. Such distinctive features, as whether p_m and p are in or out of phase, allow the possibility of distinguishing between the fast and slow modes. This has found application in the magnetosphere (see, for example, Siscoe (1983), Southwood and Hughes (1983)).

It is of interest to determine the velocity field in the fast and slow modes. Equation (3.5) shows that $v_y/v_x = k_y/k_x$, unless $\omega^2 = k_z^2 v_A^2$ (the Alfvén wave). Thus, on using Eqs (3.15), (3.16) and (3.22), we find that the velocity fields in the fast and slow modes are proportional to

$$\underset{\sim}{v}^{(\mathrm{slow})} = \frac{v_x}{k_x}(k_x, k_y, \lambda^{(-)} k_z), \quad \underset{\sim}{v}^{(\mathrm{fast})} = \frac{v_x}{k_x}(k_x, k_y, \lambda^{(+)} k_z), \tag{3.24}$$

where

$$\lambda^{(\pm)} = \frac{c_s^2(k_x^2 + k_y^2)}{(\omega^2 - k_z^2 c_s^2)}$$

is determined once ω^2 is specified by Eq (3.21). By forming the scalar product of $\underset{\sim}{v}^{(\mathrm{slow})}$ and $\underset{\sim}{v}^{(\mathrm{fast})}$ we may readily confirm that they are mutually orthogonal and, furthermore, are orthogonal to $\underset{\sim}{v}^{(\mathrm{Alfvén})}$ (Shercliff, 1965; Goedbloed, 1983). Thus, the three hydromagnetic modes generate an orthogonal triad of velocity vectors. Any velocity field, then, may be written as the sum of these three vectors.

We turn now to a consideration of the phase-speed ω/k and group velocity $\underset{\sim}{c}_g = \partial\omega/\partial\underset{\sim}{k}$ for the magnetoacoustic waves (e.g. Hughes and Young, 1966). These are best presented in graphical form. In Figure 3.2 we have sketched ω/k and $\underset{\sim}{c}_g$ in polar form for the case $v_A > c_s$. The Alfvén wave is also included. The diagram illustrates vividly the anisotropic nature of the Alfvén and slow waves; the fast mode is almost isotropic in form. The slow wave is immediately seen to be the most singular of the three waves. It is amusing to contemplate how different a favourite symphony might sound to the ear of a listener to an orchestra playing in a perfectly conducting plasma, and how different it would sound depending upon where one was seated relative to an applied magnetic field! Seats perpendicular to the field might be the most favoured, while listeners seated along the field would have to allow for both fast and slow sounds! Alfvén waves are silent, of course.

Finally, we note the behaviour of magnetoacoustic waves in the limits of high and low plasma β. A *high* β plasma corresponds to an *incompressible* fluid ($c_s >> v_A$), for which the fast wave disappears ($\omega \to \infty$) and the slow wave yields

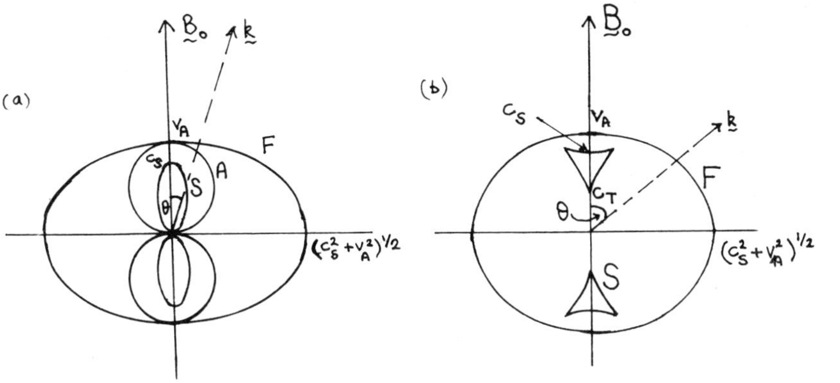

Figure 3.2 The (a) phase-speed and (b) group velocity of magnetohydrodynamic waves in the case $v_A > c_s$.

$$\omega^2 \sim k_z^2\, c_T^2 \sim k_z^2\, v_A^2. \tag{3.25}$$

This is the same dispersion relation as for an Alfvén wave (eqn (3.18)) but the polarisation of fluid motions, $\underset{\sim}{v}^{(slow)}$ and $\underset{\sim}{v}^{(Alfvén)}$, are not the same. Thus, in an incompressible ($c_s \to \infty$) fluid motions are governed by the dispersion relation $\omega^2 = k_z^2 v_A^2$, but may be Alfvén waves (if $v_z = p = 0$) or compressive (slow) modes (if v_z, $p \neq 0$). However, this distinction between the different modes is not always made.

Turning to a consideration of the *low* β case, for which $c_s << v_A$ and the magnetic field dominates, we find that (3.21) yields

$$\omega^2 \sim k_z^2\, c_T^2 \sim k_z^2\, c_s^2 \tag{3.26}$$

from the slow mode branch, and

$$\omega^2 \sim k^2(c_s^2 + v_A^2) \sim k^2 v_A^2 \tag{3.27}$$

from the fast mode. The fast wave, then, propagates isotropically; it is sometimes referred to as a 'compressional Alfvén wave' but we should note that it is quite distinct from an Alfvén wave. By contrast to the fast wave, the slow wave is highly anisotropic. Indeed, we see from (3.26) that sound waves in a strong magnetic field are propagated *one-dimensionally* (with predominantly longitudinal motions) along the essentially rigid field lines, a property in sharp contrast to the isotropic behaviour of sound waves in a non-magnetic atmosphere.

The properties of the three mhd waves are summarized in the following Table.

Wave	Characteristics
Alfvén	Driven by tension forces; no pressure or density variations; motions transverse to both the applied magnetic field and the direction of propagation. Anisotropic, unable to propagate across the field; energy flows along the field at the Alfvén speed.
slow magnetoacoustic	Driven by tension and pressure forces; involves pressure and density variations; gas and magnetic pressure variations are out of phase. Anisotropic, unable to propagate across the field; energy flow confined to near the magnetic field.
fast magnetoacoustic	Driven by tension and pressure forces; involves pressure and density variations; gas and magnetic pressure variations are in phase. Roughly isotropic, propagating fastest across the field. Forms an orthogonal triad with the other two modes.

Table 3.I : Properties of the Alfvén and magnetoacoustic waves.

3.4 WAVES IN DISCRETELY STRUCTURED MEDIA

The general properties of hydromagnetic waves in a uniform medium provides a partial guide to how those waves might behave in an astrophysical plasma, such as the Sun's atmosphere or the Earth's magnetosphere. However, the assumption of uniformity is a severe one and not generally met in astrophysical plasmas. Indeed the opposite is more frequently the case, as we have remarked earlier. It is thus necessary to consider how magnetic structuring modifies the general conclusions of the previous Section.

Two types of discretely structured media are of obvious interest, namely a medium possessing a single interface separating two regions of differing plasma properties, and a medium in which there are two interfaces combining to form a slab or tube of magnetic field which is in some sense distinct from its environment. The *single interface* model has applications to the lower penumbral magnetic field in sunspots or to the boundary of the magnetopause. The *slab* or *flux tube* model has applications to photospheric intense tubes, to sunspots, and to coronal loops.

We begin our discussion by investigating waves in an incompressible medium ($c_s \to \infty$). This circumstance has limited application to astrophysical plasmas but it is simpler to treat than the compressible case and thus is a useful preliminary to elucidating the structure of wave motions in the complicated problem. Nevertheless, it is not a substitute for a compressible treatment. Compressibility allows a far richer spectrum of wave behaviour, as we shall see in the subsequent sections.

3.4.1 Incompressible medium

Consider the incompressible limit of Eq (3.12). For convenience, we will also set $k_y = 0$ and $v_y = 0$. With $c_s \to \infty$, Eq (3.12) for two-dimensional motions $(v_x, 0, v_z)$ reduces to

$$\frac{d}{dx}\left\{\rho_0(k_z^2 v_A^2 - \omega^2)\frac{dv_x}{dx}\right\} - k_z^2\rho_0(k_z^2 v_A^2 - \omega^2)v_x = 0. \qquad (3.28)$$

We suppose the unperturbed medium to be made up of a region (say $x < 0$) with field strength B_0, gas pressure p_0 and density ρ_0, all uniform, and a region ($x > 0$) with uniform field B_e, gas pressure p_e and density ρ_e. Thus in the *single interface* model we have

$$B_0(x),\ p_0(x),\ \rho_0(x) = \begin{cases} B_e,\ p_e,\ \rho_e, & x > 0, \\ B_0,\ p_0,\ \rho_0, & x < 0, \end{cases} \qquad (3.29)$$

with constancy of total pressure (eq (3.2)) giving

$$p_0 + \frac{B_0^2}{2\mu} = p_e + \frac{B_e^2}{2\mu}\ . \qquad (3.30)$$

Combined with the gas law, Eq (3.30) implies that

$$\rho_e(c_e^2 + \tfrac{1}{2}\gamma v_{Ae}^2) = \rho_0(c_0^2 + \tfrac{1}{2}\gamma v_A^2), \qquad (3.31)$$

where c_o, c_e and v_A, v_{Ae} are the sound and Alfvén speeds in the regions $x < 0$ and $x > 0$, respectively.

Either side of the interface $x = 0$ is thus a uniform medium, and so Eq (3.28) yields

$$\rho_o(k_z^2 v_A^2 - \omega^2) \left\{ \frac{d^2 v_x}{dx^2} - k_z^2 v_x \right\} = 0 \qquad (3.32)$$

in $x < 0$, with a similar result in $x > 0$. There is evidently the possibility of $\omega^2 = k_z^2 v_A^2$ (see Section 3.3), but our interest here is directed more to the vanishing of the differential operator:

$$\frac{d^2 v_x}{dx^2} - k_z^2 v_x = 0. \qquad (3.33)$$

Equation (3.33) possesses simple exponential solutions, $v_x \sim e^{-k_z x}$ and $e^{k_z x}$. In a medium stretching form $x = -\infty$ to $+\infty$ such Laplacian solutions are unacceptable, being unbounded at one or other end, but here (since $x < 0$) it is only necessary for v_x to be bounded as $x \to -\infty$. The region $x > 0$ is treated similarly. We thus take as our solution

$$v_x(x) = \begin{cases} \alpha_e e^{-k_z x}, & x > 0, \\ \alpha_o e^{k_z x}, & x < 0, \end{cases} \qquad (3.34)$$

satisfying $v_x \to 0$ as $|x| \to \infty$ for k_z chosen positive; α_o and α_e are arbitrary constants.

To determine the dispersion relation for the wave we must connect the two regions at the interface $x = 0$. We require that the normal component of the velocity $\underset{\sim}{v}$ be continuous at $x = 0$, i.e. v_x is continuous at $x = 0$. Thus $\alpha_e = \alpha_o$. Additionally, it is evident from physical considerations that the total pressure perturbation, p_T, must also be continuous across the interface, for otherwise there would be an unbalanced pressure force at $x = 0$. The form of p_T may be written down from Eq (3.10). Alternatively, we may argue directly from Eq (3.28) that, since v_x is continuous at $x = 0$, the expression in curly brackets must be continuous also (a result evident on integrating Eq (3.28) across the interface). Hence, we require that

$$p_T \sim \rho_o(x)(k_z^2 v_A^2(x) - \omega^2) \frac{dv_x}{dx} \qquad (3.35)$$

be continuous across $x = 0$. Combining this with Eq (3.34), for $\alpha_e = \alpha_o$, finally yields (Kruskal and Schwarzschild, 1954)

$$\omega^2 = k_z^2 c_k^2 = k_z^2 \left[\frac{\rho_o v_A^2 + \rho_e v_{Ae}^2}{\rho_o + \rho_e}\right]. \tag{3.36}$$

The dispersion relation (3.36) describes a *hydromagnetic surface wave*. Such a wave exists whenever density or magnetic field differences arises. The wave propagates along the interface at a speed c_k which is intermediate between the two Alfvén speeds. The wave is confined to the interface, with a characteristic penetration of order k_z^{-1} into the surrounding fluid.

Consider now the behaviour of waves in a *slab* of magnetic field, representative of the kind of field distribution observed in, for example, the solar photosphere. (A cylindrical geometry is more appropriate for a flux tube but both geometries yield similar results, with the slab being slightly easier to treat.) The equilibrium, then, is taken to be

$$B_o(x) = \begin{cases} B_o, & |x| < x_o, \\ 0, & |x| > x_o, \end{cases} \tag{3.37}$$

representing a uniform slab of magnetic field, with width $2x_o$, surrounded by a field-free plasma. Again, Eq (3.33) is applicable, with solution

$$v_x(x) = \begin{cases} \alpha_e\, e^{-k_z x}, & x > x_o, \\ \alpha_o \cosh k_z x + \beta_o \sinh k_z x, & |x| < x_o, \\ \beta_e\, e^{k_z x}, & x < -x_o, \end{cases} \tag{3.38}$$

for arbitrary constants α_e, β_e, α_o and β_o. In addition to the boundary condition $v_x \to 0$ as $|x| \to \infty$, applied in Eq (3.38) for $k_z > 0$, we require that v_x and p_T (see Eq (3.35)) are continuous across $x = \pm x_o$. Application of these conditions results in the dispersion relation (Parker, 1964)

$$\frac{k_z^2 v_A^2}{\omega^2} = 1 + \left[\frac{\rho_e}{\rho_o}\right] \begin{bmatrix}\tanh \\ \coth\end{bmatrix} k_z x_o, \tag{3.39}$$

where v_A is the Alfvén speed within the slab of gas density ρ_o surrounded by a field-free medium of gas density ρ_e.

The dispersion relation (3.39) describes the propagation of slab waves. There are two normal modes, depending geometrically on whether the slab is vibrated symmetrically or asymmetrically. See Figure 3.3.

A symmetric disturbance corresponds to the 'tanh' function in Eq (3.39); in this mode, commonly called a *sausage* wave, the slab pulsates like a blood vessel, with the axis of symmetry (the z-axis) remaining undisturbed. In the asymmetric mode, corresponding to the 'coth' function, the slab's axis is moved back and forth during the wave motion; this mode is commonly referred to as a *kink* wave. The phase-speed, ω/k_z, of both the sausage and kink waves is reduced below the Alfvén speed v_A, a consequence of the inertia in the slab's surroundings.

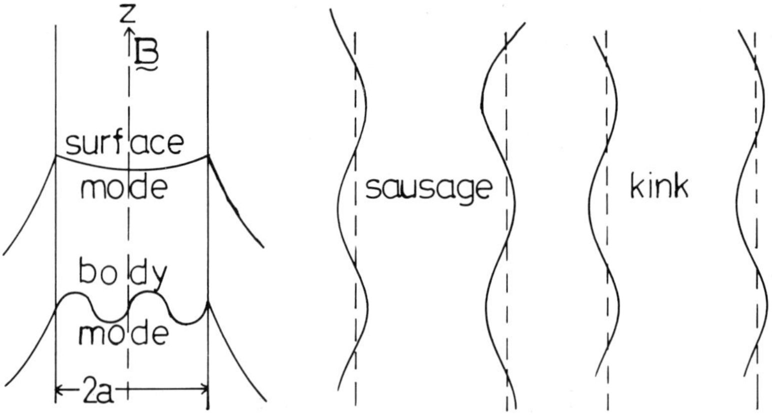

Figure 3.3 Surface and body waves in a slab or cylinder, exhibiting either sausage or kink geometrical oscillations

In contrast to the hydromagnetic surface wave (Eq (3.36)), slab waves are *dispersive*, i.e. their phase-speed is a function of wavenumber. This is because in a slab (and similarly in a cylindrical tube) the equilibrium state provides us with a length-scale (the slab width, say) by which a wave may be measured; no such length-scale arises for surface waves on a single interface or for waves in a uniform, unbounded medium. In the limit of a *thin* slab, corresponding to $k_z x_o \ll 1$, long wavelength disturbances propagate with phase-speed given by

$$\frac{\omega}{k_z} \sim \begin{cases} v_A, & \text{sausage mode,} \\ \left[\dfrac{\rho_o}{\rho_e}\right]^{\frac{1}{2}} (k_z x_o)^{\frac{1}{2}}, & \text{kink mode.} \end{cases} \qquad (3.40)$$

In the opposite extreme of a very wide slab, $k_z x_o \gg 1$, both modes have phase-speed

$$\frac{\omega}{k_z} \sim v_A \left[1 + \frac{\rho_e}{\rho_o}\right]^{-\frac{1}{2}}, \tag{3.41}$$

showing the expected surface wave behaviour at a single interface one side of which is field-free (Eq (3.36) with $v_{Ae} = 0$).

3.4.2 Compressible medium

The analysis for a compressible medium proceeds along much the same lines as in Section 3.4.1 but there are a number of complications, not least because a magnetic atmosphere can sustain two (the fast and slow) magnetoacoustic modes. Restricting attention to two-dimensional motions $(v_x, 0, v_z)$, Eq (3.12) yields

$$\frac{d}{dx}\left\{\frac{\rho_o(x)(c_s^2(x) + v_A^2(x))(k_z^2 c_T^2(x) - \omega^2)}{(k_z^2 c_s^2(x) - \omega^2)} \frac{dv_x}{dx}\right\}$$

$$- \rho_o(x)(k_z^2 v_A^2(x) - \omega^2)v_x = 0, \tag{3.42}$$

with p_T related to v_x through (see Eq (3.10))

$$p_T = \frac{i\rho_o(x)}{\omega} (c_s^2(x) + v_A^2(x)) \frac{(k_z^2 c_T^2(x) - \omega^2)}{(k_z^2 c_s^2(x) - \omega^2)} \frac{dv_x}{dx}. \tag{3.43}$$

In a *uniform* medium Eq (3.42) yields

$$\frac{d^2 v_x}{dx^2} - m_o^2 v_x = 0, \tag{3.44}$$

where

$$m_o^2 = \frac{(k_z^2 c_s^2 - \omega^2)(k_z^2 v_A^2 - \omega^2)}{(c_s^2 + v_A^2)(k_z^2 c_T^2 - \omega^2)} \tag{3.45}$$

replaces k_z^2 in the incompressible case (Eq (3.33)).

Consider first the *single interface* model, given in Eq (3.29). Proceeding much as in the incompressible problem, requiring that $v_x \to 0$ as $|x| \to \infty$ and that v_x and p_T are continuous across $x = 0$, yields (Wentzel, 1979; Roberts, 1981)

$$\frac{\omega^2}{k_z^2} = v_A^2 - \left[\frac{R}{R+1}\right] (v_A^2 - v_{Ae}^2), \tag{3.46}$$

where $R = \rho_e m_o / (\rho_o m_e)$. Here m_e is the value of m_o in the region $x > 0$.

The derivation of Eq (3.46) assumes that m_o and m_e are positive.

The dispersion relation (3.46) governs the behaviour of *magnetoacoustic surface waves*, propagating along the interface $x = 0$ with motions confined to the x-z plane. The surface waves penetrate a distance of the order of m_o^{-1} into the region $x < 0$, and a distance m_e^{-1} in $x > 0$. Notice that Eq (3.46) is transcendental, since R is a function of ω. It therefore may possess more than one solution; indeed, we may anticipate the possibility of two solutions, corresponding to fast and slow magnetoacoustic surface waves. In general, the phase-speeds will lie between the two Alfvén speeds, v_A and v_{Ae}. In the incompressible limit m_o and m_e approach k_z and so Eq (3.46) reduces to the earlier result (Eq (3.36)).

If one side of the interface is field-free, say $B_e = 0$, then Eq (3.46) reduces to

$$(k_z^2 v_A^2 - \omega^2) m_e = \left[\frac{\rho_e}{\rho_o}\right] \omega^2 m_o, \tag{3.47}$$

where now

$$m_e^2 = k_z^2 - \frac{\omega^2}{c_e^2}.$$

The solutions of Eq (3.47), the dispersion relation for magnetoacoustic surface waves at an interface one side of which is field-free, must be such that $\omega^2 < k_z^2 \min(c_e^2, v_A^2)$. There is always a *slow surface wave* and its phase-speed is below the minimum of c_T and c_e. If $c_e > c_o$ and $v_A > c_o$ then, in addition to the slow surface wave, there is a *fast surface wave*, which propagates with phase-speed between c_o and $\min(c_e, v_A)$. A more extensive discussion of these modes is given in Roberts (1981).

We turn now to a consideration of the compressible *slab* modes. The analysis (for equilibrium (3.37)) proceeds much as before so we will content ourselves with merely summarizing the main results. The governing dispersion relation for *magnetoacoustic slab waves* is (Cram and Wilson, 1975; Roberts, 1981)

$$(k_z^2 v_A^2 - \omega^2) m_e = \left[\frac{\rho_e}{\rho_o}\right] \omega^2 m_o \begin{bmatrix}\tanh\\ \coth\end{bmatrix} m_o x_o. \tag{3.48}$$

In contrast to the single interface case, m_o^2 is no longer restricted to positive values. However, we still require $m_e > 0$ and so solutions of Eq (3.48) must be consistent with $\omega^2 < k_z^2 c_e^2$. Waves with $m_o^2 > 0$ are referred to as *surface* waves; those with $m_o^2 < 0$ as *body* waves. See Figure 3.3(b).

As with the incompressible case there are both sausage and kink modes (Fig 3(a)), corresponding in the above to the 'tanh' and 'coth' functions respectively. Motions inside the slab ($|x| < x_o$) are of the form

$$v_x \sim \begin{cases} \sinh m_o x, & \text{sausage mode,} \\ \cosh m_o x, & \text{kink mode.} \end{cases}$$

(The hyperbolic functions become trigonometrical functions for body ($m_o^2 < 0$) waves, and then m_o is replaced by $|m_o|$.)

The complicated transcendental nature of Eq (3.48) means that we have to resort to numerical or graphical means for its solution. However, a number of features can be seen very readily. For example, in the incompressible limit ($c_o, c_e \to \infty$, so $m_o, m_e \to k_z$) Eq (3.48) reduces to Eq (3.39), giving a sausage surface wave and a kink surface wave. Also, in the wide slab limit ($k_z x_o \gg 1$), replacing $\tanh m_o x_o$ and $\coth m_o x_o$ by unity immediately recovers Eq (3.47), showing the existence of a slow surface wave and, possibly, a fast surface wave.

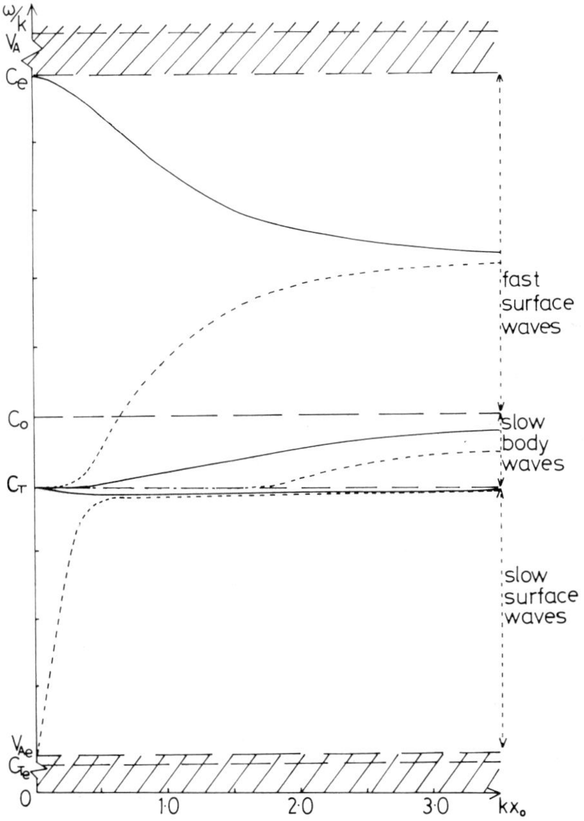

Figure 3.4 The phase-speed ω/k_z as a function of $k_z x_o$ for the case $v_A > c_e > c_o$ (after Edwin and Roberts, 1982)

In Figure 3.4 we have sketched the solutions of Eq (3.48), giving the phase-speed, ω/k_z, as a function of (dimensionless) wavenumber $k_z x_o$. The general features we have pointed out above may readily be identified. Of particular interest are the slow surface waves in a thin tube ($k_z x_o \ll 1$). The slow sausage wave in the long wavelength limit gives

$$\omega \sim k_z c_T; \tag{3.49}$$

the kink wave has the same approximate behaviour as in the incompressible case (see Eq (3.40)). We will return to these observations in later sections.

Before leaving relation (3.48) it is of interest to note that it has no *unstable* solutions. An instability corresponds to solutions with $\omega^2 < 0$; see Chapters 1 and 4. Physically, it is evident here that there cannot be any instabilities but it is amusing to observe how this conclusion may also be reached directly from the dispersion relation. For suppose $\omega^2 < 0$. Then $m_o^2 > 0$, and so the left-hand side of Eq (3.48) is positive. So also is $m_o \tanh m_o x_o$. Thus the right-hand side of Eq (3.48) is negative, leading to a contradiction. Therefore, there are no modes with $\omega^2 < 0$.

3.5 OSCILLATIONS IN A LOW-β GAS

In many astrophysical circumstances the plasma-β is small. The solar corona provides a good example of such a plasma. There the sound speed is typically around 200 km s^{-1} whereas the Alfvén speed may be of order 10^3km s^{-1}, thus producing $\beta \sim 0.05$. The magnetic field dominates in such a gas: sound waves propagate longitudinally along almost rigid field lines, while fast waves are isotropic and propagate at the Alfvén speed. In this section we examine the behaviour of the fast magnetoacoustic waves. For sake of clarity we shall take $\beta \equiv 0$, that is, $p_o = c_s = 0$. Only a fast wave (and an Alfvén wave) can then exist. (A treatment of the $\beta \neq 0$ case reveals simply the existence of guided sound waves, with the fast mode modified only slightly.)

The equilibrium is again taken to be a unidirectional magnetic field $\underset{\sim}{B}_o = B_o\hat{\underset{\sim}{z}}$, but now B_o is everywhere constant (in keeping with constancy of total pressure, since $p_o = 0$). However, non-uniformities may arise through the density distribution ρ_o. Thus the Alfvén speed is non-uniform. Both slab and cylinder geometries are investigated.

3.5.1 Slab inhomogeneities

Consider first the slab geometry. The general equation is again (3.12), with now $c_s = 0$. We restrict attention to two-dimensional motions, $(v_x, 0, v_z)$. Then, inside the slab ($|x| < x_o$) we again have Eq (3.44):

$$\frac{d^2 v_x}{dx^2} - m_o^2 v_x = 0, \tag{3.50}$$

where now

$$m_o^2 = (k_z^2 v_A^2 - \omega^2)/v_A^2.$$

Outside the slab, in $|x| > x_o$, Eq (3.50) applies with m_o^2 replaced by m_e^2, where

$$m_e^2 = (k_z^2 v_{Ae}^2 - \omega^2)/v_{Ae}^2 \tag{3.51}$$

and $v_{Ae}^2 = B_o^2/\mu\rho_e = \rho_o v_A^2/\rho_e$ is the square of the Alfvén speed in the slab's environment.

The analysis now proceeds much as in Section 3.4.2, resulting in the dispersion relations (Edwin and Roberts, 1982)

$$\tan n_o x_o = -n_o/m_e \tag{3.52}$$

for the sausage waves, and

$$\tan n_o x_o = m_e/n_o \tag{3.53}$$

for the kink waves. In the above we have written $n_o^2 = -m_o^2$, so that

$$n_o^2 = (\omega^2 - k_z^2 v_A^2)/v_A^2. \tag{3.54}$$

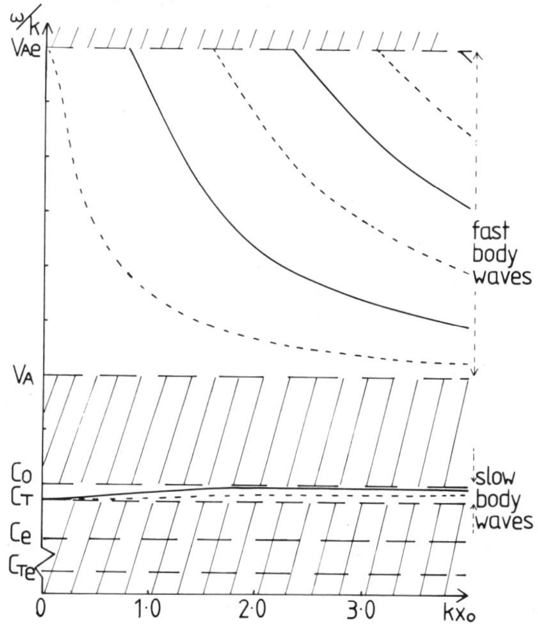

Figure 3.5 The phase-speed ω/k_z as a function of $k_z x_o$ in a coronal slab inhomogeneity with $v_{Ae} > v_A$ (after Edwin and Roberts, 1982,1983)

As in the previous discussion, we have supposed that $m_e > 0$ in deriving Eqs (3.52) and (3.53); then $v_x \to 0$ as $|x| \to \infty$. However, no assumption as to the sign of n_o^2 has yet been made. In fact, for real ω and k_z, n_o^2 must be positive. To see this, suppose $n_o^2 < 0$ and write $n_o = i\bar{n}_o$. Then $(\tan n_o x_o)/n_o$ becomes $(\tanh \bar{n}_o x_o)/\bar{n}_o$, which is positive – in contradiction to Eq (3.52). A similar argument applies to Eq (3.53). Hence, we are lead to the conclusion that sausage and kink *body* (because $m_o^2 < 0$) waves in a density inhomogeneity exist only for $v_{Ae} > v_A$, and then $k_z^2 v_A^2 < \omega^2 < k_z^2 v_{Ae}^2$. Thus, *it is regions of low Alfvén speed which permit ducted fast waves to occur.* If one tried to propagate a fast wave in a region of high Alfvén speed it would simply leak from the sides of the region, resulting in a complex ω (or k_z). No trapped wave would be possible.

Dispersion relations (3.52) and (3.53) have analogies with other fields. Eq (3.52), for the sausage modes, is equivalent to *Pekeris waves* in oceanography (Pekeris, 1948). Eq (3.53), for the kink waves, is equivalent to *Love waves* in seismology (Love, 1911).

In Figure 3.5 we have sketched the behaviour of ducted fast (body) waves, actually calculated from the more complicated dispersion relations that replace Eqs (3.52) and (3.53) when non-zero β terms are included (see Edwin and Roberts (1982) for details). We reserve our comments on the results summarized by this figure until we have discussed the cylindrical geometry below.

3.5.2 Cylindrical inhomogeneities

The cylindrical, flux tube appearance of many magnetic structures in the low-β plasmas of the magnetosphere and more especially the solar corona encourages an investigation of propagation in cylindrical geometries. The results are much as those outlined above for a slab geometry, with the occasional difference.

A convenient starting point is the general equation for mhd wave propagation in a cylindrical magnetic field $\underset{\sim}{B}_o = B_o(r)\hat{\underset{\sim}{z}}$. As with the slab geometry we suppose that both pressure $p_o(r)$ and gas density $\rho_o(r)$ are also spatially structured, here in terms of the radial coordinate r. In equilibrium, total pressure constancy (in r) holds (cf. Eq (3.2)). The general governing equation is

$$\frac{d}{dr}\left\{\frac{\rho_o(r)(k_z^2 v_A^2(r) - \omega^2)}{\left[m_o^2(r) + \frac{n^2}{r^2}\right]}\frac{1}{r}\frac{d}{dr}(r v_r)\right\}$$

$$- \rho_o(r)(k_z^2 v_A^2(r) - \omega^2)v_r = 0, \tag{3.55}$$

where we have written

$$v_r(r,\theta,z,t) \sim v_r(r)e^{i(\omega t + n\theta - k_z z)}.$$

Equation (3.55) is the twist-free *Hain-Lüst equation* (Hain and Lüst,

1958). It is the cylindrical equivalent of Eq (3.12). (The more general case of a twisted equilibrium field is of considerable interest in stability analyses.)

Associated with Eq (3.55) is the expression for the total pressure perturbation, p_T:

$$\left[m_o^2(r) + \frac{n^2}{r^2}\right]\omega p_T(r) = i\rho_o(r)(k_z^2 v_A^2(r) - \omega^2)\ \frac{1}{r}\frac{d}{dr}(rv_r). \quad (3.56)$$

The expression for m_o^2 is as before (Eq (3.45)).

The simplest case to treat is that of a zero β (cold) plasma, in which $c_s = 0$, $p_o = 0$ and B_o is everywhere uniform. For a gas density ρ_o inside a cylinder of radius $r = a$ and a density ρ_e outside, the resulting dispersion relation is (Edwin and Roberts, 1983)

$$\frac{J_n'(n_o a)}{J_n(n_o a)}\ \frac{K_n(m_e a)}{K_n'(m_e a)} = -\ \frac{n_o}{m_e}\ , \quad (3.57)$$

where J_n and K_n are Bessel functions of order n, possessing derivatives J_n' and K_n'. The expressions for m_e and n_o are as in the slab geometry (Eqs (3.51) and (3.54)). The case $n = 0$ corresponds to the sausage mode and is clearly similar to the slab relation (3.52); the case $n = 1$ corresponds to the kink mode (cf. Eq (3.53)). Equation (3.57) has analogies with guided modes in fibre optics.

The behaviour of the fast body waves described by Eq (3.57) is illustrated in Figure 3.6 where, as with slab geometry, we have included the modifications introduced by a small non-zero β. The results (Figs 3.5 and 3.6) for the two geometries should be compared. Both figures illustrate the division of the modes into two, widely separated, classes of body waves. The slow waves are essentially sound waves constrained to propagate one-dimensionally along the strong magnetic field lines. The fast waves are ducted by the inhomogeneity in the Alfvén speed. Regions of *low* Alfvén speed act as wave guides for the fast modes. The fast waves are strongly *dispersive* and possess *cut-offs*. Only the principal kink mode is able to propagate at all wavenumbers. In the thin tube case ($k_z a \ll 1$), the principal kink mode has a phase speed given by

$$\frac{\omega}{k_z} \sim \left[\frac{\rho_o v_A^2 + \rho_e v_{Ae}^2}{\rho_o + \rho_e}\right]^{1/2}, \quad (3.58)$$

which is the speed c_k of a hydromagnetic surface wave (Eq (3.36)). This is different from the slab case, where $\omega \sim k_z v_{Ae}$ for $k_z x_o \ll 1$.

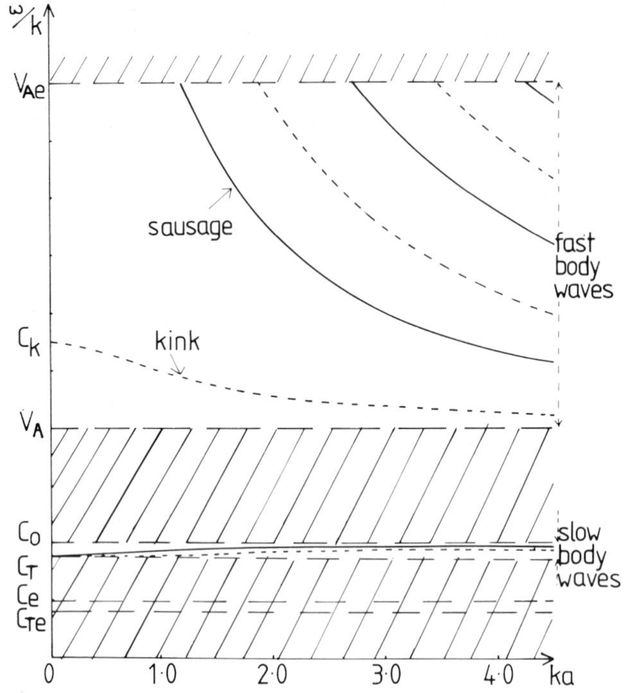

Figure 3.6 The phase-speed ω/k_z as a function of $k_z a$ in a cylinder with $v_{Ae} > v_A$ (after Edwin and Roberts, 1983)

3.5.3 Impulsively generated fast waves

In astrophysical plasmas waves are often impulsively excited. For example, in the solar corona flares (see Chapter 1) are likely to act as impulsive generators for mhd waves, just as explosions or earthquakes may impulsively excite various waves in the Earth. Since the ducted fast waves of a slab or cylinder are highly dispersive, an impulsive source may be expected to generate wave packets. What, then, is the nature of such wave packets? This question has recently been considered in the solar context by Roberts, Edwin and Benz (1983, 1984) and in the magnetospheric case by Edwin, Roberts and Hughes (1985).

The evolution of an impulsively excited wave depends critically on the form of the group velocity c_g ($= d\omega/dk_z$) of the wave. In Figure 3.7 we have sketched c_g as a function of frequency ω for the principal sausage mode in a cylinder. The sausage mode in a slab gives a similar behaviour. The important point to notice is the presence of a minimum, c_g^{min}, in the group velocity, occurring at a frequency ω^{min} slightly in excess of the mode cut-off frequency ω_c.

The frequency ω^{min} must be determined numerically from the governing dispersion relation. However, the cut-off frequency ω_c is readily determined analytically.

Inspection of Figures 3.5 and 3.6 reveals that the cut-off frequencies arise when $\omega = \omega_c = k_z v_{Ae}$. In the slab, with $\beta = 0$, Equation (3.52) gives

$$\omega_c = \frac{\pi v_{Ae}}{2x_0}\left[\frac{v_A^2}{v_{Ae}^2 - v_A^2}\right]^{\frac{1}{2}} . \tag{3.59}$$

In the cylinder, Equation (3.57) for the n = 0 (sausage) mode gives

$$\omega_c = \frac{j_0 v_{Ae}}{a}\left[\frac{v_A^2}{v_{Ae}^2 - v_A^2}\right]^{\frac{1}{2}} , \tag{3.60}$$

where j_0 ($\approx$ 2.40) is the first zero of the Bessel function J_0.

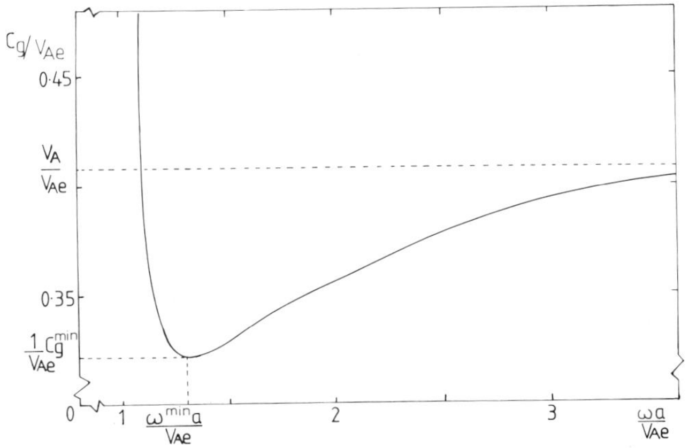

Figure 3.7 The behaviour of the group velocity $c_g = d\omega/dk_z$ as a function of frequency ω for the principal sausage mode in a cylinder with $v_{Ae} > v_A$ (after Roberts, Edwin and Benz, 1984)

It is sometimes convenient to express the mode cut-offs in terms of period $\tau_c \equiv 2\pi/\omega_c$. The cut-off period for the principal sausage mode is given as

$$\tau_c = \begin{cases} \dfrac{4x_o}{v_{Ae}}\left[\dfrac{\rho_o}{\rho_e} - 1\right]^{\frac{1}{2}}, & \text{slab,} \\ \\ \dfrac{2\pi a}{j_o v_{Ae}}\left[\dfrac{\rho_o}{\rho_e} - 1\right]^{\frac{1}{2}}, & \text{cylinder.} \end{cases} \tag{3.61}$$

Evidentally, τ_c is slightly smaller in a cylinder tube than in an equivalent slab geometry; in the extreme of a very dense inhomogeneity, $\rho_o >> \rho_e$, we see that $\tau_c \sim 4x_o/v_A$ in a slab compared with $\tau_c \sim 2.6a/v_A$ in a cylinder.

Both the cut-off frequency ω_c and the frequency ω^{min} at which the group velocity is at a minimum are important in the temporal evolution of an impulsively generated signal. Suppose, at time $t = 0$, that a ducted fast sausage wave is excited impulsively at the location $z = 0$. Then the behaviour of the disturbance at a location $z = h$ ($>> x_o, a$) may be determined by Fourier analysis, integrating the group velocity over frequency ω. Such an analysis was carried out by Pekeris (1948) in his original investigation into ducted sound waves in the ocean. In Figure 3.8 we have sketched the temporal behaviour determined by Pekeris's analysis. Strictly, this result applies to the slab geometry only, but the similarities between the slab and cylinder geometries suggest that it will also apply to a cylinder.

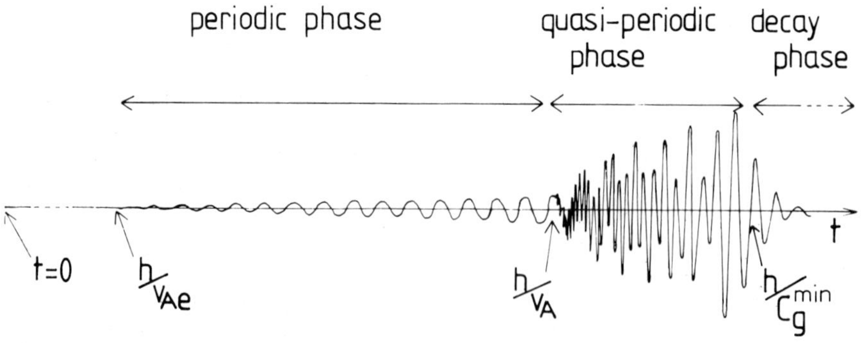

Figure 3.8 The signature at z=h of an impulsively generated (at z=0) fast mhd wave in a density duct ($v_A < v_{Ae}$); after Pekeris (1948) and Roberts, Edwin and Benz (1983, 1984)

There are three phases in the temporal behaviour. Firstly, a detector located at $z = h$ would register a small amplitude almost periodic oscillation. This begins at a time $t = h/v_{Ae}$ and continues until $t = h/v_A$. This is the *periodic phase*. The period of oscillations in this phase is given by τ_c (eq (3.61)). Suddenly, at time $t = h/v_A$, the *quasi-periodic phase* begins; it persists until a time $t = h/c_g^{min}$. Amplitudes are much larger during the quasi-periodic phase; the typical frequency of oscillation during this phase is ω^{min}. Finally, at time $t = h/c_g^{min}$ the *decay (or Airy) phase* begins, during which amplitudes decline exponentially fast (though continuing to oscillate with frequency ω^{min}).

A detailed analysis of the behaviour of c_g^{min}, ω^{min}, etc with the various parameters (e.g. ρ_o, ρ_e) in a coronal or magnetospheric plasma would be inappropriate here (see Roberts *et al.* (1984), Edwin *et al.* (1985)). Suffice it to say that the period τ_c, given in Eq (3.61), provides a useful guide to the expected periodicities. The period τ_c is, in fact, the longest timescale in the impulsively generated signal (Fig 8). Timescales in the quasi-periodic phase may be substantially smaller.

We conclude this section by giving an estimate of τ_c. For the corona we choose $v_{Ae} = 10^3 km\ s^{-1}$ and consider a cylindrical tube of radius $a = 10^3 km$. Then Eq (3.61) gives $\tau_c = 2.6s$ for an internal density of twice that in the environment (i.e. $\rho_o = 2\rho_e$), and $\tau_c = 7.8s$ for a density $\rho_o = 10\rho_e$. Interestingly enough timescales of this order have been associated with pulsations in radio wavelengths (e.g. Rosenberg, 1970).

3.6 DAMPED ALFVÉN WAVES

So far in our discussion we have ignored the effects of wave damping, as introduced by viscosity, finite resistivity, thermal conductivity, etc. Such effects are likely to be important in many circumstances. In particular, the heating of the solar coronal gas (at temperatures $\sim 10^6$K compared with 6000K in the photospheric layers) remains an unresolved problem, now that it is clear that sound waves do not propagate (with sufficient energy flux) to the higher layers of the Sun's atmosphere. Interest is therefore centred on, among other possible mechanisms, the damping of mhd waves in an inhomogeneous medium. A full discussion of this problem is not presently available. However, by confining attention to the Alfvén mode some progress may be made.

Consider, then, the propagation of an Alfvén wave in an inhomogeneous magnetic field; the equilibrium state is as before (Eq (3.1)). Following Heyvaerts and Priest (1983), we write

$$\left.\begin{aligned} \underset{\sim}{v} &= v(x,z,t)\hat{\underset{\sim}{y}}, \\ \underset{\sim}{B} &= B_o(x)\hat{\underset{\sim}{z}} + B(x,z,t)\hat{\underset{\sim}{y}}, \end{aligned}\right\} \qquad (3.62)$$

so that perturbations are assumed to be independent of y. The governing linearized equations are the momentum equation (1.2), with the inclusion of viscosity, and the induction equation (1.1) for finite conductivity, viz.

$$\rho_0(x) \frac{\partial v}{\partial t} = \frac{B_0(x)}{\mu_0} \frac{\partial B}{\partial z} + \rho_0(x)\nu \left[\frac{\partial^2 v}{\partial x^2} + \frac{\partial^2 v}{\partial z^2}\right] \tag{3.63}$$

and

$$\frac{\partial B}{\partial t} = B_0(x) \frac{\partial v}{\partial z} + \eta\left[\frac{\partial^2}{\partial x^2} + \frac{\partial^2}{\partial z^2}\right]B, \tag{3.64}$$

where ν is the coefficient of kinematic viscosity and η is the resistivity.

If either ν or η is zero (or simply small enough for their products to be neglected), then Eqs (3.63) and (3.64) may be combined to yield

$$\frac{\partial^2 v}{\partial t^2} = v_A(x) \frac{\partial^2 v}{\partial z^2} + \bar{\nu} \left[\frac{\partial^2}{\partial x^2} + \frac{\partial^2}{\partial z^2}\right]\left[\frac{\partial v}{\partial t}\right], \tag{3.65}$$

where $\bar{\nu} = \nu + \eta$.

In the absence of dissipation ($\bar{\nu} = 0$), Equation (3.65) reduces to

$$\frac{\partial^2 v}{\partial t^2} = v_A^2(x) \frac{\partial^2 v}{\partial z^2} . \tag{3.66}$$

This is simply the wave equation and shows that Alfvén waves may propagate along each field line with the local Alfvén speed of that field line. The general solution of Eq (3.66) is evidently

$$v(x,z,t) = f_-(z - v_A(x)t,x) + f_+(z + v_A(x)t,x), \tag{3.67}$$

where $f_\pm$ are arbitrary functions of two variables.

It is of interest to consider Eq (3.66) from the stand-point of Fourier analysis, for it is in such a form that the equation (or an allied form) frequently arises (cf. Eq (3.12)). In terms of Fourier components (see Eq (3.8)), Equation (3.66) yields

$$(\omega^2 - k_z^2 v_A^2(x))v(x) = 0. \tag{3.68}$$

Aside from the trivial solution $v = 0$, Equation (3.68) must be solved in terms of generalized functions:

$$v = \delta(\omega - k_z v_A(x)), \quad \delta(\omega + k_z v_A(x)), \tag{3.69}$$

where δ is the Dirac delta function. (We should not object to solutions in terms of generalized functions, since v in Eqs (3.68),

(3.69) refers to the Fourier transformed variable and not to the physical variable.) In physical variables, Eq (3.69) implies the D'Alembert solutions in Eq (3.67).

Returning now to the dissipative case, Eq (3.65), we note that the dominant contribution to dissipation is expected to arise from the x-inhomogeneities, for which $|\partial^2 v/\partial x^2| \gg |\partial^2 v/\partial z^2|$. In such circumstances, Eq (3.65) reduces to

$$\frac{\partial^2 v}{\partial t^2} = v_A^2(x)\frac{\partial^2 v}{\partial z^2} + \bar{\nu}\frac{\partial^3 v}{\partial x^2 \partial t}. \tag{3.70}$$

Ignoring, for the moment, variations in $v_A(x)$ we see that Eq (3.70) generates the dispersion relation

$$\omega^2 = k_z^2 v_A^2 + i\omega k_x^2 \bar{\nu}. \tag{3.71}$$

In the limit of *weak* damping ($|\omega\bar{\nu}k_x^2| \ll |k_z^2 v_A^2|$, $|\omega^2|$), Eq (3.71) may be solved for the imaginary part of k_z (regarding ω as given) to yield a damping length

$$L = 2v_A/(k_x^2\bar{\nu}). \tag{3.72}$$

The amplitude of the wave decays by a factor of e in a distance $z = L$. Under coronal conditions the damping length L is generally very large, implying that Alfvén waves damp very little in their propagation through the lower part of the corona (Osterbrock, 1961).

However, it is clear from Eq (3.68) that L may be substantially reduced if k_x is sufficiently large, corresponding to small horizontal wavelengths. Such small horizontal wavelengths are likely to arise in regions of strong inhomogeneities in $v_A(x)$. Thus, in a *non-uniform* medium the damping of Alfvén waves is expected to be strongly enhanced and becomes a candidate for coronal heating (Heyvaerts and Priest, 1983). An extensive discussion of the solutions of Eq (3.66) would take us too far afield and so we must be content with referring the reader to the original literature (Heyvaerts and Priest, 1983; Nocera, Leroy and Priest, 1984).

3.7 WAVES IN STRATIFIED ATMOSPHERES

Up to this point in our discussion we have ignored the effect of stratification, brought about by the presence of gravity. In fact, in many astrophysical circumstances the effects of gravity are important and so it is the combined influences of both magnetic structuring and gravitational stratification that we must consider. We can obtain a glimpse of the expected behaviour of mhd waves in a structured and stratified atmosphere by considering the general features that these two aspects introduce.

The presence of a magnetic field implies the existence of a force in the atmosphere, the $\underset{\sim}{j} \times \underset{\sim}{B}$ force, and so we may expect that force to drive waves distinct from those waves that arise in a non-magnetic atmosphere. Thus the occurrence of Alfvén and magnetoacoustic waves, superceding sound waves. A magnetic field introduces a *preferred direction* and so we may expect wave propagation to reflect that preference for one direction over another, and so be *anisotropic*, as indeed is the case. Alfvén and slow waves are highly anisotropic, fast waves only mildly so. Inhomogeneities in the magnetic field allow additional features to arise, such as the surface waves and ducted, highly dispersive, fast waves.

In a similar fashion we may anticipate some of the effects of gravity, ignoring magnetic fields. Gravity too introduces both a force (buoyancy) and a preferred direction. Thus, in addition to sound waves we may expect a *gravity* mode and also anisotropic propagation; gravity waves are unable to propagate vertically. Gravity stratifies an atmosphere and so a wave propagating vertically must gain in amplitude in order to conserve energy, given that it is propagating into a more tenuous medium. This increase in amplitude may lead to the formation of shocks. Finally, stratification, like magnetic structuring, introduces frequency cut- off and dispersion.

Of general interest is the question of how these two features, stratification and magnetic structuring, couple together to modify wave propagation. Unfortunately, our knowledge of this coupling is far from complete and only a few special cases, of particular geometries or physicaly conditions have been adequately investigated. We outline below some of these cases. But to begin it is necessary to examine the non-magnetic case first.

3.7.1 Sound waves

In the absence of a magnetic field a gas under gravity is hydrostatically stratified:

$$p_0'(z) = -\rho_0(z)g, \tag{3.73}$$

where gravity acts in the negative z-direction. A dash denotes differentiation with respect to z. Combined with the gas law, this equation may be "integrated" to yield (see Chapter 1)

$$p_0(z) = p_0(0) \exp\left[-\int_0^z \frac{dz}{\Lambda_0}\right], \tag{3.74}$$

where $\Lambda_0(z) = k_B T_0(z)/mg$ is the pressure scale-height in an atmosphere with temperature $T_0(z)$. The point $z = 0$ is an arbitrary reference level. In terms of the sound speed $c_s(z) = (\gamma p_0(z)/\rho_0(z))^{1/2}$, Eq (3.73) may be written as

$$(c_s^2)' = -\gamma g - c_s^2(\rho_0'/\rho_0). \tag{3.75}$$

Perturbations about the equilibrium (3.73) are readily investigated in the usual way. Equations (3.3) apply with the term ρg

added to the right-hand side of the momentum equation, and $\underset{\sim}{B} \equiv 0$. Following Lamb (1932), it is convenient to introduce the variable

$$\Delta = \text{div } \underset{\sim}{v}, \tag{3.76}$$

and Fourier analyse thus

$$\underset{\sim}{v} = (v_x(z), 0, v_z(z)) \exp i(\omega t - k_x x). \tag{3.77}$$

After some algebra we obtain the governing ordinary differential equation (Lamb, 1932, p 551)

$$\frac{d^2\Delta}{dz^2} + \left[\frac{c_s^{2\prime}}{c_s^2} - \frac{\gamma g}{c_s^2}\right]\frac{d\Delta}{dz} + \left\{\frac{\omega^2 - k_x^2 c_s^2}{c_s^2} - \frac{g k_x^2}{\omega^2}\left[\frac{g}{c_s^2} + \frac{\rho_0'}{\rho_0}\right]\right\}\Delta = 0. \tag{3.78}$$

By writing

$$Q = \rho_0^{\frac{1}{2}} c_s^2 \Delta \tag{3.79}$$

we may reduce Eq (3.78) to the canonical form (e.g Deubner and Gough, 1984)

$$\frac{d^2 Q}{dz^2} + \kappa^2(z) Q = 0, \tag{3.80}$$

where

$$\kappa^2(z) = \frac{\omega^2 - \omega_a^2}{c_s^2} + k_x^2\left[\frac{\omega_g^2}{\omega^2} - 1\right] \tag{3.81}$$

and

$$\omega_a^2 = \frac{c_s^2}{4H^2}(1 - 2H'), \quad \omega_g^2 = g\left[\frac{1}{H} - \frac{g}{c_s^2}\right], \quad H = -\frac{\rho_0}{\rho_0'} . \tag{3.82}$$

In the above, H is the density scale-height, ω_g^2 is the square of the Brunt-Väisälä(buoyancy) frequency, and ω_a^2 is the square of the (generalized) acoustic frequency.

It is sometimes convenient to work in terms of the velocity component v_z. The governing equation for v_z is

$$\frac{d}{dz}\left[\frac{\rho_0 c_s^2}{\omega^2 - k_x^2 c_s^2}\frac{dv_z}{dz}\right]$$

$$+ \left\{\rho_0 + \frac{g^2 k_x^2 \rho_0}{\omega^2(k_x^2 c_s^2 - \omega^2)} + \frac{k_x^2 g}{\omega^2}\left[\frac{\rho_0 c_s^2}{k_x^2 c_s^2 - \omega^2}\right]'\right\} v_z = 0. \tag{3.83}$$

Equations (3.78), (3.80) and (3.83) possess a complex structure which is not unexpected in view of the fact that the medium is compressible and stratified both in temperature (i.e. in c_s^2) and in density. In an *isothermal* atmosphere (c_s^2 = constant), the frequencies ω_a and ω_g are constants and so $\kappa^2(z)$ becomes a constant. Consequently, Eq (3.80) yields a simple dispersion relation (for $Q \sim e^{-ik_z z}$), namely

$$k_z^2 = \frac{\omega^2 - \omega_a^2}{c_s^2} + k_x^2 \left[\frac{\omega_g^2}{\omega^2} - 1\right], \tag{3.84}$$

where now the frequencies ω_a, ω_g and the density scale-height H take on their isothermal values:

$$\omega_a = \frac{c_s}{2\Lambda_o} = \frac{\gamma g}{2c_s}, \quad \omega_g = (\gamma - 1)^{\frac{1}{2}} \frac{g}{c_s}, \quad H = \Lambda_o. \tag{3.85}$$

The dispersion relation (3.84) for propagation in an isothermal atmosphere may be rearranged into the form

$$\omega^4 - (\omega_a^2 + k^2 c_s^2)\omega^2 + c_s^2 \omega_g^2 k^2 \sin^2\theta = 0, \tag{3.86}$$

where θ is the angle that the wavevector $\underline{k} = (k_x, 0, k_z)$ makes with the z-axis, and $k^2 = k_x^2 + k_z^2$. Equation (3.86) makes it clear that there are two modes, an *acoustic* wave and a *gravity* (or internal) wave, the propagation characteristics of which depend upon the parameters θ and $k\Lambda_o$.

If $k\Lambda_o \gg 1$, so that waves are much shorter than the pressure scale-height, then Eq (3.86) yields

$$\omega^2 \sim k^2 c_s^2, \quad \text{and} \quad \omega^2 \sim \omega_g^2 \sin^2\theta, \tag{3.87}$$

showing that the sound waves and the gravity waves are almost completely decoupled from one another: each mode propagates almost independently of the presence of the other. In the opposite extreme of very long wavelengths, $k\Lambda_o \ll 1$, Eq (3.86) yields

$$\omega^2 \sim \frac{c_s^2}{4\Lambda_o^2}, \quad \text{and} \quad \omega^2 \sim 4\left[\frac{\gamma - 1}{\gamma^2}\right] k^2 c_s^2 \sin^2\theta. \tag{3.88}$$

Returning to Eq (3.84), we may see that not all frequencies may propagate. Non-propagation (or evanescence) in the z-direction arises whenever $k_z^2 < 0$; regions of $k_z^2 > 0$ correspond to propagation. In Figure 3.9 we have sketched the regions in the ω-k_x plane that correspond to $k_z^2 > 0$ or $k_z^2 < 0$, found from Eq (3.84) by setting $k_z^2 = 0$. This is the so-called *diagnostic diagram*. It has been used extensively in solar studies (e.g. Stein and Leibacher, 1974; Thomas, 1983).

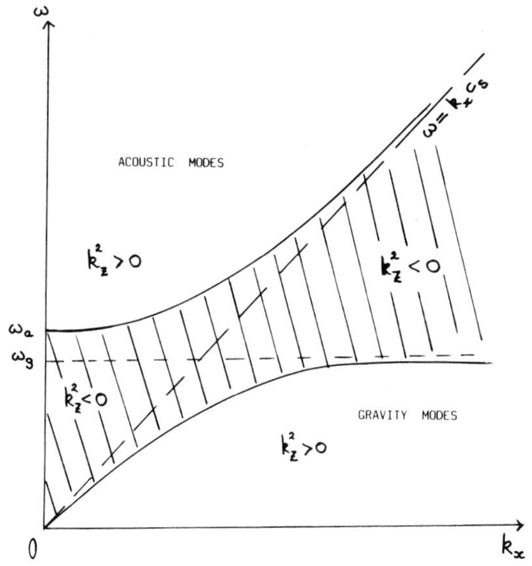

Figure 3.9 The diagnostic diagram for acoustic-gravity waves, dividing the ω-k_z plane into regions of vertical propagation ($k_z^2 > 0$) and evanescence ($k_z^2 < 0$)

It is evident from Eq (3.86) and Figure 3.9 that gravity waves are unable to propagate vertically (i.e. as $\theta \to 0$, $\omega^2 \to 0$ for the gravity mode). Indeed, for *vertical propagation* ($\theta = 0$), sound waves in an isothermal atmosphere satisfy the dispersion relation

$$\omega^2 = k_z^2 c_s^2 + \omega_a^2. \tag{3.89}$$

The partial differential equation equivalent to this dispersion relation is

$$\frac{\partial^2 Q}{\partial t^2} - c_s^2 \frac{\partial^2 Q}{\partial z^2} + \omega_a^2 Q = 0. \tag{3.90}$$

This is the *Klein-Gordon* equation (Morse and Feshbach, 1953); it also arises in describing flux tube waves (see Section 3.8).

Finally, we consider briefly the case of a *non-isothermal* atmosphere. The governing equations are Eqns (3.78) and (3.83). The special case of a *polytropic* atmosphere, in which $c_s^2(z) = -\alpha z$, $z < 0$, was investigated by Lamb (1932), who showed that Eq (3.78) has solution

$$\Delta = e^{k_x z} M(\lambda, m+2, -2k_x z), \tag{3.91}$$

where M is a confluent hypergeometric function and m is the polytropic index, related to the gradient α by

$$m = \frac{\gamma g}{\alpha} - 1. \tag{3.92}$$

The parameter λ is given by

$$2\lambda = m+2 - \frac{\omega^2}{\gamma g k_x}(m+1) - \frac{k_x g}{\omega^2}\left[\frac{(\gamma-1)g}{\alpha} - 1\right]. \tag{3.93}$$

The solution (3.91) satisfies the condition that v_z and Δ are finite at the top of the atmosphere ($z = 0$). However, to comply with the requirement that v_z is bounded as $z \to -\infty$ we require λ to be a negative integer, thus reducing M to a (Laguerre) polynomial. This provides us with the governing dispersion relation determining $\omega = \omega_n$ as a function of $k_x > 0$ and integer n (Spiegel and Unno, 1962; Christensen-Dalsgaard, 1980; Deubner and Gough, 1984):

$$(m+1)\left[\frac{\omega^2}{gk_x}\right]^2 - \gamma(m+2n)\left[\frac{\omega^2}{gk_x}\right] + [(\gamma-1)m - 1] = 0, \tag{3.94}$$

for $n = 1,2,3,\ldots$. The dispersion relation is quadratic in ω^2, corresponding to the two modes (the acoustic and gravity waves). Inspection of this relation reveals that $\omega^2 \sim k_x$. More specifically, for large n we find that

$$\omega^2 \sim \begin{cases} 2\gamma g k_x\,(n + \tfrac{1}{2}m)/(m+1), & \text{acoustic waves,} \\ \gamma g k_x\left[m - \dfrac{m+1}{\gamma}\right]/(n + \tfrac{1}{2}m), & \text{gravity waves.} \end{cases} \tag{3.95}$$

The acoustic waves form a Sturmian sequence accumulating at infinity ($\omega_1^2 < \omega_2^2 < \omega_3^2 < \ldots < \omega_n^2$, $\omega_n^2 \to \infty$ as $n \to \infty$), the gravity waves an anti-Sturmian sequence accumulating at the origin ($\omega_1^2 > \omega_2^2 > \omega_3^2 > \ldots > \omega_n^2$, $\omega_n^2 \to 0$ as $n \to \infty$).

Remarkably, the parabolic dependence of ω on k_x, illustrated in Eq (3.95), is in agreement with solar observations (e.g. Deubner, 1975), revealing that the solar interior traps sound waves of 5-minute period. Such an agreement between theory and observation has promoted the rapid development of a new subject - helioseismology - by which a study of the Sun's oscillations may provide a diagnostic tool for probing physical conditions within the solar interior (see the review by Deubner and Gough (1984)).

3.7.2 The influence of a horizontal magnetic field

In many astrophysical circumstances sound wave propagation is influenced by a horizontal magnetic field. The penumbral structure of a sunspot provides a possible illustration. In the presence of a horizontal magnetic field $\underset{\sim}{B}_0 = B_0(z)\hat{\underset{\sim}{x}}$ the equilibrium is one of magnetohydrostatic balance:

$$\frac{d}{dz}\left[p_o + \frac{B_o^2}{2\mu}\right] = -\rho_o(z)g. \tag{3.96}$$

Linear perturbations are then described by the equation

$$\frac{d}{dz}\left\{\frac{\rho_o(c_s^2 + v_A^2)(\omega^2 - k_x^2c_T^2)}{(\omega^2 - k_x^2c_s^2)}\frac{dv_z}{dz}\right\}$$

$$+\left\{\rho_o(\omega^2 - k_x^2v_A^2) + \frac{g^2k_x^2\rho_o}{(k_x^2c_s^2 - \omega^2)} + gk_x^2\left[\frac{\rho_o\ c_s^2}{k_x^2c_s^2 - \omega^2}\right]'\right\}v_z = 0. \tag{3.97}$$

This rather complicated equation generalizes Eq (3.83), to which it reduces when $v_A = 0$, and Eq (3.42), to which it reduces when $g = 0$. Its mathematical structure is evidently akin to that of Eq (3.42). However, elucidating the general properties of Eq (3.97) is presently not possible so we will content ourselves with examining the special case of an incompressible fluid. Compressible studies (e.g. Nye and Thomas, 1976; Small and Roberts, 1984) have been carried out, though, and the results applied to the problem of explaining the outward propagation of waves observed in the penumbral magnetic fields of sunspots - the so called *running penumbral waves* (see the observational review by Moore (1981)).

The *incompressible* ($c_s \to \infty$) limit of Eq (3.97) yields

$$\frac{d}{dz}\left\{\rho_o(\omega^2 - k_x^2v_A^2)\frac{dv_z}{dz}\right\} - \left\{k_x^2\rho_o(\omega^2 - k_x^2v_A^2) + k_x^2g\rho_o'\right\}v_z = 0. \tag{3.98}$$

To illustrate the solution of this equation consider the *Rayleigh-Taylor* instability of a dense fluid superposed on a lighter fluid (see also Chapter 4). For simplicity we take the horizontal field to be uniform, so $B_o(z) = B_o$, a constant. The density $\rho_o(x)$ is taken to be

$$\rho_o(z) = \begin{cases}\rho_e, & z > 0,\\ \rho_o, & z < 0,\end{cases} \tag{3.99}$$

for constants ρ_o and ρ_e.

For such an equilibrium the coefficients of Eq (3.98) are constants in the regions $z > 0$ and $z < 0$. Thus, aside from an Alfvén wave ($\omega^2 = k_x^2v_A^2$), we must consider the equation

$$\frac{d^2 v_z}{dz^2} - k_x^2 v_z = 0, \tag{3.100}$$

applying in both z < 0 and z > 0. Hence

$$v_z(x) = \begin{cases} \alpha_e e^{-|k_x|z}, & z > 0, \\ \alpha_o e^{|k_x|z}, & z < 0, \end{cases} \tag{3.101}$$

satisfying the condition that v_z be bounded at $\pm\infty$.

The solutions (3.101) are to be matched across the interface z = 0. We require that the normal component of velocity be continuous across z = 0, i.e. $\alpha_o = \alpha_e$. Also, integrating Eq (3.98) across a layer near z = 0 in which $\rho_o(x)$ is considered to change very rapidly yields the requirement that

$$\rho_o(z)(\omega^2 - k_x^2 v_A^2)\frac{dv_z}{dz} - k_x^2 g\rho_o(z)v_z$$

be continuous across z = 0. This requirement implies the dispersion relation

$$\omega^2 = k_x^2 c_k^2 - g|k_x|\left[\frac{\rho_e - \rho_o}{\rho_e + \rho_o}\right], \tag{3.102}$$

where c_k is the mean Alfvén speed for the two layers (see Eq (3.36)).

Thus we see the appearance of the surface wave speed c_k and the destabilising effect of gravity if $\rho_e > \rho_o$.

3.8 SLENDER FLUX TUBES

We end this chapter with a discussion of wave propagation in isolated magnetic flux tubes of the type generally supposed to be presented in the photospheric layers (and below) of the solar atmosphere. The observational evidence indicates field strengths of about 1.5 kG, which corresponds to a magnetic pressure ($B^2/2\mu \rightarrow B^2/8\pi$ in cgs units) of about 10^5 dynes cm^{-2}. This is comparable with the gas pressure at the Sun's surface and so the plasma β within an intense flux tube is close to unity. Also stratification effects are strong in the photospheric layers, where the pressure scale-height is about 150 km. The inferred diameter (of 100–300 km) of intense tubes is close to the pressure scale-height, though below the photospheric surface the tubes are expected to narrow whereas the pressure scale-height increases. Intense flux tubes, then, provide a good example of a strongly stratified plasma with β of order unity. In other words, both stratification and magnetic structuring are important in the equilibrium and dynamics of photospheric flux tubes.

The propagation of waves in such an atmosphere is evidently a complicated problem. Progress has come about by assuming that the tubes are *slender* (or thin). This amounts to supposing that wavelengths are much greater than tube diameters or pressure scale-heights. In the absence of gravity, this is simply that $k_z x_0$ (or $k_z a$) is small. We consider the sausage and kink modes separately.

3.8.1 The slender flux tube equations: sausage modes

We examine vertical motions in a slender flux tube that is slowly diverging with height z and has a radius much smaller than the local pressure scale-height. The governing equations are those of continuity, vertical momentum, transverse momentum, and isentropic energy:

$$\frac{\partial}{\partial t}\rho A + \frac{\partial}{\partial z}\rho v A = 0, \tag{3.103}$$

$$\rho\left[\frac{\partial v}{\partial t} + v\frac{\partial v}{\partial z}\right] = -\frac{\partial p}{\partial z} - \rho g, \tag{3.104}$$

$$p + \frac{B^2}{2\mu} = p_e, \tag{3.105}$$

$$\frac{\partial p}{\partial t} + v\frac{\partial p}{\partial z} = \frac{\gamma p}{\rho}\left[\frac{\partial \rho}{\partial t} + v\frac{\partial \rho}{\partial z}\right], \tag{3.106}$$

$$BA = \text{constant}. \tag{3.107}$$

Equations (3.103)–(3.107) govern the nonlinear behaviour of longitudinal, isentropic motions $v(z,t)$ of a gas of density $\rho(z,t)$ and pressure $p(z,t)$ confined within an elastic tube of cross-sectional area $A(z,t)$. Equations (3.103), (3.104) and (3.106) apply to *any* elastic tube in which motions are predominantly along the tube (cf. Lighthill, 1978). The specification that the elastic tube is in fact magnetic comes about through Eq (3.105), which relates gas pressure variations to magnetic pressure variations; it is required that the total pressure is maintained equal to the confining gas pressure $p_e(z,t)$, which must either be specified or calculated from consideration of the external medium. Equations (3.103)–(3.107) are commonly referred to as the *slender (or thin) flux tube equations* (for the sausage mode). They have been derived from the full mhd equations by an expansion about the axis of the tube (Roberts and Webb, 1978), somewhat akin to shallow water-wave theory (see, for example, Whitham 1974). The incompressible case of the slender flux tube equations was first discussed by Parker (1974). Defouw (1976) was first to examine the case of an isothermal gas, showing the existence of propagation cut-off. Non-isothermal effects were considered by Roberts and Webb (1978), Webb and Roberts (1978), Spruit and Zweibel (1979), and Unno and Ando (1979). A careful examination of the approximation in the unstratified ($g = 0$) case is given in Roberts and Webb (1979). A general discussion may be found in Spruit (1981a).

The *equilibrium* configuration of the tube is given by setting $v = 0$ and $\partial/\partial t = 0$ in the above. Supposing that the external medium is in hydrostatic equilibrium, with the same temperature and scale-height inside and outside the tube, Eqs (3.103)-(3.107) yield

$$p_0(z) = p_0(0)e^{-n(z)}, \quad \rho_0(z) = \rho_0(0)\,\frac{A_0(0)}{A_0(z)}\,e^{-n(z)},$$
$$A_0(z) = A_0(0)e^{\frac{1}{2}n(z)}, \tag{3.108}$$

where $n(z) = \int_0^z dz/\Lambda_0(z)$, and $\Lambda_0(z)$ is the pressure scale-height. It follows from Eq (3.108) that the radius $r_0(z)$ of the diverging flux tube (see Figure 3.10) expands like $p_e^{-\frac{1}{4}}(z)$; in an isothermal atmosphere this gives $r_0(z) \sim \exp(z/4\Lambda_0)$, showing that the tube's radius e-folds once in four scale-heights (about 500-600 km in the upper photosphere).

Linear perturbations about the equilibrium (3.108) may be investigated in the usual way. To begin with, we consider the linearized forms of the equations of continuity (3.103), longitudinal momentum (3.104) and isentropic flow (3.106). These may be combined to yield (Rae and Roberts, 1982)

$$\frac{\partial^2 Q}{\partial t^2} - c^2(z)\,\frac{\partial^2 Q}{\partial z^2} + \omega_v^2(z)Q = 0, \tag{3.109}$$

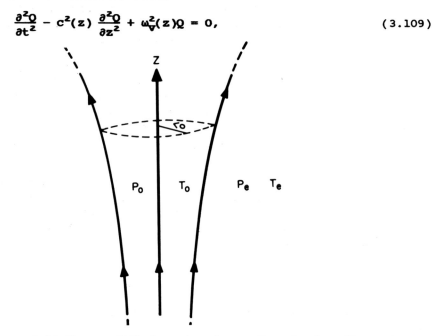

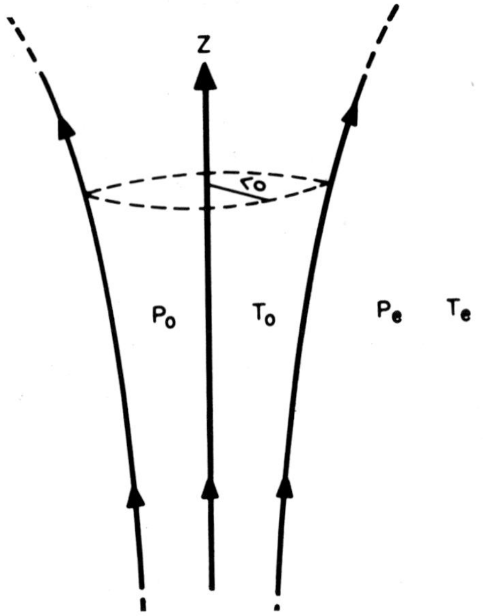

Figure 3.10 The equilibrium structure of a slender flux tube

where the speed c(z) is defined by

$$\frac{1}{c^2} = \frac{1}{c_s^2(z)} + \frac{\rho_0(z)}{A_0(z)}\left[\frac{\partial A}{\partial p}\right]_{p=0} \tag{3.110}$$

and the frequency $\omega_v(z)$ by

$$\omega_v^2 = \omega_g^2 + c^2\left\{\frac{1}{2}\left[\frac{\rho_0'}{\rho_0} + \frac{A_0'}{A_0} + \frac{c^{2\prime}}{c^2}\right]' + \left[\frac{q}{c_s^2} - \frac{A_0'}{A_0}\right]'\right.$$

$$\left. + \frac{1}{4}\left[\frac{\rho_0'}{\rho_0} + \frac{A_0'}{A_0} + \frac{c^{2\prime}}{c^2}\right]^2 + \left[\frac{q}{c_s^2} - \frac{A_0'}{A_0}\right]\left[\frac{\rho_0'}{\rho_0} + \frac{c^{2\prime}}{c^2} + \frac{q}{c_s^2}\right]\right\}. \tag{3.111}$$

Q(z,t) is related to the velocity v(z,t) through

$$Q(z,t) = \left[\frac{\rho_0(z)A_0(z)c^2(z)}{\rho_0(0)A_o(0)c^2(0)}\right]^{1/2} v(z,t). \tag{3.112}$$

So far we have made no use of Eqs (3.105) and (3.107). Thus Eq (3.109) applies to any longitudinal linear motion in an elastic tube. Notice that Eq (3.109) is of the Klein–Gordon type.

To illustrate the nature of our governing equation we consider a number of special cases. We suppose that the atmosphere is *isothermal*. As a first example, consider a *straight* ($A_0(z) = A_0(0)$) and *rigid* ($\partial A/\partial p = 0$) tube. We find, from Eq (3.110), that $c = c_s$, the sound speed, and that $\omega_v = c_s/2\Lambda_0 \equiv \omega_a$, the acoustic cut-off frequency. We have therefore recovered Eq (3.92), the equation for vertical propagation of acoustic-gravity waves. As a second example, suppose again that the tube is *rigid* but now of varying cross-sectional area $A_o(z) = A_o(0)\exp(\alpha z/\Lambda_o)$, for expansion parameter α and (isothermal) scale-height Λ_0. Then $c = c_s$, as before, but now

$$\omega_v^2 = \left[\alpha^2 - \left[\frac{4}{\gamma} - 2\right]\alpha + 1\right]\omega_a^2. \tag{3.113}$$

The case $\alpha = 0$ recovers the straight tube example. Equation (3.113) shows that ω_v has a minimum value when $\alpha = 2/\gamma - 1$; for $\gamma = 5/3$, this is $\alpha = 1/5$. The value $\alpha = 1/2$ is of special interest since it corresponds to the equilibrium profile (3.108) of a magnetic tube. With $\alpha = 1/2$, we obtain

$$\omega_v^2 = \left[\frac{9}{4} - \frac{2}{\gamma}\right]\omega_a^2. \tag{3.114}$$

The significance of the speed c and frequency ω_v will be made clear shortly.

Consider, finally, the case of a *magnetic flux tube*. To do this we bring in the linearized forms of Eqs (3.105) and (3.107), supposing that p_e is simply the undisturbed external gas pressure. The atmosphere is again supposed isothermal. Application of Eqs (3.110)–(3.112) reveals that

$$c = c_T, \qquad v = e^{z/4\Lambda_0}Q, \tag{3.115}$$

and

$$\omega_v^2 = \left[\frac{9}{4} - \frac{2}{\gamma}\right]\omega_a^2 - \left[\frac{3}{2} - \frac{2}{\gamma}\right]\left[\frac{\beta}{\beta + \frac{2}{\gamma}}\right]\omega_a^2. \tag{3.116}$$

Thus the speed c is simply the cusp speed c_T, which is both sub-sonic and sub-Alfvénic, and the velocity v e-folds in four scale-heights (for oscillatory Q). The squared tube frequency, ω_v^2, is evidently made up of two parts: the first part depends purely on the equilibrium *geometry* of the tube (cf. Eq (3.114)), and the second part on the tube's *elasticity* (as measured by the plasma $\beta \equiv 2\mu p_0(z)/B_0^2(z)$, which is a constant).

3.8.2 Pulse propagation

To bring out the significance of the speed c and the frequency ω_v, notice that in each of the above illustrations (rigid and elastic tubes) the governing equation is the Klein-Gordon one (Eq (3.109)) with constant coefficients. It therefore has the associated dispersion relation

$$\omega^2 = k_z^2c^2 + \omega_v^2 \tag{3.117}$$

for Fourier dependence of the form

$$Q(z) = Q_0 \exp i(\omega t - k_z z).$$

The dispersion relation (3.117) makes clear that only those frequencies ω above the cut-off ω_v are able to propagate; motions with $\omega < \omega_v$ are *evanescent* (non-propagating). In the case of vertically propagating sound waves, for which $\omega_v = \omega_a$, the acoustic cut-off ω_a is about 0.03 s^{-1} in the solar photosphere, corresponding to a period of around 200s. This may be contrasted with its value of about 0.015 s^{-1} (corresponding to about 7 minutes) for air on a warm summer's day; see Lighthill (1978).

What effect does propagation cut-off have on an impulsively generated wave? To answer this, we may write the general solution of our Klein-Gordon equation as (see, for example, Whitham (1974))

$$Q(z,t) = \int_{-\infty}^{\infty} f_+(k_z)e^{i(k_z z+\omega t)}dk_z + \int_{-\infty}^{\infty} f_-(k_z)e^{i(k_z z-\omega t)}dk_z, \tag{3.118}$$

where $\omega = \omega(k_z)$ satisfies Eq (3.117). Suppose that $Q(z,t=0) = 0$ and $\partial Q/\partial t\ (z,t=0) \propto \delta(z/\Lambda_0)$, representing an initial pressure excitation concentrated at $z = 0$. These initial conditions permit us to determine the arbitrary functions $f_\pm(k_z)$, giving

$$Q(z,t) \propto \int_0^\infty \frac{1}{\omega(k_z)} \cos k_z z \sin \omega(k_z) t \, dk_z. \tag{3.119}$$

The above integral may be evaluated to give

$$Q(z,t) \propto \begin{cases} J_0\left[\frac{\omega_v}{c}(c^2t^2 - z^2)^{\frac{1}{2}}\right], & 0 < z < ct, \\ \\ 0, & z > ct. \end{cases} \tag{3.120}$$

Relation (3.120) shows the propagation of a wavefront with speed c (= c_s in a rigid tube, c_T in a magnetic tube) behind which is trailed a *wake* oscillating with frequency ω_v (Rae and Roberts, 1982). This behaviour is illustrated in Figure 3.11. An interesting question is what happens to such a wavefront and its wake when nonlinearities are allowed for. Hollweg (1982) has investigated this question numerically and, indeed, argued that such motions may well explain the curious jet-like dynamical phenomena called *spicules* that are observed to rise with speeds of 20–30 km s^{-1} through the solar chromosphere.

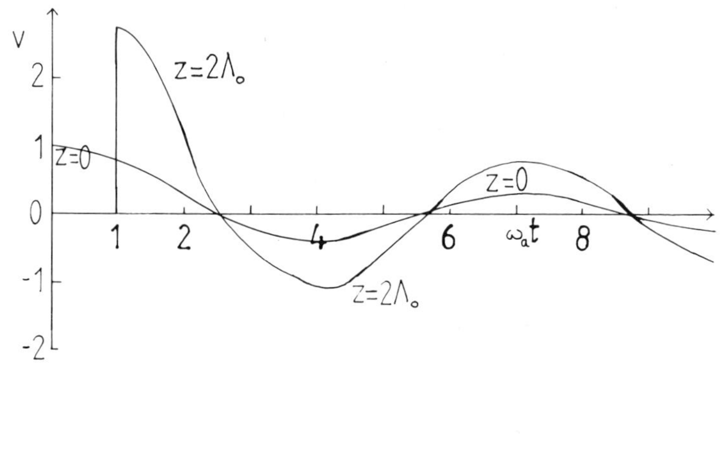

Figure 3.11. The wavefront and wake of an impulsively generated wave in a slender flux tube (after Roberts, 1984b)

3.8.3 Kink modes

The derivation of Equation (3.109) is based upon the slender flux tube equations (3.103)–(3.107). In particular, it pertains to the sausage modes of a flux tube. Spruit (1981b) has shown how to incorporate the kink modes of a slender tube. He finds that linear *transverse* motions $v_\perp$ are governed by the equation

$$\frac{\partial^2 v_\perp}{\partial t^2} = g\left[\frac{\rho_o - \rho_e}{\rho_o + \rho_e}\right]\frac{\partial v_\perp}{\partial z} + \left[\frac{\rho_o}{\rho_o + \rho_e}\right]v_A^2\,\frac{\partial^2 v_\perp}{\partial z^2}\,. \qquad (3.121)$$

The first term on the right-hand side represents the effect of buoyancy on the isolated flux tube, the second term the elastic restoring force of magnetic tension in the field.

Just as for the sausage mode, the kink wave in a slender tube may be linked to the Klein-Gordon equation. If we write

$$v_\perp(z,t) = e^{z/4\Lambda_o}\,Q_\perp(z,t), \qquad (3.122)$$

for an isothermal atmosphere, then Eq (3.121) becomes the Klein-Gordon equation for $Q_\perp$, with now

$$c = \left[\frac{\rho_o}{\rho_o + \rho_e}\right]^{\frac{1}{2}} v_A = \left[\frac{\beta}{1 + 2\beta}\right]^{\frac{1}{2}} v_A, \text{ and } \omega_v = \frac{c}{4\Lambda_o}. \qquad (3.123)$$

Thus we obtain the mean Alfvén speed c_k (see Eq (3.58)) for the speed c.

In several respects the behaviour of the kink mode is similar to that found earlier for the sausage wave: both modes e-fold in four scale-heights of an isothermal atmosphere and their speeds of propagation are of similar magnitude (for $\beta = 1$ we obtain $c_k = 0.58\ v_A = 0.63\ c_s$ ($= 6.3$ km s^{-1} for $c_s = 10$ km s^{-1}) and $c_T = 0.74\ c_s$). However, they differ significantly in one respect: their cut-off frequencies are different. The sausage mode has a cut-off of about 0.03 s^{-1}, whereas the kink mode has a cut-off of 0.009 s^{-1} (with corresponding period of about 700s).

This large difference in the cut-offs of the two modes has led Spruit (1981b) to argue that kink modes, generated by turbulence in the upper convection zone, may be more likely to propagate up to the chromosphere, without significant reflection, than sausage modes. Certainly, the transversal oscillations detected in H_α fibrils are likely to have been generated in the photosphere as the kink modes of a slender tube; but spicules involve longitudinal motions and so are likely to have arisen from the sausage modes of a photospheric flux tube (Roberts, 1984b).

3.8.4 Instabilities in tubes

Much of the above discussion of waves in slender flux tubes has concentrated upon the circumstances of an isothermal atmosphere. However, the theory of motions –both sausage and kink – in a slender tube allows for the possibility of non-isothermal atmospheres. Of particular interest is the behaviour of the sausage mode in the convectively unstable atmosphere of the convection zone. Indeed, it transpires that the concentration of flux tubes to kilogauss field strengths is to be seen as a consequence of the convectively unstable atmosphere in which the tube resides.

Consider the general Klein–Gordon equation (3.109). For a magnetic flux tube the speed c is $c_T(z)$. Returning to the vertical velocity component v(z,t) as primary variable and writing

$$v(z,t) = \hat{v}(z)e^{i\omega t}$$

we obtain the equation

$$\frac{B_o}{\rho_o}\frac{d}{dz}\left[\frac{\rho_o c_T^2}{B_o}\frac{d\hat{v}}{dz}\right] + \left[\omega^2 - \omega_g^2\left[\frac{c_T^2}{v_A^2} + \frac{\gamma c_T^2}{2c_s^2}\right]\right]\hat{v} = 0. \qquad (3.124)$$

Equation (3.124) is in standard Sturm-Liouville form. It follows that, supplemented by Sturmian boundary conditions (e.g. $\hat{v} = 0$ at two levels z), Equation (3.124) possesses *stable* (i.e. $\omega^2 > 0$) solutions provided

$$\omega_g^2 > 0 \qquad (3.125)$$

throughout the domain of the vertical flow. The inequality (3.125) is in fact the usual Schwarzschild criterion for stability in a non-magnetic atmosphere. Inequality (3.125) proves a *sufficient* condition for stability; if $\omega_g^2 < 0$, then instability ($\omega^2 < 0$) may or may not occur, depending upon the magnitude of the field $B_o(z)$ and the strength of the superadiabaticity (as measured by the magnitude of negative ω_g^2).

A useful guide to the behaviour of solutions of equation (3.124) is provided by the 'local approximation', in which the coefficients of our equation are assumed not to vary greatly with z. Then, requiring for illustrative purposes that $\hat{v} = 0$ at z = 0 and z = –d, we may write the solution of Eq (3.124) as

$$\hat{v} = e^{z/4\Lambda_o}\sin\left[\frac{n\pi z}{d}\right] \qquad (3.126)$$

for integer n. The associated dispersion relation is (Webb and Roberts, 1978)

$$\frac{\omega^2}{c_T^2} = \frac{n^2\pi^2}{d^2} + \frac{1}{16\Lambda_o^2} + \omega_g^2\left[\frac{1}{v_A^2} + \frac{\gamma}{2c_s^2}\right]. \qquad (3.127)$$

Of particular interest is the condition of *marginal stability*, which corresponds to $\omega^2 = 0$. Equation (3.127) shows that marginal stability occurs for sufficiently strong magnetic field. In terms of the plasma β, Eq (3.127) with $\omega^2 = 0$ and d sufficiently large yields

$$\beta = \frac{1}{8\left|\Lambda_0' + \frac{\gamma-1}{\gamma}\right|} - 1, \qquad (3.128)$$

given $\omega_g^2 < 0$. For example, with $\gamma = 1.2$ and $\Lambda_0' = -0.20$ (both illustrative of conditions in the upper part of the convection zone), we find a critical value of $\beta = 2.75$; a slightly stronger super-adiabaticity, say $\Lambda_0' = -0.25$, requires a stronger magnetic field, to reach a critical β of 0.5. Of course, we should not push the results of a local analysis too far but it does demonstrate that a critical β of order unity (corresponding to an Alfvén speed close to the sound speed) is necessary to ensure stability of the tube. A calculation of the marginal stability value of β for a model convection zone atmosphere yields $\beta = 1.8$ (Spruit and Zweibel, 1979). The general conclusion from these and similar analyses (Parker, 1978, 1979; Webb and Roberts, 1978; Spruit and Zweibel, 1979; Spruit, 1979; Unno and Ando, 1979) is that field strengths of 1–2 kG are necessary to stabilize an intense flux tube against convective instability. This is, of course, entirely consistent with the observational record – see Parker (1979) for a survey.

REFERENCES

Adam, J A (1981) *Astrophysics and Sp. Sci.* **78**, 293.
Adam, J A (1982) *Phys. Rep.* **86**, 217.
Alfvén, H (1942) *Nature* **150**, 405.
Anzer, U and Galloway, D J (1983) *Mon. Not. Roy. Astron. Soc.* **203**, 637.
Appert, K, Gruber, R and Vaclavik, J (1974) *Phys. Fluids* **17**, 1471.
Christensen-Dalsgaard, J (1980) *Mon. Not. Roy. Astron. Soc.* **190**, 765.
Cram, L E and Thomas J H (1981), editors, *The Physics of Sunspots*, Sacramento Peak Observatory.
Defouw, R J (1976) *Astrophys. J.* **209**, 266.
Deubner, F-L (1975) *Astron. Astrophys.* **44**, 371.
Deubner, F-L and Gough, D (1984) *Ann. Rev. Astron. Astrophys.* **22**, 593.
Edwin, P M and Roberts, B (1982) *Solar Phys.* **76**, 239.
Edwin, P M and Roberts, B (1983) *Solar Phys.* **88**, 179.
Edwin, P M, Roberts, B and Hughes, W J (1985) *J. Geophys. Res.*, submitted.
Giovanelli, R G (1980) *Solar Phys.* **68**, 49.
Giovanelli, R G and Jones, H P (1982) *Solar Phys.* **79**, 247.
Goedbloed, J P (1983) *Lecture Notes on Ideal Magnetohydrodynamics*, Rijnhuizen Rep 83-145, Nieuwegein.
Hain, K and Lüst, R (1958) *Z. Naturforsch* **13a**, 936.
Heyvaerts, J and Priest, E R (1983) *Astron. Astrophys.* **117**, 220.
Hollweg, J V (1982) *Astrophys. J.* **257**, 345.

Hughes, W F and Young, F J (1966) *The Electromagnetodynamics of Fluids*, Wiley, New York.
Kruskal, M and Schwarzschild, M (1954) *Proc. Roy. Soc. Lond.* **A223**, 348.
Lamb, H (1932) *Hydrodynamics*, Cambridge Univ Press.
Lighthill, J (1978) *Waves in Fluids*, Cambridge Univ Press.
Love, A E H (1908) *Some Problems of Geodynamics*, Cambridge Univ Press.
Moore, R (1981) in *The Physics of Sunspots* (Ed. L E Cram and J H Thomas), Sacramento Peak Observatory, pp 259–311.
Morse, P M and Feshbach, H (1953) *Methods of Theoretical Physics*, McGraw-Hill, New York.
Nocera, L, Leroy, B and Priest, E R (1984) *Astron. Astrophys.* **133**, 387.
Nye, A H and Thomas, J H (1976) *Astrophys. J.* **204**, 582.
Osterbrock, D E (1961) *Astrophys. J.* **134**, 347.
Parker, E N (1964) *Astrophys. J.* **139**, 690.
Parker, E N (1974) *Astrophys. J.* **189**, 563.
Parker, E N (1978) *Astrophys. J.* **221**, 368.
Parker, E N (1979) *Cosmical Magnetic Fields*, Oxford Univ Press.
Pekeris, C L (1948) *Geol. Soc. Amer. Mem.* **27**.
Rae, I C and Roberts, B (1982) *Astrophys. J.* **256**, 761.
Roberts, B (1981) *Solar Phys.* **69**, 27.
Roberts, B (1984a) in *The Hydromagnetics of the Sun*, Noordwijkerhout, ESA SP-220, pp 137–145.
Roberts, B (1984b) in *Adv. Space Res.* **4**, 17.
Roberts, B, Edwin P M and Benz, A O (1983) *Nature* **305**, 688.
Roberts, B, Edwin P M and Benz, A O (1984) *Astrophys. J.* **279**, 857.
Roberts, B and Webb, A R (1978) *Solar Phys.* **56**, 5.
Roberts, B and Webb, A R (1979) *Solar Phys.* **59**, 249.
Rosenberg, H (1970) *Astron, Astrophys.* **9**, 159.
Shercliff, J A (1965) *A Textbook of Magnetohydrodynamics*, Pergamon.
Siscoe, G L (1983) in *Solar-Terrestrial Physics* (Ed. R Carovillano and J Forbes), Dordrecht, Reidel, pp 11–92.
Small, L M and Roberts, B (1984) in *The Hydromagnetics of the Sun*, Noordwijkerhout, ESA SP-220, p 257.
Southwood, D J and Hughes, W J (1983) *Space Sci. Rev.* **35**, 301.
Spiegel, E A and Unno, W (1962) *Publ. Astron. Soc. Jpn.* **14**, 28.
Spruit, H C (1981a) in *The Sun as a Star* (Ed. S Jordan), NASA SP-450, pp 385–412.
Spruit, H C (1981b) *Astron. Astrophys.* **98**, 155.
Spruit, H C and Roberts, B (1983) *Nature* **304**, 401.
Spruit, H C and Zweibel, E G (1979) *Solar Phys.* **62**, 15.
Stein, R F and Leibacher, J W (1974) *Ann. Rev. Astron. Astrophys.* **12**, 407.
Thomas, J H (1983) *Ann. Rev. Fluid Mech.* **15**, 321.
Unno, W and Ando, H (1979) *Geophys. Astrophys. Fluid Dyn.* **12**, 107.
Webb, A R and Roberts, B (1978) *Solar Phys.* **59**, 249.
Wentzel, D G (1979) *Astron. Astrophys.* **76**, 20.
Whitham, G B (1974) *Linear and Nonlinear Waves*, Wiley, New York.

CHAPTER 4

MHD INSTABILITIES

A W Hood
Applied Mathematics Department
The University
St Andrews KY16 9SS

4.1 EQUILIBRIUM SOLUTIONS

4.1.1 Introduction

In a course of this length, it is impossible to cover all aspects of Magnetohydrodynamic (MHD) Instabilities and so an elementary introduction is given with references to the original and more advanced publications.

It is assumed that the plasma may be described by the MHD equations, (1.1) – (1.6). In the energy equation, (1.4), the right hand side, $\mathcal{L}$ is the energy loss function. When there is no exchange of heat between the plasma and its surroundings, $\mathcal{L} = 0$ and the plasma is said to be adiabatic. In studying magnetic instabilities $\mathcal{L}$ will be neglected but it is important when studying purely thermal instabilities. (See Priest, 1982, Chap 2 for alternative forms of equation (1.4)).

The simple form of Ohm's law,

$$\mathbf{j} = \sigma(\mathbf{E} + \mathbf{v} \times \mathbf{B}) \tag{4.1}$$

has been used to derive the induction equation but extra terms, e.g. the Hall term, may also be important in certain circumstances. (See Boyd and Sanderson (1969), Chap 3 and Priest (1982), Chap 2).

4.1.2 Energetics

In this section we will manipulate the MHD equations to obtain an energy equation containing the different types of energy, e.g. heat, magnetic and mechanical energy.

Defining a gravitational potential, Φ, such that $\mathbf{g} = -\nabla\Phi$, and combining $v^2/2 \times$ (1.3) and the dot product of $\mathbf{v}$ and (1.2) gives an energy equation for the kinetic, magnetic and gravitational energies.

$$\tfrac{1}{2} v^2 \frac{\partial \rho}{\partial t} + \rho \mathbf{v} \cdot \frac{\partial \mathbf{v}}{\partial t} = \frac{\partial}{\partial t} (\tfrac{1}{2} \rho v^2),$$

$$\tfrac{1}{2}\, v^2\nabla.(\rho\mathbf{v}) + \mathbf{v}.(\rho\mathbf{v}.\nabla\mathbf{v}) = \nabla.(\tfrac{1}{2}\, \rho v^2\mathbf{v}),$$

$$-\mathbf{v}.\nabla p = -\nabla.(p\mathbf{v}) + p\nabla.\mathbf{v},$$

$$\mathbf{v}.(\mathbf{j}\times\mathbf{B}) = -\mathbf{j}.(\mathbf{v}\times\mathbf{B}) = -\frac{j^2}{\sigma} + \mathbf{j}.\mathbf{E}, \quad \text{using (4.1)},$$

$$= -\frac{j^2}{\sigma} - \nabla.\frac{(\mathbf{E}\times\mathbf{B})}{\mu} - \frac{\partial}{\partial t}\left[\frac{B^2}{2\mu}\right], \text{using } \nabla\times\mathbf{E} = -\frac{\partial\mathbf{B}}{\partial t}.$$

$$-\mathbf{v}.\rho\nabla\Phi = -\nabla.(\rho\Phi\mathbf{v}) + \Phi\nabla.(\rho\mathbf{v})$$

$$= -\nabla.(\rho\Phi\mathbf{v}) - \Phi\frac{\partial\rho}{\partial t}, \quad \text{using (1.3)},$$

$$= -\nabla.(\rho\Phi\mathbf{v}) - \frac{\partial}{\partial t}(\rho\Phi), \quad \text{since } \frac{\partial\Phi}{\partial t} = 0.$$

Combining these results gives

$$\frac{\partial}{\partial t}\left[\tfrac{1}{2}\rho v^2 + \rho\Phi + \frac{B^2}{2\mu}\right] + \nabla.\left[\left[\tfrac{1}{2}\rho v^2 + \rho\Phi + p\right]\mathbf{v} + \frac{\mathbf{E}\times\mathbf{B}}{\mu}\right]$$

$$= p\nabla.\mathbf{v} - j^2/\sigma.$$

An expression for $p\nabla.\mathbf{v}$ can be obtained from equation (1.4), using (1.3),

$$\frac{\partial p}{\partial t} + \mathbf{v}.\nabla p = -\gamma p\nabla.\mathbf{v} - \mathcal{L},$$

$$-(\gamma-1)p\nabla.\mathbf{v} = \frac{\partial p}{\partial t} + \nabla.(p\mathbf{v}) + \mathcal{L}.$$

Finally, the full energy equation may be written as

$$\frac{\partial}{\partial t}\left[\tfrac{1}{2}\rho v^2 + \rho\Phi + \frac{p}{\gamma-1} + \frac{B^2}{2\mu}\right] + \nabla.\left[\left[\tfrac{1}{2}\rho v^2 + \rho\Phi + \frac{\gamma p}{\gamma-1}\right]\mathbf{v} + \frac{\mathbf{E}\times\mathbf{B}}{\mu}\right]$$

$$= -\frac{j^2}{\sigma} - \frac{\mathcal{L}}{\gamma-1}. \tag{4.2}$$

The first term represents the rate of change of kinetic, gravitational, internal and magnetic energies. The second term represents the flux of these quantities. $\gamma p/\gamma-1$ is the enthalpy per unit mass and $\mathbf{E}\times\mathbf{B}/\mu$ is the Poynting flux. In ideal MHD, the right hand side is neglected.

This equation will be used later.

4.1.3 The Lorentz Force

The Lorentz force, $(\nabla\times\mathbf{B})\times\mathbf{B}/\mu$, may be written as

$$\frac{(\nabla\times\mathbf{B})\times\mathbf{B}}{\mu} = \frac{(\mathbf{B}.\nabla)\mathbf{B}}{\mu} - \nabla\left[\frac{1}{2}\frac{B^2}{\mu}\right].$$

The first term is non-zero if $\mathbf{B}$ varies along the direction of $\mathbf{B}$ and represents the effect of a tension parallel to $\mathbf{B}$ of magnitude B^2/μ per unit area. The second term represents the effect of a pressure of magnitude $B^2/2\mu$.

The Lorentz force, therefore, has two effects. It acts to shorten the field lines through the tension term and to compress the plasma through the pressure term.

4.1.4 Magnetohydrostatic (MHS) Equilibria

Before studying MHD instabilities, it is essential to discuss the equilibrium state. This is obtained by setting, $\partial/\partial t = \mathbf{v} = 0$ in equations (1.1) - (1.6), giving

$$\nabla p = (\nabla\times\mathbf{B})\times\mathbf{B}/\mu - \rho g\hat{\mathbf{z}}, \tag{4.3}$$

$$p = \frac{\mathfrak{R}}{\tilde{\mu}}\rho T = \frac{k_B}{m}\rho T, \tag{4.4}$$

$$\nabla.\mathbf{B} = 0, \tag{4.5}$$

where $\mathfrak{R}$ is the gas constant and $\tilde{\mu}$ is the mean molecular weight. In general, the temperature satisfies an energy equation of the form

$$\mathcal{L} = 0,$$

but, in the following description, this will be neglected and the temperature is assumed known. (A description of thermal equilibria is given in Priest (1982), Chap 11.) In addition, the equilibrium is assumed ideal, i.e. $\eta = 0$.

Comparing the size of the gas pressure, p, to the magnetic pressure, $B^2/2\mu$, introduces a parameter, called the plasma beta, $\beta = p/(B^2/2\mu)$. If $\beta \gg 1$, then the magnetic field is dominated by the gas pressure and the Lorentz force is neglected in (4.3). On the other hand, if $\beta << 1$, then the magnetic field dominates and p may be neglected.

Comparing terms 1 and 3 in (4.3) gives

$$\frac{p}{\ell} \sim \rho g = \frac{p g \tilde{\mu}}{\Re T} = \frac{p}{\Lambda},$$

where ℓ is a typical length scale over which pressure variations occur. $\Lambda = \Re T/\mu g$ is the pressure scale height, the height over which the pressure falls by a factor of e due to gravity. If $\ell << \Lambda$ then gravity may be neglected.

The simplest form of (4.3) occurs for the situation when $\ell << \Lambda$, and $\beta << 1$. Then (4.3) reduces to

$$(\nabla \times \mathbf{B}) \times \mathbf{B} = 0, \tag{4.6}$$

and the magnetic field is said to be *force-free*. The "solution" is that the current is parallel to the magnetic field.

$$\nabla \times \mathbf{B} = \alpha \mathbf{B}, \tag{4.7}$$

where α may be a scalar function of position.

This seemingly straightforward equation, (4.7), is proving exceptionally hard to solve analytically except in particular cases. If α is constant everywhere, the equation is linear and now may be solved by standard methods.

$$\nabla^2 \mathbf{B} + \alpha^2 \mathbf{B} = 0 \tag{4.8}$$

is the constant-α field equation.

4.1.5 Cylindrically Symmetric Magnetic Fields

Consider a cylindrically symmetric flux tube with magnetic field components

$$\mathbf{B} = (0, B_\theta(r), B_z(r)), \tag{4.9}$$

and gas pressure, $p = p(r)$.

(4.9) automatically satisfies $\nabla.\mathbf{B} = 0$ and, neglecting gravity, the MHS equation reduces to

$$\underbrace{\frac{d}{dr}\left[p + \frac{B_\theta^2 + B_z^2}{2\mu}\right]}_{\text{pressure}} + \underbrace{\frac{B_\theta^2}{\mu r}}_{\text{tension}} = 0. \tag{4.10}$$

All the other components are identically zero. There is only one equation for the 3 unknowns and, hence, two may be specified and the third deduced.

(a) An easy way to obtain solutions to (4.10) is to choose

$$p + \frac{B_\theta^2 + B_z^2}{2\mu} = f(r),$$

where f(r), the generating function, is a prescribed function of r. (4.10) then implies that

$$\frac{B_\theta^2}{\mu r} = -\frac{df}{dr},$$

$$p + \frac{B_z^2}{2\mu} = f - \frac{1}{2} r \frac{df}{dr}. \qquad (4.11)$$

(b) Alternatively, B_θ and B_z may be prescribed and p deduced. If the field is force-free then only one of B_θ and B_z can be chosen.

e.g. $$B_\theta = \frac{r}{1+r^2}, \quad B_z = \frac{1}{1+r^2}, \qquad (4.12)$$

$$\nabla \times \mathbf{B} = (0, -B_z', \frac{1}{r}(rB_\theta)') = \left[0, \frac{2r}{(1+r^2)^2}, \frac{2}{(1+r^2)^2}\right] = \alpha\mathbf{B}.$$

This implies $\alpha(r) = 2/(1+r^2)$.

(4.12) gives a field with uniform twist, where the twist is defined as $\Phi = LB_\theta/(rB_z)$ (and L is the length of the flux tube).

(c) Instead of the above approaches, $\alpha(r)$ could be prescribed and the force free field deduced. Consider the constant-α field, (4.8) and (4.7) can be written as

$$\frac{1}{r}(rB_z')' + \alpha^2 B_z = 0, \qquad (4.13)$$

and

$$B_\theta = -\frac{1}{\alpha} B_z'. \qquad (4.14)$$

Solving (4.13) and (4.14) gives

$$B_\theta = J_1(\alpha r), \quad B_z = J_0(\alpha r),$$

where J_0 and J_1 are Bessel functions of the first kind of order 0 and 1.

4.1.6 2-Dimensional Magnetic Fields

Assume the magnetic field only depends on two space coordinates. For example, this may model an arcade of magnetic flux tubes in which there is no variation along the y direction.

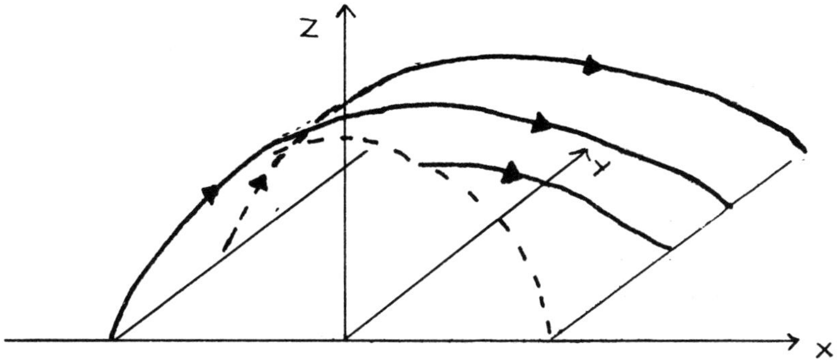

Figure 4.1 An arcade of magnetic field lines

The vector magnetic potential is

$$\mathbf{A} = (A_x(x,z),\ A_y(x,z),\ A_z(x,z))$$

and

$$\mathbf{B} = \nabla\times\mathbf{A} = \left[-\frac{\partial A_y}{\partial z},\ \frac{\partial A_x}{\partial z} - \frac{\partial A_z}{\partial x},\ \frac{\partial A_y}{\partial x}\right].$$

Set $A_y = A$ and $B_y = \frac{\partial A_x}{\partial z} - \frac{\partial A_z}{\partial x}$. Hence, the magnetic field may be expressed as

$$\mathbf{B} = \left[-\frac{\partial A}{\partial z},\ B_y,\ \frac{\partial A}{\partial x}\right], \tag{4.15}$$

and $\nabla.\mathbf{B} = 0$ is automatically satisfied.

$$\mathbf{j} = \left[-\frac{\partial B_y}{\partial z},\ -\nabla^2 A,\ \frac{\partial B_y}{\partial x}\right]/\mu,$$

where $\nabla^2 A = \frac{\partial^2 A}{\partial x^2} + \frac{\partial^2 A}{\partial z^2}$.

Assume that the plasma is *isothermal* (T = const), then

$$\nabla p = \mathbf{j}\times\mathbf{B} - \frac{p}{\Lambda}\,\hat{\mathbf{z}},$$

is satisfied if,

$$e^{-z/\Lambda}\frac{\partial}{\partial x}\left[pe^{z/\Lambda}\right] = -\frac{\partial A}{\partial x}\nabla^2 A - \frac{\partial}{\partial x}\left[\tfrac{1}{2}B_y^2\right], \tag{4.16}$$

$$e^{-z/\Lambda}\frac{\partial}{\partial z}\left[pe^{z/\Lambda}\right] = -\frac{\partial A}{\partial z}\nabla^2 A - \frac{\partial}{\partial z}\left[\tfrac{1}{2}B_y^2\right], \tag{4.17}$$

$$\frac{\partial A}{\partial x}\frac{\partial B_y}{\partial z} - \frac{\partial A}{\partial z}\frac{\partial B_y}{\partial x} \equiv J(A,B_y) = 0. \tag{4.18}$$

The most general solution to (4.18) is

$$B_y = B_y(A). \tag{4.19}$$

Taking $\frac{\partial A}{\partial z}\times$ (4.16) and $\frac{\partial A}{\partial x}\times$ (4.17) gives

$$\frac{\partial A}{\partial z}\frac{\partial}{\partial x}\left[pe^{z/\Lambda}\right] - \frac{\partial A}{\partial x}\frac{\partial}{\partial z}\left[pe^{z/\Lambda}\right] \equiv J\left[A,pe^{z/\Lambda}\right] = 0 ,$$

and so

$$p = P(A)e^{-z/\Lambda}. \tag{4.20}$$

Using (4.19) and (4.20), the MHS equation reduces to

$$\nabla^2 A = -\frac{\partial}{\partial A}\left\{P(A)e^{-z/\Lambda} + \tfrac{1}{2}B_y^2(A)\right\}. \tag{4.21}$$

(4.21) is normally solved by choosing forms for $P(A)$ and $B_y(A)$, although the actual forms will depend on the boundary conditions, see Low (1982) and Melville *et al* (1983,1984) and Bateman (1980) Chap 4, for the Grad-Shafranov equation.

4.2 PHYSICAL DESCRIPTION OF MHD INSTABILITIES

Consider the linear pinch, in which a *uniform* axial current flows along a plasma cylinder. This induces an azimuthal magnetic field, B_θ, and

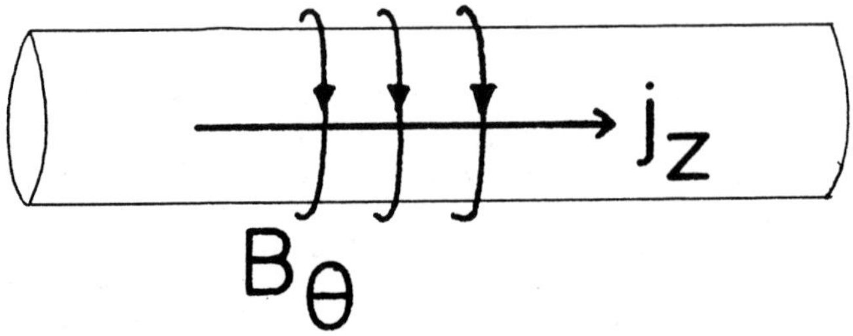

Figure 4.2 **Linear Pinch**

the $\mathbf{j}\times\mathbf{B}$ force acts radially inwards compressing the plasma. This is balanced by an outward pressure gradient. Now from (1.5)

$$\mathbf{j} = \begin{cases} j\hat{\mathbf{z}} = \dfrac{1}{\mu r}\dfrac{d}{dr}(rB_\theta)\hat{\mathbf{z}} & , \quad r < r_o, \\ 0 & , \quad r > r_o, \end{cases}$$

and so

$$B_\theta = \begin{cases} \dfrac{\mu}{2}\, jr & , \quad r < r_o, \\ \dfrac{B_1}{r} & , \quad r > r_o, \end{cases}$$

and from (4.10)

$$p = \begin{cases} p_o - \dfrac{\mu j^2 r^2}{4} & , \quad r < r_o, \\ 0 & , \quad r > r_o, \end{cases}$$

where $p_o = \mu j^2 r_o^2/4$ is the pressure on the axis and $B_1 = \mu j r_o^2/2$, (using continuity of total pressure). The total current flowing in the plasma is $J = \int jr\, dr d\theta = \pi j r_o^2$ and remains constant.

(i) Now assume that the plasma is subjected to a perturbation of the type shown. At the constrictions the radius is r_1. The old field is $B_\theta = \mu/2\ Jr_1/\pi r_o^2$ whereas the new field is $B_1/r_1 = \mu/2\pi\ J/r_1$. Hence the magnetic pressure is increased,

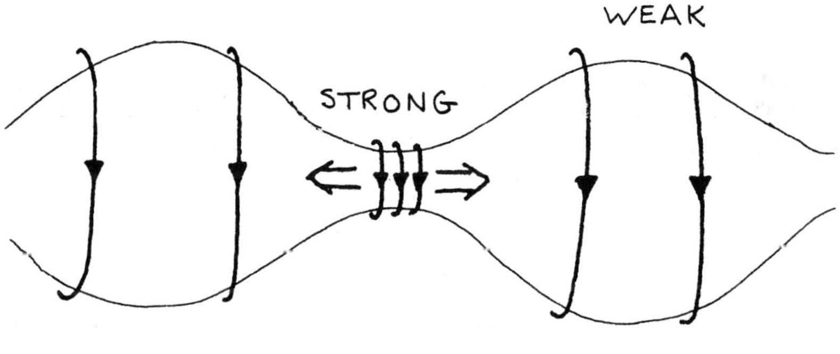

Figure 4.3 **Sausage Instability**

while at the bulges it is decreased. Therefore, the distortion increases and the plasma is unstable. This sausage instability can be stabilised by imposing an axial field, B_z, inside the plasma. The effect of this extra field is to provide an increase in the internal magnetic pressure at the constrictions and a reduction at the bulges.

(ii) The kink instability occurs when the plasma is distorted as shown in Fig 4. By a similar argument, the magnetic pressure is larger on the inside of the curve than on the outside. The resulting

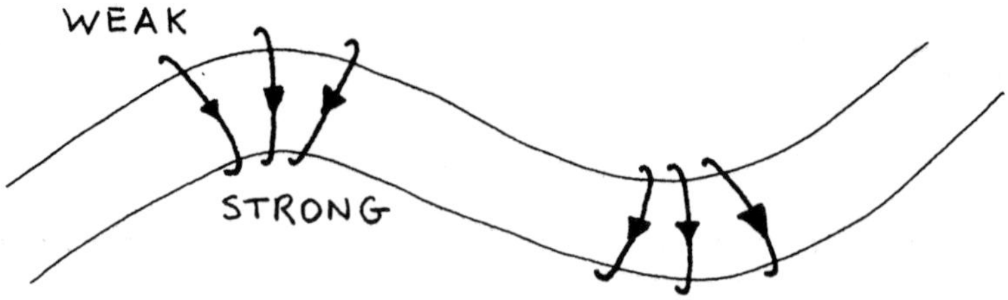

Figure 4.4 **Kink Instability**

pressure difference will increase the displacement and again the plasma is unstable. This particular lateral kink is also stabilised by including an axial field since the kinks increase the internal magnetic tension.

While the above description is intuitively simple to understand, more complicated equilibria require a more rigorous treatment. This is developed in the following sections.

4.3 LINEARISED MHD EQUATIONS

Assuming that the equilibrium satisfies the MHD equations, (4.3) - (4.5), all variables are expressed as their equilibrium value plus a small perturbed amount so that

$$p = p_0(\mathbf{r}) + p_1(\mathbf{r},t),$$

$$\rho = \rho_0(\mathbf{r}) + \rho_1(\mathbf{r},t),$$

$$\mathbf{B} = \mathbf{B}_0(\mathbf{r}) + \mathbf{B}_1(\mathbf{r},t), \qquad (4.22)$$

$$\mathbf{v} = 0 + \mathbf{v}_1(\mathbf{r},t),$$

etc.

Substituting into the MHD equations and neglecting products of perturbed quantities produces a linearised system of equations:

$$\frac{\partial \rho_1}{\partial t} + \nabla.(\rho_0 \mathbf{v}_1) = 0, \qquad (4.23)$$

$$\rho_0 \frac{\partial \mathbf{v}_1}{\partial t} = -\nabla p_1 + (\nabla \times \mathbf{B}_1) \times \mathbf{B}_0/\mu + (\nabla \times \mathbf{B}_0) \times \mathbf{B}_1/\mu + \rho_1 \mathbf{g}, \qquad (4.24)$$

$$\frac{\partial p_1}{\partial t} + \mathbf{v}_1.\nabla p_0 = -\gamma p_0 \nabla.\mathbf{v}_1, \qquad (4.25)$$

$$\frac{\partial \mathbf{B}_1}{\partial t} = \nabla \times (\mathbf{v}_1 \times \mathbf{B}_0), \tag{4.26}$$

$$\nabla . \mathbf{B}_1 = 0. \tag{4.27}$$

It is assumed, for the moment, that the plasma is evolving *adiabatically* and *resistive effects are neglected.*

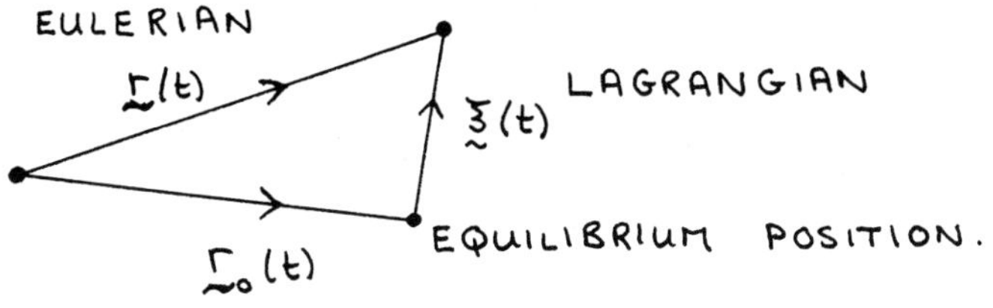

Figure 4.5 **Eulerian and Lagrangian Variables**

The perturbed velocity vector is related to the Eulerian position vector by

$$\mathbf{v}_1 = \frac{D\mathbf{r}}{Dt}. \tag{4.28}$$

If the Lagrangian equilibrium position vector is $\mathbf{r}_0$, the displacement, $\boldsymbol{\xi}(\mathbf{r}_0, t)$ is just $\boldsymbol{\xi} = \mathbf{r} - \mathbf{r}_0$. Hence,

$$\mathbf{v}_1 = \frac{\partial \boldsymbol{\xi}}{\partial t}. \tag{4.29}$$

It is frequently convenient to reduce (4.23) – (4.27) to one equation in $\boldsymbol{\xi}$ and to change from Eulerian to Lagrangian variables (although these are identical to first order). Assuming that the plasma is initially in equilibrium, so that all perturbed variables are zero at t = 0, equations (4.23), (4.25) and (4.26) may be integrated in time, using (4.29). The equation of motion (4.24) is now,

$$\rho_0 \frac{\partial^2 \boldsymbol{\xi}}{\partial t^2} = \mathbf{F}(\boldsymbol{\xi}), \tag{4.30}$$

where the force function is

$$\mathbf{F}(\boldsymbol{\xi}) = \nabla(\boldsymbol{\xi} . \nabla p_0 + \gamma p_0 \nabla . \boldsymbol{\xi}) - \nabla . (\rho_0 \boldsymbol{\xi}) \mathbf{g}$$
$$+ (\nabla \times \nabla \times (\boldsymbol{\xi} \times \mathbf{B}_0)) \times \mathbf{B}_0 / \mu + (\nabla \times \mathbf{B}_0) \times \nabla \times (\boldsymbol{\xi} \times \mathbf{B}_0)/\mu. \tag{4.31}$$

The spatial derivatives are evaluated at the equilibrium position and not at the displaced position. (An expansion would produce 2nd order terms which would then be neglected.) Equations (4.30) and (4.31) are the starting point for the *ideal MHD stability analysis.*

Before proceeding, it is, perhaps, worthwhile drawing an analogy with one-dimensional motion. If V(x) is the potential then the equation of motion of a particle in this potential is

$$m\ddot{x} = -\frac{\partial V}{\partial x}, \tag{4.32}$$

and the equilibrium conditions are

$$\dot{x} = 0 \quad \text{and} \quad \frac{\partial V}{\partial x} = 0. \tag{4.33}$$

There are 3 possible potential variations depending on the sign of $\partial^2 V/\partial x^2$.

If the equilibrium position is x_0, then $\left.\frac{\partial V}{\partial x}\right|_{x_0} = 0$. At time t,

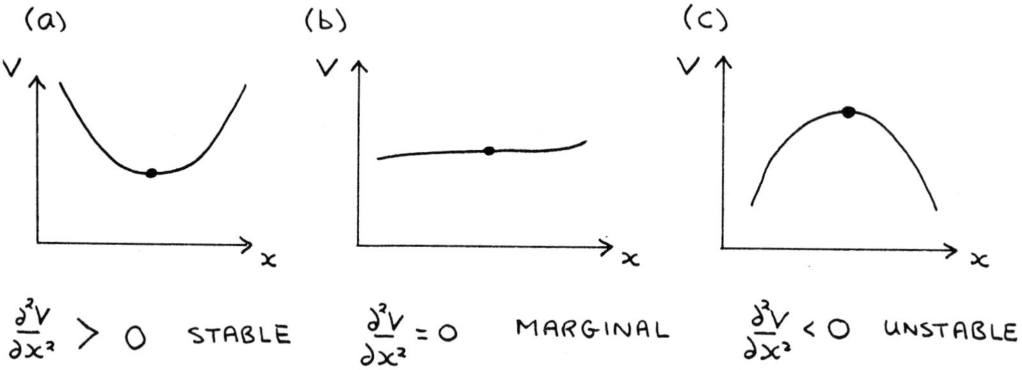

Figure 4.6 **Possible Variations in V(x)**

the particle's position is $x = x_0 + \xi(t)$ and expanding the potential gives

$$V(x) = V(x_0) + \left.\frac{\partial V}{\partial x}\right|_{x_0} \xi + \left.\frac{\partial^2 V}{\partial x^2}\right|_{x_0} \frac{\xi^2}{2} + \dots .$$

The equation of motion is

$$m\ddot{\xi} = -\left.\frac{\partial^2 V}{\partial x^2}\right|_{x_0} \xi ,$$

with solution $\xi = \xi_0 e^{i\omega t}$, where $\omega^2 = \frac{1}{m}\frac{\partial^2 V}{\partial x^2}$.

(a) If $\left.\frac{\partial^2 V}{\partial x^2}\right|_{x_0} > 0$, then ω is real and the particle oscillates about the equilibrium position (STABLE).

(b) If $\left.\frac{\partial^2 V}{\partial x^2}\right|_{x_0} < 0$, then ω is imaginary and there is one exponentially growing solution, $e^{\gamma t}$, with $\gamma = \sqrt{-\frac{1}{m}\left.\frac{\partial^2 V}{\partial x^2}\right|_{x_0}}$ (UNSTABLE).

4.4 NORMAL MODES METHOD

The linearised equation of motion is equation (4.30), where the force function does not explicitly depend on time t. Hence, it is possible to look for normal modes and assume that

$$\boldsymbol{\xi}(\mathbf{r},t) = \boldsymbol{\xi}(\mathbf{r})e^{i\omega t}. \tag{4.34}$$

Equation (4.30) reduces to the eigenvalue problem

$$-\rho_0\omega^2\boldsymbol{\xi} = \mathbf{F}(\boldsymbol{\xi}). \tag{4.35}$$

Equation (4.35) is then solved, with the appropriate boundary conditions, and the eigenvalues ω^2 are determined. The sign of ω^2 determines the stability of the plasma in that

if *all* $\omega^2 > 0$ then the plasma is stable.

if *one* $\omega^2 < 0$ then the plasma is unstable.

For a plasma in a finite region the ω's normally form a complete discrete set but in an infinite or semi-infinite region they may form a discrete set and/or continuous spectrum.

One advantage of the Normal Mode method is that the solution to (4.35) explicitly determines the dispersion relation, which allows a calculation of the fastest growing mode. (This is *possibly* the most important mode.) However, for non-uniform equilibria, the determination of ω^2 can become an extremely formidable task. In such cases, the question of stability or instability can often be answered by using an Energy Method.

4.5 ENERGY (OR VARIATIONAL) METHOD

In Section 4.1.2, an energy equation, (4.2), was derived. Assuming, as above, that the plasma is an *ideal MHD plasma*, then the loss terms on the right-hand side of (4.2) are neglected. Integrating over the volume of the plasma gives

$$\frac{\partial}{\partial t}\left[\int_V \tfrac{1}{2}\rho v^2 d^3\mathbf{x} + \int_V \rho\Phi + \frac{p}{\gamma-1} + \frac{B^2}{2\mu}\, d^3\mathbf{x}\right]$$

$$+ \int_S \hat{\mathbf{n}} \cdot \left[\tfrac{1}{2}\rho v^2 + \rho\Phi + \frac{\gamma p}{\gamma-1}\right]\mathbf{v} + \hat{\mathbf{n}}.(\mathbf{E}\times\mathbf{B})/\mu d^3\mathbf{x} = 0, \qquad (4.36)$$

where the surface integral represents the flux of quantities across the bounding surface. If the boundary conditions are such that the surface integral vanishes, then the total energy of the system is constant.

This is obvious from the time integral of (4.36),

$$\int_V \tfrac{1}{2}\rho v^2 d^3\mathbf{x} + \int_V \rho\Phi + \frac{p}{\gamma-1} + \frac{B^2}{2\mu}\, d^3\mathbf{x} = \text{constant} \qquad (4.37)$$

or

Kinetic Energy (K) + Potential Energy (W) = constant.

At $t = 0$, the plasma is initially in equilibrium, and stability or instability is determined by whether the potential energy is at a minimum or maximum value.

If W has a minimum at equilibrium, then any motion will cause W to increase and so K must decrease (eventually to zero). This corresponds to stability since the perturbation cannot grow indefinitely.

However, if W has a maximum initially, then any motion will cause W to decrease and so K increases. This corresponds to instability since the perturbations can grow (exponentially) in time.

Taking the potential energy,

$$W = \int \rho\Phi + \frac{p}{\gamma-1} + \frac{B^2}{2\mu}\, d^3\mathbf{x}, \qquad (4.38)$$

all variables are written as their equilibrium value plus a small perturbation, where the perturbations satisfy the constraints

$$\rho_1 = -\nabla.(\rho_0\boldsymbol{\xi}), \qquad (4.39)$$

$$p_1 = -\boldsymbol{\xi}.\nabla p_0 - \gamma p_0\nabla.\boldsymbol{\xi}, \qquad (4.40)$$

$$\mathbf{B}_1 = \nabla \times (\boldsymbol{\xi}\times\mathbf{B}_0). \qquad (4.41)$$

These are obtained from (4.23)–(4.26). Using (4.39)–(4.41), (4.38) may be written as

$$W = W_0 + \delta_1 W + \delta_2 W, \tag{4.42}$$

where W_0 is the potential energy at the equilibrium value,

$$\delta_1 W = \int \boldsymbol{\xi}.(\nabla p_0 - (\nabla \times \mathbf{B}_0) \times \mathbf{B}_0/\mu + \rho_0 \nabla\Phi) d^3x = 0,$$

since this involves the MHS equation, and

$$\delta_2 W = \frac{1}{2} \int \left\{ \frac{1}{\mu} |\nabla \times (\boldsymbol{\xi} \times \mathbf{B}_0)|^2 + \frac{\nabla \times \mathbf{B}_0}{\mu} . \left[\boldsymbol{\xi} \times (\nabla \times (\boldsymbol{\xi} \times \mathbf{B}_0)) \right] \right.$$

$$\left. + (\boldsymbol{\xi}.\nabla p_0) \nabla.\boldsymbol{\xi} + \gamma p_0 (\nabla.\boldsymbol{\xi})^2 - (\boldsymbol{\xi}.\nabla\Phi)\nabla.(\rho_0 \boldsymbol{\xi}) \right\} d^3x. \tag{4.43}$$

A detailed derivation of (4.43) is given in Roberts (1967).

An alternative approach to deriving (4.43) is to use the linearised equation of motion, (4.30). Multiplying by $\partial\boldsymbol{\xi}/\partial t$ and integrating over all space gives

$$\int \frac{1}{2} \rho_0 \frac{\partial \boldsymbol{\xi}}{\partial t} . \frac{\partial^2 \boldsymbol{\xi}}{\partial t} d^3x = \int \frac{\partial \boldsymbol{\xi}}{\partial t} . \mathbf{F}(\boldsymbol{\xi}) d^3x . \tag{4.44}$$

Now conservation of energy requires that $\mathbf{F}$ is a self-adjoint operator, so that

$$\int \boldsymbol{\eta}.\mathbf{F}(\boldsymbol{\xi}) d^3x = \int \boldsymbol{\xi}.\mathbf{F}(\boldsymbol{\eta}) d^3x, \tag{4.45}$$

where $\boldsymbol{\eta}$ and $\boldsymbol{\xi}$ satisfy the same boundary conditions. (See Kadomstev (1966) and Kulsrud (1962) for a proof of self-adjointness without gravity.) Using (4.45), with $\boldsymbol{\eta} = \partial\boldsymbol{\xi}/\partial t$, (4.44) becomes

$$\frac{\partial}{\partial t} \left[\frac{1}{2} \int \rho_0 \frac{\partial \boldsymbol{\xi}}{\partial t} \frac{\partial \boldsymbol{\xi}}{\partial t} d^3x - \frac{1}{2} \int \boldsymbol{\xi}.\mathbf{F}(\boldsymbol{\xi}) d^3x \right] = 0, \tag{4.46}$$

showing the conservation of energy. The first term is the kinetic energy and the second term is the second-order change in potential energy,

$$\delta_2 W = -\frac{1}{2} \int \boldsymbol{\xi}.\mathbf{F}(\boldsymbol{\xi}) d^3x. \tag{4.47}$$

To show that (4.43) and (4.47) are identical requires the use of some vector identities

$$-\boldsymbol{\xi}.[\nabla\times\nabla\times(\boldsymbol{\xi}\times\mathbf{B}_0)]\times\mathbf{B}_0 = \nabla\times\nabla\times(\boldsymbol{\xi}\times\mathbf{B}_0).(\boldsymbol{\xi}\times\mathbf{B}_0)$$

$$= \nabla\times(\boldsymbol{\xi}\times\mathbf{B}_0).\nabla\times(\boldsymbol{\xi}\times\mathbf{B}_0) + \nabla.(\nabla\times(\boldsymbol{\xi}\times\mathbf{B}_0)\times\boldsymbol{\xi}\times\mathbf{B}_0),$$

$$-\boldsymbol{\xi}.\nabla(\gamma p_0\nabla.\boldsymbol{\xi} + \boldsymbol{\xi}.\nabla p_0) = \gamma p_0(\nabla.\boldsymbol{\xi})^2 + \boldsymbol{\xi}.\nabla p_0\nabla.\boldsymbol{\xi}$$

$$- \nabla.[(\gamma p_0\nabla.\boldsymbol{\xi} + \boldsymbol{\xi}.\nabla p_0)\boldsymbol{\xi}].$$

Integrating the divergence terms, the resulting surface integrals are zero if, for example, $\hat{\mathbf{n}}.\boldsymbol{\xi} = 0$. Then it can be seen that (4.43) and (4.47) are the same.

The stability of the system is now governed by the sign of $\delta_2 W$. The advantage of the energy method is that trial functions may be used to prove instability or $\delta_2 W$ may be shown, under certain circumstances, to be positive, thus proving stability.

Summarising

$$\begin{aligned}&\text{If } \delta_2 W > 0 \quad \textit{for all } \boldsymbol{\xi} \text{ then STABLE,}\\ &\text{If } \delta_2 W < 0 \quad \textit{for one } \boldsymbol{\xi} \text{ then UNSTABLE.}\end{aligned} \tag{4.48}$$

(A proof of the necessity of the energy method is given by Laval *et al.* (1964).)

Thus, the question of stability or instability may be settled without resorting to a detailed solution of the linear equation of motion. *An estimate* of the size of the growth rate, $-\omega^2$, may be obtained from the expression

$$-\omega^2_{est} = \frac{-\delta_2 W}{\frac{1}{2}\int \rho_0\boldsymbol{\xi}.\boldsymbol{\xi}d^3x} \tag{4.49}$$

using a trial function $\boldsymbol{\xi}$.

4.6 THE RAYLEIGH-TAYLOR INSTABILITY

Following Wesson (1981), the normal mode and energy methods will be used to investigate the Rayleigh-Taylor instability. Three equilibria are discussed, namely two fluids, a continuous fluid and a magnetic fluid.

4.6.1 Normal Modes - Two Fluids

Since the fluid is uniform in x and y, all variables may be expressed

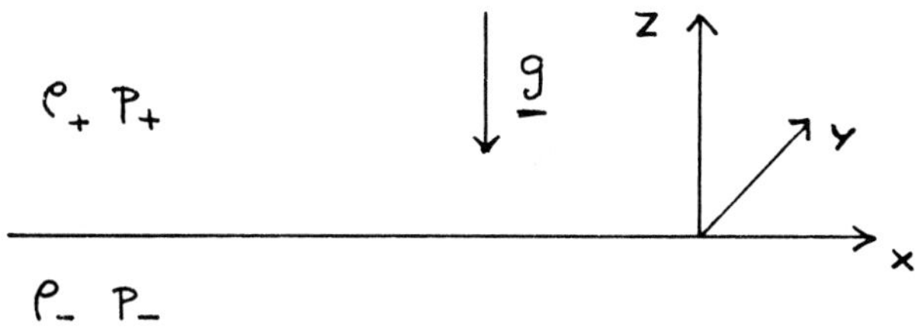

Figure 4.7 **Two Fluids**

in terms of the Fourier component

$$f(z)e^{i(kx-\omega t)}. \tag{4.50}$$

Substituting into the linearised equations of motion, and assuming the plasma is incompressible, $\nabla.\mathbf{v}_1 = 0$, instead of (4.25) gives

$$-i\omega\rho_1 = -v_z \frac{d\rho_0}{dz}, \tag{4.51}$$

$$-\rho_0 i\omega v_x = -ikp_1, \tag{4.52}$$

$$-\rho_0 i\omega v_z = -\frac{dp_1}{dz} - \rho_1 g. \tag{4.53}$$

Eliminating variables in favour of v_z gives

$$\frac{d}{dz}\left[\rho_0 \frac{dv_z}{dz}\right] - k^2\left[\rho_0 + \frac{g}{\omega^2}\frac{d\rho_0}{dz}\right] v_z = 0. \tag{4.54}$$

Now inside each fluid the densities are constant and so the upper and lower solutions to (4.54) are

$$v_{z\pm} = v_z(0)e^{\mp kz}, \tag{4.55}$$

where continuity of normal velocity has been used to match the solutions across the interface. The other joining condition is obtained by integrating (4.54) from a small distance below z = 0 to a small distance above, i.e. from $-\epsilon$ to $+\epsilon$ and letting $\epsilon \to 0$

$$\rho_0 \frac{dv_z}{dz}\bigg|_{+\epsilon} - \frac{k^2 g}{\omega^2}\rho_0 v_z\bigg|_{+\epsilon} = \rho_0 \frac{dv_z}{dz}\bigg|_{-\epsilon} - \frac{k^2 g}{\omega^2}\rho_0 v_z\bigg|_{-\epsilon}. \tag{4.56}$$

Using the solution (4.55) in each region, (4.56) defines the dispersion relation

$$\omega^2 = kg\,\frac{(\rho_- - \rho_+)}{(\rho_- + \rho_+)}. \tag{4.57}$$

Obviously, the equilibrium is stable if the lighter fluid rests on top of the heavier, i.e. $\rho_- > \rho_+$. However, if $\rho_+ > \rho_-$, then $\omega^2 < 0$ and there is an exponentially growing solution in time and so the equilibrium is unstable. Therefore the necessary and sufficient condition for stability is that $\rho_+ \leqslant \rho_-$.

The fastest growing instability occurs for very short wavelengths in the x-direction ($k \to \infty$). This mode is highly localised about $z = 0$.

4.6.2 Normal Modes - Continuous Fluid

When the density is a function of z, (4.54) must be solved to obtain the eigenfunctions $v_z(z)$ and the eigenvalues ω^2. As an illustration, Wesson (1981) assumes

$$\rho_0 = \rho_0(0)e^{z/\lambda}, \tag{4.58}$$

where λ is the density scale height and the boundary conditions are taken as

$$v_z = 0 \quad \text{at} \quad z = 0 \text{ and } h.$$

The solution to (4.54) is

$$v_z(z) = \sin(n\pi z/h)e^{-z/2\lambda},$$

where n is an integer and the dispersion relation is

$$\omega^2 = \frac{-gk^2}{\lambda\left[\left[\frac{1}{2\lambda}\right]^2 + \left[\frac{n\pi}{h}\right]^2 + k^2\right]}. \tag{4.59}$$

Obviously, the plasma is unstable but, in this case, there is a limiting finite growth rate as $k \to \infty$ for finite λ, in contrast to the sharp boundary. (However, the previous result can be obtained from (4.59) by setting $k = + 1/2\lambda$, and $\rho_-/\rho_+ \to 0$ as $k \to \infty$.)

4.6.3 Simple Energy Method - Two Fluids

Assume that the equilibrium, Fig 8 (i), is perturbed in the manner shown in Fig 8 (ii) in such a way that the area A is equal to the area B. Let $\pm d/2$ be the height/depth of the centres of mass from the equilibrium position. The upper fluid has lost the potential energy, $\rho_+ gAd$, whereas the lower fluid has gained the potential energy, $\rho_- gAd$.

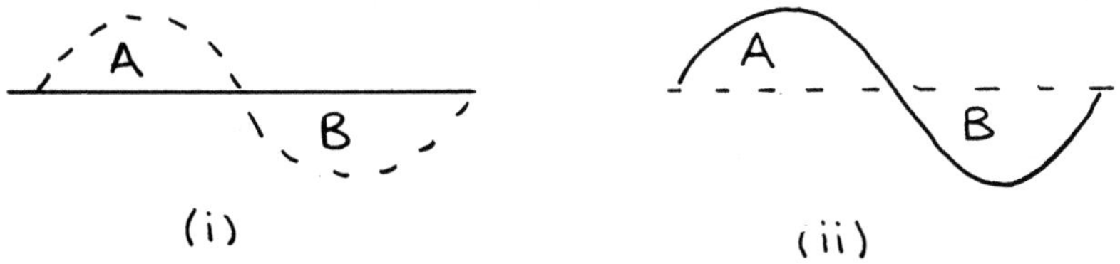

Figure 4.8 **Two Fluids – Energy Approach**

Hence, the total change in potential energy is $\delta W = (\rho_- - \rho_+)gAd$. Therefore,

$$\rho_- > \rho_+ \Rightarrow \delta W > 0 \qquad \text{STABLE,}$$

$$\rho_- < \rho_+ \Rightarrow \delta W < 0 \qquad \text{UNSTABLE.}$$

(The same conclusion can be reached by setting $\nabla.\boldsymbol{\xi} = 0$ in (4.43) and using the trial function

$$\xi_z = \begin{cases} 1 & |z| < d/2. \\ 0 & |z| > d/2, \end{cases} \tag{4.60}$$

in $\delta_z W$.)

4.6.4 Energy Method – Continuous Fluid

Using the incompressible assumption, $\nabla.\boldsymbol{\xi} = 0$, and $\nabla\Phi = -g\hat{\mathbf{z}}$, (4.43) reduces to

$$\delta_z W = -\tfrac{1}{2}\int g\xi_z^2 \frac{d\rho_0}{dz}\, d^3x. \tag{4.61}$$

Clearly $\delta_z W$ can be made negative whenever $d\rho_0/dz$ is positive, i.e. whenever heavier fluid overlies lighter fluid, by choosing a trial function similar to (4.60).

So far the perturbations have been assumed incompressible. Compressibility has been included by, for example, Bateman, (1980) Chap 3. The plasma is now unstable if at some height $de/dz > 0$, where $e = p/\rho^\gamma$ is the entropy of the fluid. (See Bateman for details.)

4.6.5 MHD Incompressible Rayleigh-Taylor Instability

In this section the effect of a magnetic field, with shear, is included. Magnetic shear means that the magnetic field vector points in different directions at different heights. Therefore,

$$\mathbf{B}_0 = (B_{ox}(z),\ B_{oy}(z),\ 0), \tag{4.62}$$

and the equilibrium satisfies,

$$\frac{d}{dz}\left[p_0 + \frac{B_{ox}^2 + B_{oy}^2}{2\mu}\right] = -\rho_0 g. \tag{4.63}$$

Assuming the plasma is incompressible, again for illustration, and taking Fourier components,

$$\boldsymbol{\xi} = \boldsymbol{\xi}(z)e^{i(kz+\ell y)}, \tag{4.64}$$

the energy integral (4.43) may be reduced to

$$\delta_2 W = \frac{1}{2}\int\left\{-g\frac{d\rho_0}{dz}\,\xi_z^2 + \frac{(\mathbf{k}.\mathbf{B}_0)^2}{\mu}\left[\xi_z^2 + \frac{1}{k^2+\ell^2}\left[\frac{d\xi_z}{dz}\right]^2\right]\right\}dz, \tag{4.65}$$

where, using algebraic manipulation, $\delta_2 W$ has been minimised by choosing

$$\xi_x B_{oy} - \xi_y B_{ox} = \frac{(\ell B_{ox} - kB_{oy})}{k^2+\ell^2}\,\frac{d\xi_z}{dz}, \tag{4.66}$$

and

$$k\xi_x + \ell\xi_y = -\frac{d\xi_z}{dz}. \tag{4.67}$$

From (4.65), the magnetic terms are positive and help to stabilise the plasma *except* near the region where the wave vector is perpendicular to the magnetic field, i.e. where

$$\mathbf{k}.\mathbf{B}_0 = kB_{ox} + \ell B_{oy} = 0. \tag{4.68}$$

This region defines a singular layer (or resonant surface) and corresponds to a neighbourhood in which the displacement ξ_z moves the magnetic field *without* producing any bending. So magnetic tension is unable to stabilise the instability.

Completely minimising $\delta_2 W$ gives rise to the Euler-Lagrange equation

$$\frac{d}{dz}\left[\frac{(\mathbf{k}.\mathbf{B}_0)^2}{\mu}\,\frac{d\xi_z}{dz}\right] + \left[(k^2+\ell^2)\,g\frac{d\rho_0}{dz} - \frac{(\mathbf{k}.\mathbf{B}_0)^2}{\mu}\right]\xi_z = 0. \tag{4.69}$$

Equation (4.69) becomes singular at the resonant surface. The importance of this resonant surface will be discussed later. However, it is worth noting that a mathematical singularity frequently suggests that some physics is missing.

4.7 THE SHARP PINCH - NORMAL MODES

As a final example of the Normal Modes method, the physical description of the sausage mode instability, introduced in Section 4.2, is studied in more detail. The equilibrium is defined by

$$\left.\begin{array}{l} p = p_0 \\ \rho = \rho_0 \\ \mathbf{B} = (0,0,B_0) \end{array}\right\} \quad r \leq a,$$

$$\left.\begin{array}{l} p = 0 \\ \rho = 0 \\ \mathbf{B} = \left[0, \frac{B_1 a}{r}, 0\right] \end{array}\right\} \quad r > a. \qquad (4.70)$$

The plasma is therefore surrounded by a vacuum, although the same result applies for a low-density, low-pressure external region.

Figure 4.9 The Sharp Pinch

Using continuity of total pressure, at r = a, defines B_1 as

$$\frac{B_1^2}{2\mu} = \frac{B_0^2}{2\mu} + p_0. \qquad (4.71)$$

The current is now restricted to a "thin" skin on the surface r = a. Because of the cylindrical nature of the problem, cylindrical coordinates are used and, since t, θ and z do not explicitly appear in the coefficients of the linearised equation of motion, Fourier components,

$$\boldsymbol{\xi} = \boldsymbol{\xi}(r)e^{i(m\theta+kz-\omega t)}, \qquad (4.72)$$

are chosen. The value of the azimuthal mode number, m, determines the nature and nomenclature of the instability. For example,

$$\begin{array}{ll} m = 0 & \text{sausage mode,} \\ m = 1 & \text{kink mode,} \\ m \to \infty & \text{interchange mode.} \end{array} \tag{4.73}$$

Although the general m mode can be solved analytically, (see Kruskal and Tuck, 1958), the algebra is simpler if attention is focussed on the sausage mode, m = 0. (See Boyd and Sanderson, (1969) Chap 4.) The normal mode equation, (4.35), is solved separately in the inner and outer regions and then matched, using continuity of normal velocity and continuity of total pressure *at the perturbed boundary* ($r = a + \xi(a)e^{i(kz-\omega t)}$).

4.7.1 Inner Solution r < a

Using the uniformity of B_z, p_o (and ρ_o), (4.35) reduces to

$$-\rho_o\omega^2\xi_r = \gamma p_o \frac{d\eta}{dr} + \frac{B_o^2}{\mu}\left[\frac{d}{dr}\left[\frac{1}{r}\frac{d}{dr}(r\xi_r)\right] - k^2\xi_r\right],$$

$$-\rho_o\omega^2\xi_\theta = -\frac{k^2B_o^2}{\mu}\xi_\theta,$$

$$-\rho_o\omega^2\xi_z = ik\gamma p_o\eta,$$

where $\eta = \frac{1}{r}\frac{d}{dr}(r\xi_r) + ik\xi_z$. Defining the sound speed by $c_s^2 = \gamma p_o/\rho_o$

and the Alfven speed by $v_A^2 = B_o^2/\mu\rho_o$, and using the definition of η, these equations may be rewritten as

$$(k^2v_A^2 - \omega^2)\xi_r = (c_s^2 + v_A^2)\frac{d}{dr}\left[\frac{1}{r}\frac{d\xi_r}{dr}\right] + ikc_s^2\xi_z, \tag{4.74}$$

$$(k^2v_A^2 - \omega^2)\xi_\theta = 0, \tag{4.75}$$

$$(k^2c_s^2 - \omega^2)\xi_z = ikc_s^2\left[\frac{1}{r}\frac{d\xi_r}{dr}\right]. \tag{4.76}$$

Equation (4.75) corresponds either to a torsional Alfven wave or to $\xi_\theta = 0$.

ξ_r may be eliminated, by taking $\frac{1}{r}\frac{d}{dr}$ of (4.74), and rearranging the resulting equation for ξ_z leaves

$$\frac{d^2\xi_z}{dr^2} + \frac{1}{r}\frac{d\xi_z}{dr} - \frac{(k^2c_s^2 - \omega^2)(k^2v_A^2 - \omega^2)}{k^2c_s^2v_A^2 - \omega^2(c_s^2 + v_A^2)}\xi_z = 0. \tag{4.77}$$

The solution to (4.77) is a modified Bessel function of the first kind and order 0,

$$\xi_z = I_0(Kr), \tag{4.78}$$

where

$$K^2 = k^2\left[1 + \frac{(\omega/k)^4}{c_s^2v_A^2 - (\omega/k)^2(c_s^2 + v_A^2)}\right], \tag{4.79}$$

(see Boyd and Sanderson, (1969) Chap 4). Then using (4.78) in either (4.74) or (4.76), gives

$$\xi_r = \frac{Kr}{ik}\left[\frac{c_s^2(k^2v_A^2 - \omega^2) - \omega^2v_A^2}{c_s^2(k^2v_A^2 - \omega^2)}\right]I_0'(Kr). \tag{4.80}$$

4.7.2 Outer Solution r > a

In the vacuum no currents can flow and so

$$\nabla \times \nabla \times (\boldsymbol{\xi} \times \mathbf{B}_0) = 0.$$

The radial component implies that in the outer region

$$ik\xi_z = -r\frac{d}{dr}\left[\frac{1}{r}\xi_r\right]. \tag{4.81}$$

This, in turn, implies that the perturbed magnetic field in the vacuum vanishes.

4.7.3 Matching Conditions at r = a

The matching conditions are obtained from continuity of the normal displacement, i.e.

$$\xi_r^{inner} = \xi_r^{outer}, \tag{4.82}$$

and from continuity of total pressure

$$\left[p + \frac{B^2}{2\mu}\right]^{inner} = \left[p + \frac{B^2}{2\mu}\right]^{outer} \tag{4.83}$$

where the variables are evaluated on the perturbed boundary at $r = a + \xi_r(a)$. (4.82), to first order, remains the same but evaluated at $r = a$. However, (4.83) requires some explanation. Remember, that the pressure is, for a general cylindrical pinch,

$$p = p_0(r) + p_1(r),$$

where $p_1 = -\boldsymbol{\xi}.\nabla p_0 - \gamma p_0 \nabla.\boldsymbol{\xi}$. Evaluating at $\mathbf{r} = a\hat{\mathbf{r}} + \boldsymbol{\xi}(a)$ and expanding up to first order gives

$$p(a\hat{\mathbf{r}} + \boldsymbol{\xi}) = p_0(a) + \boldsymbol{\xi}.\nabla p_0\Big|_a - (\boldsymbol{\xi}.\nabla p_0 + \gamma p_0 \nabla.\boldsymbol{\xi})\Big|_a,$$

$$= p_o(a) \qquad - \gamma p_o \nabla.\boldsymbol{\xi}\Big|_a.$$

ZERO FIRST ORDER

Similarly,

$$\tfrac{1}{2}B^2 = \tfrac{1}{2}\mathbf{B}.\mathbf{B} = \tfrac{1}{2}\mathbf{B}_0.\mathbf{B}_0\Big|_a + \mathbf{B}_0.\mathbf{B}_1\Big|_a + \mathbf{B}_0.(\boldsymbol{\xi}.\nabla\mathbf{B}_0)\Big|_a.$$

Using the given equilibrium and that $\mathbf{B}_1$ is zero in the outer region, (4.83) reduces to

$$\left[-\left[\gamma p_0 + \frac{B_0^2}{\mu}\right]\frac{1}{r}\frac{d}{dr}(r\xi_r) - ik\gamma p_0 \xi_z\right]^{\text{inner}}_{r=a} = \left[-\frac{B_1^2 a^2}{\mu r^3}\,\xi_r\right]^{\text{outer}}_{r=a} \tag{4.84}$$

From the definition of ξ_r and ξ_z in the inner region and using (4.82), (4.84) can be written as (see Boyd and Sanderson, (1969) Chap 4)

$$\omega^2 = \frac{k^2 B_0^2}{\mu \rho_0} - \frac{B_1^2}{\mu \rho_0 a^2}\left[\frac{KaI_0'(Ka)}{I_0(Ka)}\right] \tag{4.85}$$

However, the dispersion relation is not as simple as it at first appears since K depends on ω^2 and in general ω^2 must be obtained by numerical methods.

Two important limits can be studied now. Firstly, when the internal axial field is zero, $B_0 = v_A = 0$, then (4.85) becomes

$$\omega^2 = -\frac{B_1^2}{\mu \rho_0 a^2}\left[\frac{KaI_0'(Ka)}{I_0(Ka)}\right],$$

where $K^2 = k^2(1 - (\omega/kc_s)^2)$.

The dispersion relation is roughly sketched in Figure 10 and the plasma is obviously unstable.

As an illustration of the usefulness of the energy principle, the general cylindrical pinch is discussed. The first detailed analysis was performed by Newcomb (1960) and a good description is presented by Wesson (1981). The equilibrium satisfies (4.10), i.e.

$$\frac{d}{dr}\left[p + \frac{B_\theta^2 + B_z^2}{2\mu}\right] = -\frac{B_\theta^2}{\mu r} . \tag{4.86}$$

Taking Fourier components

$$\boldsymbol{\xi} = \boldsymbol{\xi}(r)e^{i(m\theta+kz)},$$

$\delta_z W$ may be rearranged as (see Newcomb)

$$\delta_z W = \frac{\pi}{2}\int\left\{\Lambda\left[\xi, \frac{d\xi}{dr}\right] + \gamma p_0(\nabla.\boldsymbol{\xi})^2 + \frac{m^2 + k^2r^2}{\mu r^2}\right.$$

$$\left.\times\left[\zeta - \zeta_o\left[\xi, \frac{d\xi}{dr}\right]\right]^2\right\} r dr , \tag{4.87}$$

where $\xi = \xi_r$, $\zeta = i(\xi_\theta B_z - \xi_z B_\theta)$ and

$$\zeta_0 = \frac{r}{m^2 + k^2r^2}\left[(k_r B_\theta - mB_z)\frac{d\xi}{dr} - (krB_\theta + mB_z)\frac{\xi}{r}\right].$$

Since ξ_θ and ξ_z do not involve derivatives with respect to r, $\delta_z W$ may be minimised by purely algebraic means if the second and third terms are zero. Thus,

$$\delta_z W = \frac{\pi}{2\mu}\int\left\{\frac{1}{m^2 + k^2r^2}\left[(mB_\theta + krB_z)\frac{d\xi}{dr} - (mB_\theta - krB_z)\frac{\xi}{r}\right]^2\right.$$

$$\left. + \left[(mB_\theta + krB_z)^2 - 2B_\theta\frac{d}{dr}(rB_\theta)\right]\frac{\xi^2}{r^2}\right\} r dr. \tag{4.88}$$

Immediately, (4.88) allows the derivation of a simple sufficient condition for stability. The only possible negative, and hence destabilising, term is the last. This is positive (implying $\delta_z W > 0$, for all ξ) if

$$B_\theta \frac{d}{dr}(rB_\theta) < 0. \tag{4.89}$$

However, in most astrophysical situations this condition will not be met, particularly near r = 0. (4.89) may be satisfied if there is a current-carrying conductor inside the plasma.

In deriving (4.88), (4.87) was minimised by choosing displacements that are incompressible *but it does not imply* that the most unstable, in the sense of the fastest growing, displacement is also incompressible.

Expanding the square in (4.88), using integration by parts to change the term $\xi d\xi/dr$ to a term in ξ^2, and using the equilibrium condition (4.86), allows the potential energy to be expressed as

$$\delta_2 W = \frac{\pi}{2\mu} \int \left[f \left[\frac{d\xi}{dr}\right]^2 + g\xi^2 \right] dr, \tag{4.90}$$

where

$$f = \frac{r(mB_\theta + krB_z)^2}{m^2 + k^2r^2}, \tag{4.91}$$

and

$$g = \frac{2\mu k^2r^2}{m^2 + k^2r^2}\frac{dp}{dr} + \frac{m^2 + k^2r^2 - 1}{m^2 + k^2r^2}\frac{1}{r}(mB_\theta + krB_z)^2$$

$$+ \frac{2k^2r}{(m^2 + k^2r^2)^2}(k^2r^2B_z^2 - m^2B_\theta^2). \tag{4.92}$$

Obviously a stricter sufficient condition for stability is

$$g > 0 \quad \text{for all } r. \tag{4.93}$$

The ξ that minimises (4.90) is the solution to the Euler-Lagrange equation

$$\frac{d}{dr}\left[f\frac{d\xi}{dr}\right] = g\xi. \tag{4.94}$$

However, from (4.91), the equation has a singularity at the mode rational surfaces

$$\mathbf{k}.\mathbf{B}_0 = mB_\theta + krB_z = 0. \tag{4.95}$$

As pointed out by Wesson, it is not possible immediately to use the solution of (4.94) to determine the sign of $\delta_2 W$, since it will not, in general, satisfy the appropriate boundary conditions. Before discussing the formalism of Newcomb in a later section, it is interesting to note that minimising (4.90), with the constraint,

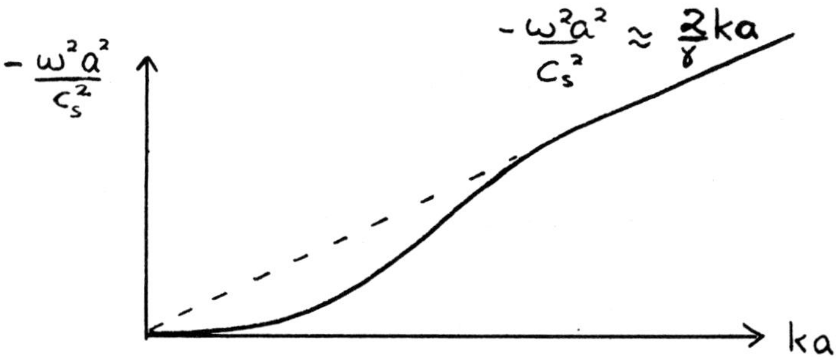

Figure 4.10 **The Dispersion Relation**

Secondly, the magnitude of the axial field required to stabilise the sausage mode instability is obtained by setting $\omega^2 = 0$ in (4.85). Thus,

$$B_0^2 = \frac{B_1^2}{(ka)} \frac{I_0'(ka)}{I_0(ka)} .$$

This condition gives the minimum value of B_0 required for stability

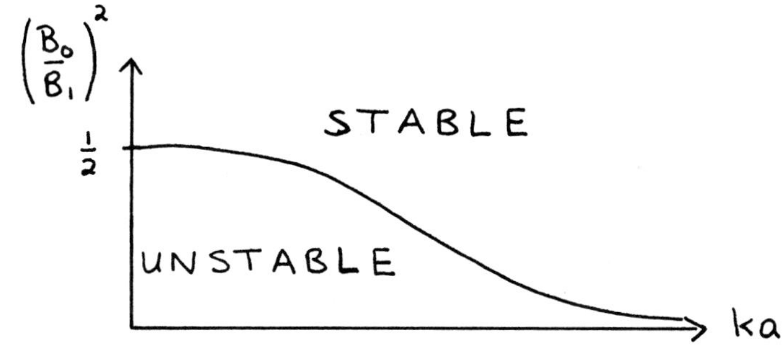

Figure 4.11 **Minimum value of B_0 for stability**

but it must be remembered that there is a maximum value $B_0 < B_1$ from the pressure balance condition

$$p_0 + \frac{B_0^2}{2\mu} = \frac{B_1^2}{2\mu} .$$

Our simple intuitive picture of section 4.2 has now been backed up by a more rigorous analysis.

4.8 GENERAL CYLINDRICAL PINCH – ENERGY METHOD

4.8.1 Minimisation of $\delta_2 W$

$$\frac{\pi}{2\mu}\int \xi^2 dr = \text{constant},$$

gives the Euler–Lagrange equation

$$\frac{d}{dr}\left[f\,\frac{d\xi}{dr}\right] = (g-\lambda)\xi \ ,$$

where λ is a Lagrange multiplier. λ may be used to satisfy the boundary conditions (e.g. ξ or $d\xi/dr = 0$), and the *sign* of $\delta_2 W$ is then given by the sign of the minimum λ.

$$\delta_2 W = \frac{\pi}{2\mu}\int_a^b f\left[\frac{d\xi}{dr}\right]^2 + g\xi^2 dr$$

$$= \frac{\pi}{2\mu}\left\{\left[f\xi\,\frac{d\xi}{dr}\right]_b^a - \int_a^b \left[\frac{d}{dr}\left[f\,\frac{d\xi}{dr}\right] - g\xi\right]\xi dr\right\}$$

$$= \frac{\pi}{2\mu}\int_a^b \lambda\xi^2 dr = \lambda \times \text{constant} \ ,$$

λ, however, has the same sign but not the same magnitude as ω^2.

4.8.2 Suydam's Criterion - A Necessary Condition

One of the powerful uses of the energy principle is to prove instability by a trial function. This provides a necessary condition for stability (sufficient for instability). A particularly important condition was derived by Suydam (1958) by studying modes localised about a mode rational surface at $r = r_s$. This derivation follows Wesson (1981).

Choose, as a trial function

$$\xi = \begin{cases} \xi(x) & |x| < 1, \\ 0 & |x| > 1, \end{cases} \tag{4.96}$$

where $x = (r-r_s)/\epsilon$ and $\epsilon \ll 1$. ϵ is a measure of the localisation of the mode. Then expanding all variables in powers of ϵ

$$\begin{aligned} \xi &= \xi_0(x) + \epsilon\xi_1(x) + \dots, \\ f &= 0 + 0 + \frac{\epsilon^2 x^2}{2} f''(r_s) + \dots, \\ g &= g_s + \epsilon x g'(r_s) + \dots \ ; \end{aligned} \tag{4.97}$$

where $\frac{f''}{2}(r_s) = \frac{rB_\theta^2 B_z^2}{B^2}\left[\frac{q'}{q}\right]^2$, $q = \frac{B_\theta}{rB_z}$, $g_s = \frac{2\mu B_\theta^2}{B^2}\frac{dp}{dr}$,

and it is assumed all coefficients are evaluated at r_s. In deriving these expressions, the $\mathbf{k}.\mathbf{B}_0 = 0$ condition is used to eliminate m and k and to obtain g_s it is assumed that $m^2\epsilon^2 << 1$.

To lowest order in ϵ $\delta_2 W$ is

$$\delta_2 W = \frac{\pi}{2\mu}\int_{-1}^{1} \frac{f''(r_s)}{2} x^2\left[\frac{d\xi_0}{dx}\right]^2 + g_s\xi_0^2 dx. \qquad (4.98)$$

Consider the region $0 < x < 1$, first of all. The Euler-Lagrange equation is

$$\frac{d}{dx}\left[x^2\frac{d\xi_0}{dx}\right] + c\xi_0 = 0, \qquad (4.99)$$

where c is defined by

$$c = -2g_s/f''(r_s) = -\frac{2\mu\ dp/dr}{rB_z^2(q'/q)^2}. \qquad (4.100)$$

The solution to (4.99) satisfying $\xi_0(1) = 0$ is

$$\xi_0(x) = x^{\nu_+} - x^{\nu_-}, \qquad (4.101)$$

where ν_+ and ν_- are the roots of the indicial equation

$$\nu(\nu+1) + c = 0, \qquad (4.102)$$

namely,

$$\nu_\pm = -\tfrac{1}{2} \pm (\tfrac{1}{4} - c)^{\frac{1}{2}}. \qquad (4.103)$$

Now to show instability assume that $c > 1/4$, so that (4.101) becomes

$$\xi_0(x) = x^{-\frac{1}{2}}\sin((c-\tfrac{1}{4})^{\frac{1}{2}}\log x), \qquad (4.104)$$

ξ_0 is a rapidly oscillating function as $x \to 0$. However, the singularity at $x=0$ creates complications, so the actual trial function is taken as (4.104), for $d < x < 1$ and

$$\xi_0 = \xi_0(d) \qquad 0 \leqslant x \leqslant d. \qquad (4.105)$$

The distance d is assumed small but arbitrary. The resulting trial function looks like,

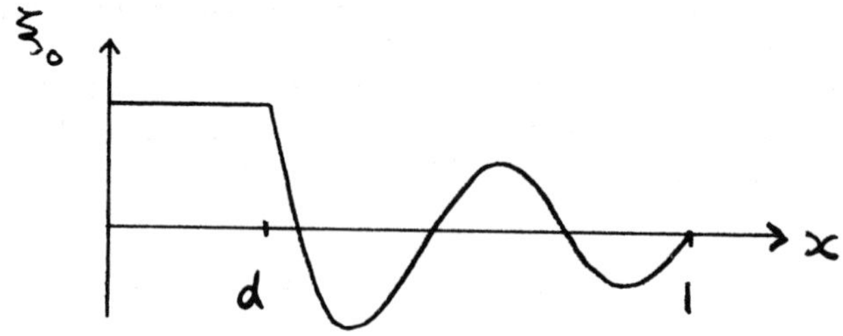

Figure 4.12 **Trial Function for Suydam's Criterion**

Substituting (4.104) and (4.105) into (4.98), for the region $0 < x < 1$, and integrating (by parts) gives

$$\delta_2 W = g_s \sin^2\theta - \frac{f''(r_s)}{2}\left[(c - \tfrac{1}{4})^{1/2}\cos\theta\sin\theta - \tfrac{1}{2}\sin^2\theta\right],$$

where $\theta = (c - \frac{1}{4})^{1/2} \log(d)$, and on rearranging, in terms of double angles,

$$\delta_2 W = -\frac{f''(r_s)}{4}\left[-(c - \tfrac{1}{2})\cos 2\theta + (c - \tfrac{1}{4})^{1/2}\sin 2\theta + c - \tfrac{1}{2}\right]$$

$$= -\frac{f''(r_s)}{4}\left[c - \tfrac{1}{2} + c\sin(2\theta + \delta)\right],$$

where $\tan\delta = -(c - \frac{1}{2})/(c - \frac{1}{4})^{1/2}$. Now, $f''(r_s) > 0$, since f has a minimum value at $r = r_s$, so $\delta_2 W$ attains its minimum value (for this trial function) when d is chosen so that $\sin(2\theta + \delta)$ is one. Therefore,

$$\delta_2 W = -\frac{f''(r_s)}{2}(c - \tfrac{1}{4}) < 0 , \qquad (4.106)$$

since

$$c > \tfrac{1}{4}. \qquad (4.107)$$

This is Suydam's criterion, or in terms of the physical variables, a *necessary condition for stability* ($c < \frac{1}{4}$) is

$$\frac{dp}{dr} + \frac{rB_z^2}{8\mu}\left[\frac{dq/dr}{q}\right]^2 > 0. \qquad (4.108)$$

If this conditon is violated anywhere in the plasma, the equilibrium is unstable. The second term shows that magnetic shear helps to stabilise the plasma but, if the pressure gradient is large enough, then instability must ensue.

If Suydam's criterion is satisfied the plasma may still be unstable to global perturbations. Then, stability can only be guaranteed by deriving necessary and sufficient conditions.

4.9 NECESSARY AND SUFFICIENT CONDITIONS - NEWCOMB'S ANALYSIS

The analysis of Newcomb is now summarised and the method for deriving necessary and sufficient conditions for ideal MHD stability is discussed. The mathematical analysis is fairly complicated so, following Wesson, only the essential results are presented.

To illustrate the general approach, assume that the wavenumbers, m and k, are such that the Euler-Lagrange equation (4.94) is regular across the plasma from $r = a$ to $r = b$. So f does not have any zeros in this region.

$$\frac{d}{dr}\left[f\frac{d\xi}{dr}\right] = g\xi. \tag{4.109}$$

For illustration, assume that the required boundary conditions are

$$\xi(a) = \xi(b) = 0. \tag{4.110}$$

Now integrating (4.109) from $r = a$ (using $d\xi/dr = 1$ as a normalising condition) will result in one of the three possible types of solution. For curve 1, the solution does *not* have a zero for $a < r < b$. For curve 2, the solution is exactly zero at $r = b$ and so both boundary conditions are satisfied. For curve 3, the first zero occurs *between* $r = a$ and $r = b$. From (4.90), δ_2W is zero for curve 2 and so this case corresponds to marginal stability.

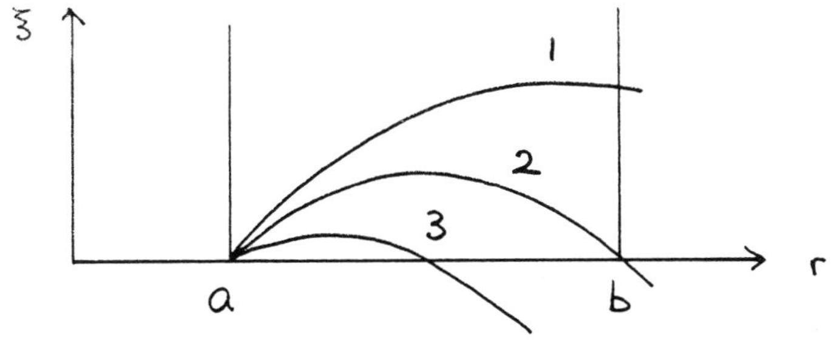

Figure 4.13 **Possible solutions for ξ**

$$\int_a^b f\left[\frac{d\xi}{dr}\right]^2 + g\xi^2 dr = \left[f\xi \frac{d\xi}{dr}\right]_a^b - \int_a^b \xi\left[\frac{d}{dr}\left[f\frac{d\xi}{dr}\right] - g\xi\right]dr = 0.$$

Curve 2, therefore, separates unstable solutions from stable solutions. It can now be shown that curve 3 corresponds to an unstable equilibrium by using the trial function shown below.

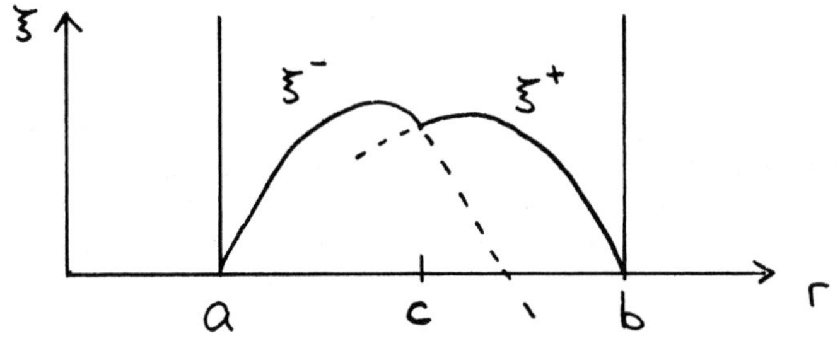

Figure 4.14 Trial function that shows $\delta_2 W < 0$

Solving the Euler Lagrange equation in the region $a < r < c$, with the appropriate boundary condition at $r = a$, gives the solution ξ^-, whereas in the region $c < r < b$ with the correct condition at $r = b$, gives ξ^+. The magnitude of the solutions can be adjusted so that they match at $r = c$. Now,

$$\delta_2 W = \int_a^b f\left[\frac{d\xi}{dr}\right]^2 + g\xi^2 dr = \int_a^c f\left[\frac{d\xi}{dr}\right]^2 + g\xi^2 dr + \int_c^b f\left[\frac{d\xi}{dr}\right]^2 + g\xi^2 dr,$$

$$= \left[f\xi^- \frac{d\xi^-}{dr}\right]\Bigg|_c - \left[f\xi^+ \frac{d\xi^+}{dr}\right]\Bigg|_c = (f\xi)_c \left[\frac{d\xi^-}{dr} - \frac{d\xi^+}{dr}\right]_{r=c} . \quad (4.111)$$

For curve 3, this expression is always negative and so the equilibrium is unstable if ξ has a zero before $r = b$. Hence, curve 1 corresponds to the stable situation.

If there are no singular points in the region $a < r < b$, the necessary and sufficient condition for stability is that the solution to the Euler-Lagrange equation which is zero at $r = a$ should not have a zero for $a < r < b$.

However, in practice, there will exist choices of m and k that make $f = 0$. In other words,

$$mB_\theta + krB_z = 0,$$

for some radius (perhaps more than one point). The region is now divided into sub-intervals on either side of the singularity or, if there are more singularities, into sub-intervals between singular points. In each sub-interval, the above type of trial function will correctly deduce the sign of $\delta_2 W$. The main difference is that the Euler-Lagrange solution near the singular point, r_s, is

$$\xi = a(|r - r_s|)^{\nu_+}, \tag{4.112}$$

where a is a constant and ν_+ is the larger root of the indicial equation (8.17), namely,

$$\nu_+ = -\tfrac{1}{2} + (\tfrac{1}{4} - c)^{\frac{1}{2}}. \tag{4.113}$$

Newcomb then showed that, for fixed m and k, a necessary and sufficient condition for stability is that in all sub-intervals (i) Suydam's criterion is satisfied and (ii) the "small" (i.e. (4.112) and (4.113) or (4.110)) Euler-Lagrange solution, at one end, has no zeros in the sub-interval.

4.10 RESISTIVE INSTABILITIES - TEARING MODES

4.10.1 Introduction

The inclusion of uniform resistivity in the induction equation,

$$\frac{\partial \mathbf{B}}{\partial t} = \nabla \times (\mathbf{v} \times \mathbf{B}) + \eta \nabla^2 \mathbf{B}, \tag{4.114}$$

allows a new class of instabilities. Three types of resistive instabilities are discussed in Furth, Killeen and Rosenbluth (1963) (FKR) namely (i) the resistive gravity mode, (ii) the rippling mode and (iii) the tearing mode which are driven by (i) gravity and an unfavourable density distribution or by curvature, (ii) gradients in the resistivity (not considered here) and (iii) the non-uniform current density.

In this section, only the tearing mode is studied and the development follows Priest (1984). First of all, an intuitive description of the tearing mode is presented and a detailed analysis is left to the next section. In the ideal limit, the field is frozen into the plasma and field lines are not allowed to break and reconnect (field lines retain their identity). Thus, the effect of pressing together oppositely directed field lines of a current sheet is just to build up the magnetic field strength at the centre. However, when η is included, the field may now diffuse relative to the fluid (as in Fig 15(c)) and reconnect to change its topology (as in Fig 15 (d)). Thus, lower energy states are now accessible and the new instabilities can take place that were absent in the ideal limit. Two qualititative descriptions are now given.

The time-scale for diffusion for a field that changes rapidly over a length l (the width of the current sheet) is l^2/η. Thus, when the field is squeezed together, as in Fig 15(b), l becomes smaller and

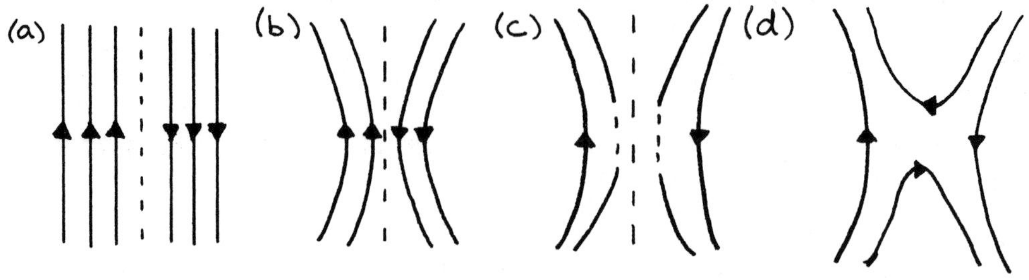

Figure 4.15 **Magnetic Reconnection**

so the field diffuses faster at the point of squeezing than elsewhere. This leads to the situation in Fig 15(c). Hence, the field lines must reconnect. The restoring force due to magnetic tension is reduced if the wavelength of the disturbance is long.

Alternatively, the current sheet may be represented by equally spaced current carrying wires. A small displacement that causes some wires to bunch together will then continue because of the mutual attraction of the wires.

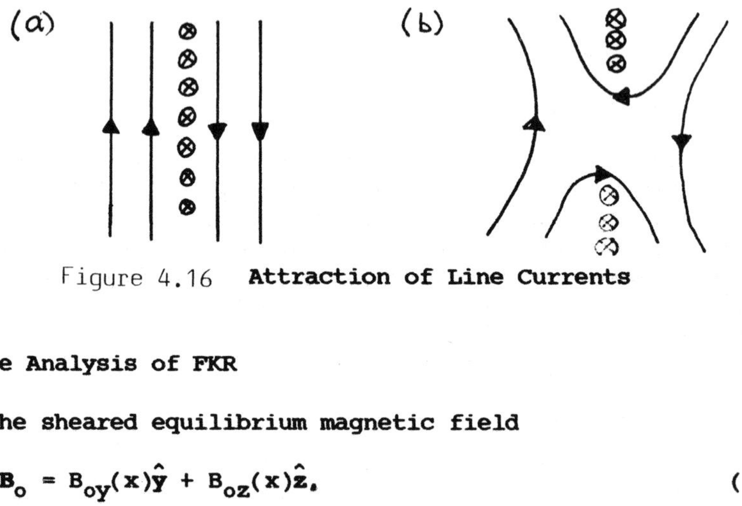

Figure 4.16 **Attraction of Line Currents**

4.10.2 The Analysis of FKR

Consider the sheared equilibrium magnetic field

$$\mathbf{B}_0 = B_{0y}(x)\hat{\mathbf{y}} + B_{0z}(x)\hat{\mathbf{z}}. \tag{4.115}$$

Then the linearised equations, for an incompressible plasma, are

$$\frac{\partial \mathbf{B}_1}{\partial t} = \nabla \times (\mathbf{v}_1 \times \mathbf{B}_0) + \eta \nabla^2 \mathbf{B}_1, \tag{4.116}$$

$$\rho_0 \frac{\partial \mathbf{v}_1}{\partial t} = -\nabla p_1 + (\nabla \times \mathbf{B}_1) \times \mathbf{B}_0/\mu + (\nabla \times \mathbf{B}_0) \times \mathbf{B}_1/\mu, \tag{4.117}$$

$$\nabla.\mathbf{B}_1 = \nabla.\mathbf{v}_1 = 0. \tag{4.118}$$

The Fourier components of the perturbed variables are

$$f(x)e^{i(k_y y + k_z z + \omega t)}$$

and so, using the x and z components of $\nabla \times$(4.117), the x component of (4.116) and (4.118), the resistive equations are

$$\omega B_{1x} = i v_{1x}(\mathbf{k}.\mathbf{B}_0) + \eta\left[\frac{d^2 B_{1x}}{dx^2} - k^2 B_{1x}\right], \quad (4.119)$$

$$\omega\left[\frac{d^2 v_{1x}}{dx^2} - k^2 v_{1x}\right] = \frac{i(\mathbf{k}.\mathbf{B}_0)}{\mu\rho_0}\left[-\frac{B_{1x}(\mathbf{k}.\mathbf{B}_0)''}{(\mathbf{k}.\mathbf{B}_0)} + \frac{d^2 B_{1x}}{dx^2} - k^2 B_{1x}\right].$$

(4.120)

The incompressible assumption is valid if the instability growth rate is much smaller than the Alfven rate and the fluid velocities are subsonic. $\mathbf{k}.\mathbf{B}_0$ and k^2 are defined as

$$\mathbf{k}.\mathbf{B}_0 = k_y B_{oy} + k_z B_{oz},$$
$$k^2 = k_y^2 + k_z^2, \quad (4.121)$$

and dashes denote derivatives with respect to x. The first term on the right-hand side of (4.119) normally dominates the second except in the neighbourhood of a singular layer, where $\mathbf{k}.\mathbf{B}_0 = 0$.

Equations (4.119) and (4.120) are nondimensionalised against a typical field strength, B_0, scale length, a, diffusion time, $\tau_d = a^2/\eta$ and Alfven travel time, $\tau_A = a/v_A$. Thus, using

$$\bar{\mathbf{B}} = \mathbf{B}/B_0, \quad \bar{\mathbf{v}}_1 = ik\tau_d \mathbf{v}_1, \quad \bar{k} = ka, \quad \bar{\omega} = \omega\tau_d, \quad \bar{x} = x/a.$$

(4.119) and (4.120) are rewritten as

$$\bar{\omega}\bar{B}_{1x} = -\bar{v}_{1x}F + (\bar{B}''_{1x} - \bar{k}^2\bar{B}_{1x}), \quad (4.122)$$

$$\bar{\omega}(\bar{v}''_{1x} - \bar{k}^2\bar{v}_{1x}) = S^2 F\left[-\bar{B}_{1x}\frac{F''}{F} + \bar{B}''_{1x} - \bar{k}^2\bar{B}_{1x}\right], \quad (4.123)$$

where $F = (\mathbf{k}.\mathbf{B}_0)/k$.

The Lundquist number, $S = \tau_d/\tau_A$, the ratio of the diffusion to Alfven travel times, is assumed to be much greater than unity.

It is assumed that the instability will grow on a hybrid time scale that is very much smaller than τ_d but very much greater than τ_A.

$$\tau_d >> \tau = \frac{1}{\omega} = \frac{\tau_d}{\bar{\omega}} >> \tau_A$$

or

$$1 << \bar{\omega} << S.$$

This can happen if $\bar{\omega}$ depends on S as

$$\bar{\omega} \sim S^c,$$

where $0 < c < 1$. $c = 0$ is the diffusive limit and $c = 1$ is the ideal limit.

Before analysing a specific equilibrium, estimates of the diffusion width (width of the boundary layer) and the tearing mode growth rate are obtained. Following Bateman (Chap 10), terms like d^2B_{1x}/dx^2 are only large in a thin boundary layer near the resonant surface, x_s. Hence, the derivative of B_{1x}, B'_{1x}, will appear to have a discontinuity here, when viewed on the large scale of the plasma as a whole. Therefore,

$$\frac{d^2B_{1x}}{dx^2} \approx \frac{B'_{1x}(x_s+\epsilon) - B'_{1x}(x_s+\epsilon)}{2\epsilon},$$

$$= \frac{\Delta' B_{1x}}{\epsilon}, \tag{4.124}$$

where, in keeping with the literature,

$$\Delta' \equiv \frac{B'_{1x}(x_s+\epsilon) - B'_{1x}(x_s-\epsilon)}{2B_{1x}}. \tag{4.125}$$

Now, inside the resistive layer all three terms of (4.119) are of similar size and so, using (4.124)

$$\omega \approx \eta\Delta'/\epsilon, \tag{4.126}$$

and

$$iv_{1x}(\mathbf{k}.\mathbf{B}_0) \approx \eta\Delta' B_{1x}/\epsilon. \tag{4.127}$$

In the resistive layer $(\mathbf{k}.\mathbf{B}_0)$ is approximately $(\mathbf{k}.\mathbf{B}_0)'\epsilon$ and the inertia term in (4.120) becomes important. Setting

$$\frac{d^2v_{1x}}{dx^2} \sim \frac{v_{1x}}{\epsilon^2},$$

and balancing with the second term on the right-hand side of (4.120), (this is the main driving term due to an increase in the sheet current) gives

$$\frac{\omega v_{1x}}{\epsilon^2} \approx \frac{i(\mathbf{k}.\mathbf{B}_o)'}{\mu\rho_o}\Delta' B_{1x}. \tag{4.128}$$

Then using (4.127) the width of the resistive layer is given in terms of the growth rate

$$\epsilon^4 \approx \omega\mu\rho_o\eta/(\mathbf{k}.\mathbf{B}_o)'^2. \tag{4.129}$$

Substituting (4.129) into (4.126) and rearranging defines the growth rate as

$$\omega \sim (\mathbf{k}.\mathbf{B}_o)'^{2/5}\eta^{3/5}\Delta'^{4/5}/(\mu\rho_o)^{1/5}.$$

This may be manipulated (see Bateman) to give

$$\omega \sim \tau_d^{-3/5}\tau_A^{-2/5}\Delta'^{4/5} \tag{4.130}$$

$\bar{\omega}$ then depends on S to the power 2/5. ($c = 2/5$). This simple-minded argument is now backed up by the analysis of a particular equilibrium namely, in dimensionless variables,

$$\bar{B}_{oy} = \begin{cases} 1 & \bar{x} > 1, \\ \bar{x} & |\bar{x}| < 1, \\ -1 & \bar{x} < -1. \end{cases} \tag{4.131}$$

Setting $k_z = 0$, $k_y = k$, then the singular layer is at $\bar{x} = 0$. The solution to (4.122) and (4.123) is obtained in the *outer* region (away from $\bar{x} = 0$) firstly, then secondly in the *inner* region (near $\bar{x} = 0$) and finally the two solutions are *matched* to obtain the dispersion relation.

In the *outer* region, the plasma is assumed to be ideal and so terms of order S^{-1} are neglected. (4.123) then becomes

$$\bar{B}''_{1x} - \bar{k}^2\bar{B}_{1x} - (\bar{B}''_{oy}/\bar{B}_{oy})\bar{B}_{1x} = 0 \tag{4.132}$$

or

$$\bar{B}''_{1x} - \bar{k}^2\bar{B}_{1x} = 0,$$

and so the solution which vanishes at large distances is

$$\bar{B}_{1x} = \begin{cases} e^{+\bar{k}\bar{x}} & \bar{x} < -1, \\ \alpha_- \sinh \bar{k}\bar{x} + \beta_- \cosh \bar{k}\bar{x} & -1 < \bar{x} < 0, \\ \alpha_+ \sinh \bar{k}\bar{x} + \beta_+ \cosh \bar{k}\bar{x} & 0 < \bar{x} < 1, \\ e^{-\bar{k}\bar{x}} & 1 < \bar{x}. \end{cases}$$

The unknown coefficients α_+, α_-, β_+ and β_- are determined by matching the solutions at $x = 1$ and $x = -1$. The matching conditions are obtained from (4.132) and are, at $x = 1$,

$$\bar{B}_{1x}(1_+) = \bar{B}_{1x}(1_-) \tag{4.133}$$

and

$$\frac{\bar{B}'_{1x}(1_+)}{\bar{B}_{1x}(1_+)} - \frac{\bar{B}'_{oy}(1_+)}{\bar{B}_{oy}(1_+)} = \frac{\bar{B}'_{1x}(1_-)}{\bar{B}_{1x}(1_-)} - \frac{\bar{B}'_{oy}(1_-)}{\bar{B}_{oy}(1_-)} . \tag{4.134}$$

These imply

$$\alpha_+ = \frac{e^{-\bar{k}}\cosh \bar{k} - 1}{\bar{k}} , \quad \beta_+ = 1 - \frac{e^{-k}\sinh \bar{k}}{\bar{k}} . \tag{4.135}$$

The conditions at $x = -1$ give

$$\alpha_- = - \alpha_+ , \quad \beta_- = \beta_+ . \tag{4.136}$$

$\bar{B}_{1x}$ is continuous at the origin but $\bar{B}'_{1x}$ is not. The jump in $\bar{B}'_{1x}/\bar{B}_{1x}$ is

$$\Delta' = \frac{2\alpha_+\bar{k}}{\beta_+} = \frac{2\bar{k}(e^{-\bar{k}} \cosh \bar{k} - \bar{k})}{(\bar{k} - e^{-\bar{k}} \sinh \bar{k})} . \tag{4.137}$$

In the *inner* region, F has become so small that inertia and diffusion are now important. Variables are now changing over a distance ϵa instead of the outer distance a, where from (4.129)

$$\epsilon^4 = \bar{\omega}/4\bar{k}^2 S^2 . \tag{4.138}$$

Stretched variables

$$X = \frac{x}{\epsilon} , \quad V_{1X} = \bar{v}_{1x}(4\epsilon/\bar{\omega}) , \quad B_{1X} = \bar{B}_{1x}$$

allow (4.122) and (4.123) to be rewritten as

$$\ddot{B}_{1X} = \epsilon^2\bar{k}B_{1X} + \epsilon^2\bar{\omega}(B_{1X} + \tfrac{1}{4} XV_{1X}),$$
$$\ddot{V}_{1X} = V_{1X}(\bar{k}^2\epsilon^2 + \tfrac{1}{4} X^2) + B_{1X}X, \qquad (4.139)$$

where dots denote differentiation with respect to X. (4.139) is normally solved numerically and Δ'_{inner} is calculated at X = 1. Assuming $\bar{k} \ll 1$, $\bar{\omega}$ (long wavelength), (4.139) can be approximated by

$$\ddot{B}_{1X} \approx \epsilon^2\bar{\omega}B_{1X} \, ,$$

$$B_{1X} \approx \cosh \sqrt{\epsilon^2\bar{\omega}}\, X \, ,$$

$$\Delta'_{inner} = 2\left[\frac{\bar{B}_{1X}}{\bar{B}_{1X}}\right]_{\bar{x}\sim\epsilon} = 2\left[\frac{\dot{B}_{1X}}{\epsilon B_{1X}}\right]_{X=1} = \frac{2\sqrt{\epsilon^2\bar{\omega}}\,\sinh\sqrt{\epsilon^2\bar{\omega}}}{\epsilon\cosh\sqrt{\epsilon^2\bar{\omega}}} \, ,$$

$$\Delta'_{inner} \sim \frac{2\epsilon^2\bar{\omega}}{\epsilon} = 2\epsilon\bar{\omega}. \qquad (4.140)$$

Equating (4.137) and (4.140), gives ($\bar{k} \ll 1$)

$$2\epsilon\bar{\omega} = \frac{2}{k}$$

and, eliminating ϵ using (4.138),

$$\bar{\omega} = \left[\frac{2S}{\bar{k}}\right]^{2/5}.$$

If B_{1X} is approximately constant in the inner region (the constant-ψ approximation where $\psi = B_{1X}$ in the FKB notation), (4.139) can be solved in terms of Hermite functions giving, when $\bar{k}\epsilon \ll 1$,

$$\Delta' = 3\epsilon\bar{\omega}.$$

The dispersion relation can be written as

$$\Delta' = \Delta'_{inner}(\bar{\omega}) \, ,$$

but, since $\Delta'_{inner}(0) = 0$, the question of stability can be answered by investigating the sign of Δ' in the outer region.

$\Delta' < 0$ STABLE,

$\Delta' > 0$ UNSTABLE.

For this example,

$$\Delta' = 0 \Rightarrow \bar{k} = e^{-\bar{k}} \cosh \bar{k}$$

or

$$\bar{k} \approx 0.64.$$

(See Priest 1984 for modifications and restrictions. Wesson gives a readable introduction to the tearing mode and the resistive gravity mode.)

4.11 APPLICATIONS OF MHD INSTABILITIES

4.11.1 Introduction

There are many astrophysical situations that can be described by MHD stability theory. For example, coronal loops and prominences exist for longer than the relevant timescales in the solar corona and, therefore, must be in a stable configuration. However, these objects sometimes erupt over the short Alfven time and presumably an instability has been triggered. Examples are solar flares, both the simple loop and two ribbon flares, and erupting prominences. The energy released during a solar flare is the order of 10^{25}J and it must be stored in the coronal magnetic field since there is no other adequate energy source. MHD stability theory has concentrated on obtaining critical conditions for the onset of a solar flare and the amount of energy stored in the coronal magnetic fields.

The main stabilising effect is due to the extremely dense photospheric material, $\rho_{corona}/\rho_{photosphere} < 10^{-8}$, which anchors the magnetic footpoints. Mathematically, this has been modelled in two different ways. If a disturbance $\boldsymbol{\xi}$ is generated in the corona then at the photospheric boundary either

$$\boldsymbol{\xi} \times \mathbf{B} = 0 \text{ and } \boldsymbol{\xi}.\mathbf{B} \text{ constrained so that energy is conserved,} \tag{4.141}$$

as discussed by Van Hoven *et al.* (1981), or

$$\boldsymbol{\xi} = 0, \tag{4.142}$$

as discussed by Rosner *et al.* (1984), where **B** is the equilibrium magnetic field. One would perhaps expect the latter to be the correct choice on physical grounds that waves would be totally reflected.

4.11.2 Ideal Kink Instability of Coronal Loops

A simple model of a coronal loop is shown below

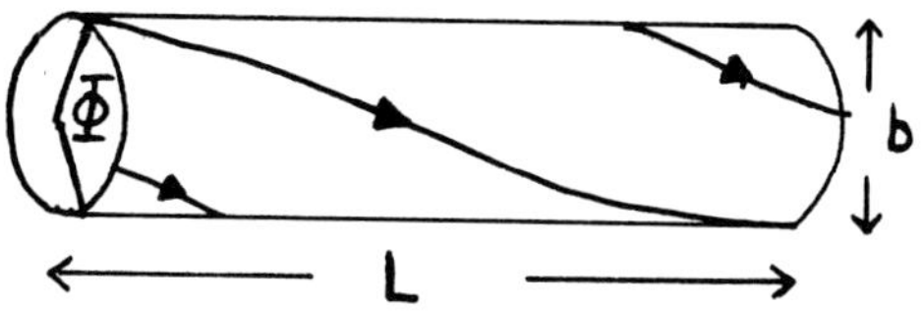

Figure 4.17 **A coronal loop**

Such a loop is remarkably stable but it may become unstable if the field is either twisted or stretched beyond a critical value. For example, the force-free field $B_z = B_o/(1 + (r/b)^2)$ $B_\theta = B_z(r/b)$ where $\Phi = L/b$ has been studied by Hood and Priest (1981) and Einaudi and Van Hoven (1983). Using the trial function $\xi = \xi(r,z)e^{im\theta}$, Hood and Priest solved the resulting Euler-Lagrange equations numerically. The critical value of twist, $\Phi_c = 2.5\pi$, was obtained. Einaudi and Van Hoven derived the same value using a truncated Fourier series

$$\xi = \sum_{n=1}^{N} \xi_n(r)\sin \frac{n\pi z}{L} e^{im\theta} \text{ with N typically 5.}$$

A more detailed account of line-tying can be found in Priest, 1982, Chapters 7 and 10.

4.11.3 Two-Ribbon Flares

The large two-ribbon flares appear to have a different geometry from the small loop flares. They occur when an active region filament and the overlying field becomes unstable and are modelled by an arcade structure. The equilibrium is 2-D and variations along the length of

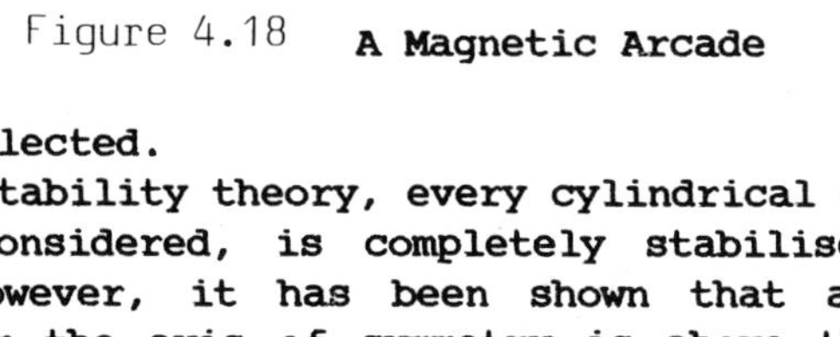

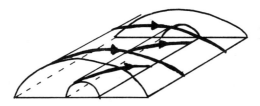

Figure 4.18 **A Magnetic Arcade**

the arcade are neglected.

Using ideal stability theory, every cylindrical force-free arcade, so far considered, is completely stabilised by line tying (Hood, 1983). However, it has been shown that an instability is possible if either the axis of symmetry is above the photosphere or pressure gradients are included (Migliuolo *et al.*, 1984, Cargill *et al.*, 1985). For the field

$$B_\theta = \frac{B_o(r/b)}{(1+r^2/b^2)}, \quad B_z = \frac{\lambda B_o}{(1+r^2/b^2)}, \quad p = \frac{B_o^2}{2\mu} \frac{(1-\lambda^2)}{(1+r^2/b^2)^2},$$

a trial function $\xi(r,\theta)e^{ikz}$ was substituted into (4.47) and the resulting Euler-Lagrange solved numerically (Cargill *et al.*, 1985). There exists a critical value of λ, λ_c say, such that for $\lambda < \lambda_c$ the field is unstable and stable for $\lambda > \lambda_c$. The value of λ_c depends on the choice of line tying boundary conditions. For (4.141) $\lambda_c = 0.48$ but for (4.142) $\lambda_c = 0.02$.

A detailed discussion of two-ribbon flares is presented in Priest, 1982, Chapter 10.

REFERENCES

Bateman, G (1980) "MHD Instabilities", MIT Press.

Boyd, T and Sanderson, J (1969) "Plasma Physics", Nelson.

Cargill, P, Hood, A W and Migliuolo, S (1985) Astrophys J submitted.

Einaudi, G and Van Hoven, G (1983) Solar Phys **88**, 168.

Furth, H, Killeen, J and Rosenbluth, M (1963) Phys Fluids **6**, 459.

Hood, A W (1983) Solar Phys **87**, 279.

Hood, A W and Priest, E R (1981) Geophys Astrophys Fluid Dynamics **17**, 297.

Kadomstev, B (1966) Rev Plasma Phys **2**, 153-199.

Kruskal, M and Tuck, J (1958) Proc Roy Soc London **A245**, 222.

Kulsrud, R (1962) in "Advanced Plasma Theory", Int School of Physics Course, **25**, Varenna.

Laval, G, Mercier, C and Pellat, R (1964) Nuclear Fusion **5**, 156.

Low, B C (1982) Rev Geophys Space Phys **20**, 145.

Melville, J, Hood, A W and Priest, E R (1983) Solar Phys **87**, 301.

Melville, J, Hood, A W And Priest, E R (1984) Solar Phys **92**, 15.

Migliuolo, S, Cargill, P and Hood, A W (1984) Astrophys J **281**, 413.

Newcomb, W (1960) Annals of Physics **10**, 232.

Priest, E R (1982) "Solar Magnetohydrodynamics", D Reidel.

Priest, E R (1984) Reports on Progress in Physics, to appear.

Roberts, P H (1967) "An Introduction to Magnetohydrodynamics", Longman Green.

Rosner, R, Low, B C and Holzer, T (1984) "Physics of the Sun", Ed P Sturrock, T Holzer, D Mihalas and R Ulrich.

Suydam, B (1958) IAEA Geneva Conference **31**, 157-159.

Van Hoven, G, Ma, S and Einaudi, G (1981) Astron. Astrophys. **97**, 232.

Wesson, J (1981) Chap 9 of "Plasma Physics and Nuclear Fusion Research", Ed. R D Gill, Academic Press.

CHAPTER 5

MAGNETIC RECONNECTION

Stanley W.H. Cowley
The Blackett Laboratory
Imperial College of Science and Technology
London SW7 2BZ

5.1 INTRODUCTION

Magnetic reconnection is a phenomenon of considerable importance in solar system plasmas. In the solar corona it results in the rapid release to the plasma of energy stored in large-scale magnetic configurations which become unstable, resulting in solar flares, while small-scale reconnection may play a role in heating the coronal plasma which leads to the outflow of the solar wind. Reconnection also results in the formation of magnetically "open" planetary magnetospheric field structures, leading to efficient coupling of solar wind momentum into the magnetospheres via magnetic stresses, as well as plasma mass exchange along the "open" flux tubes. In the extended magnetic tails formed by the solar wind interaction with solar system bodies, the onset of rapid reconnection between the tail lobes can produce large-scale dynamical plasma-field reconfigurations which are associated with auroral substorms on Earth and structure in the plasma tails of comets. Major comet tail disconnection events have also been suggested to result from dayside reconnection following changes in the direction of the solar wind magnetic field.

These examples should serve to indicate the importance of the reconnection process in a solar system plasma context. Reconnection has also been much studied in relation to laboratory devices such as tokamak fusion machines, where it can cause major disruption to the plasma confinement. The process also most probably plays an important role in astrophysical plasma systems such as accretion discs, and in various current sheet interface regions formed in interstellar and intergalactic space. For all these reasons the theory of reconnection has been actively pursued over the past thirty years, and remains an active research field to the present day. It is not the aim of the present chapter, however, to provide a detailed review of the latest developments of this subject. For these purposes the reader should consult the proceedings, edited by Hones (1984), of a recent conference devoted specifically to reconnection, or the relevant section of the NASA STP workshop report edited by Butler and Papadopoulos (1984). Rather it is our aim to provide an introduction to the theory of magnetic reconnection which emphasizes the physical principles involved, in keeping with the aims of the summer school lectures out of which this chapter evolved. The second section below thus discusses what reconnection is and why it is important in large-scale cosmic plasmas. The third and fourth sections then discuss two aspects of reconnection theory which have developed along essentially independent paths in the past i.e. theory based on MHD which is directly applicable to collisional solar coronal plasmas, and theory based on the study of single particle motion in model electromagnetic fields which is applicable to collision-free solar wind and magnetospheric plasmas. In both cases the development of detailed theory is demanding, not least because the MHD problem involves a localized but essential breakdown of the usual "frozen-in field" approximation at the reconnection site, while the single particle problem similarly involves a related violation of the usual "guiding centre" approximation for particle motion. It will be shown, however, that when self-consistency considerations are applied to the single particle approach the resulting picture is remarkably similar to that based on MHD. This conclusion should not occasion too much surprise since the major features discussed here are principally determined by conservation of mass, momentum and energy.

5.2 RECONNECTION: WHAT IT IS AND WHAT IT DOES

The aim of this section is to introduce the principal physics of the reconnection process and to indicate why it occupies a position of such importance in solar system (and other) plasma systems. In so doing we will also introduce some basic nomenclature and definitions.

For purposes of this discussion it is sufficient to consider a plasma obeying a simple Ohm's law of the form

$$\underline{E} + \underline{v} \wedge \underline{B} = \underline{j}/\sigma. \qquad (5.1)$$

The lhs of (5.1) is the electric field in the rest frame of the plasma, while the only term from the generalized Ohm's law to be retained on the rhs is a simple Ohmic current proportional to scalar conductivity σ (the latter being determined by either particle-particle or wave-particle collision times). Substituting $\underline{E}$ into Faraday's law and using Ampere's law with the displacement current neglected then yields the induction equation which determines how $\underline{B}$ varies in time

$$\frac{\partial \underline{B}}{\partial t} = \text{curl } (\underline{v} \wedge \underline{B}) + \frac{\nabla^2 \underline{B}}{\mu_0 \sigma} \qquad (5.2)$$

The field behaviour then depends on which term on the rhs of (5.2) is dominant. The first term describes convection of the field with the plasma flow, and if it dominates such that the second term may be neglected then the field and the flow are "frozen" together such that a set of fluid elements which are initially located along a given field line remain so connected in all subsequent motion. Neglect of the second term in equation (5.2) is called the perfect conductivity approximation irrespective of the actual value of σ, since it is the limit obtained by putting $\sigma \to \infty$ in that equation. The frozen-in flow which occurs in that limit may be pictured either in terms of magnetic field lines being convected along with the flow (any resulting field distortions then reacting back on the plasma motion via the $\underline{j} \wedge \underline{B}$ force), or we may consider the magnetic flux tubes themselves as moving, carrying with them the plasma they contain. Which of these equivalent pictures is more appropriate in a given situation is determined by the relative energy densities in the flow and in the magnetic field.

An opposite limit occurs if the second term on the rhs of (5.2) dominates the first. In this case the equation becomes a diffusion equation, and we may picture the field lines as diffusing through the plasma down the field gradients, such as to reduce those gradients (i.e. the current densities). There is then essentially no coupling between the field and the fluid flow.

The relative magnitude of the two terms on the rhs of (5.2) is conveniently summarized in the magnetic Reynolds number of the plasma, a dimensionless ratio given by

$$R_m = \mu_0 \sigma v L \approx \frac{|\mathrm{curl}\ (\underline{v} \wedge \underline{B})|}{\left|\frac{1}{\mu_0 \sigma} \nabla^2 \underline{B}\right|} \tag{5.3}$$

where v is a characteristic speed in the flow and L a characteristic length of the plasma system. If R_m is large convection dominates and "frozen-in flow" prevails, while if R_m is small diffusion dominates and the coupling to the plasma is weak. Now in solar system plasmas very large values of R_m are guaranteed by the very large over-all spatial scales of the plasma. We find for example $R_m \sim 10^8$ for the solar flare problem, while $R_m \gtrsim 10^{11}$ would be appropriate to solar wind and magnetospheric problems. It is to be emphasized that in deriving these R_m values the overall size of the plasma system has been used as L in (5.3) e.g. $\sim 10^4$km in the flare system, a few solar radii for the solar wind and some fraction of the cavity radius for the case of planetary magnetospheres. The resulting very large R_m values then clearly show that on these spatial scales field convection is overwhelmingly dominant and that the effects of diffusion can be entirely neglected. Evidence for the validity of this conclusion is readily seen in the magnetically ordered coronal plasma structures observed during eclipses, in the spiral structure of the interplanetary magnetic field which is carried out into the solar system frozen into the solar wind flow, and in the magnetic organization of planetary magnetospheres.

On the basis of the above discussion it seems justifiable to neglect diffusive processes in solar system plasmas, and if this is done it leads to strong constraints on the behaviour of the plasma since all cross-field mixing of plasma elements is supressed in this limit. Particles may freely mix along field lines (within any limitations imposed e.g. by magnetic mirror, gravitational or other forces), but they are completely ordered cross-field since they always remain tied to the same field line as it convects in the plasma flow.

Let us then consider the important problem of what happens when two initially separate plasma regimes come into contact with each other, as occurs, for example, in the interaction between the solar wind and a planetary magnetic field. Assuming from above that the perfect conductivity approximation is justified such that each plasma is frozen to its own magnetic field and diffusion between them is absent, then we must conclude that the two plasmas will not mix but instead that a thin boundary layer will form between them separating the two plasma populations and magnetic fields. In equilibrium the location of the boundary layer will be determined by pressure balance. Since in general the frozen fields on either side of the boundary will have differing strengths and orientations tangential to the boundary the layer must also constitute a current sheet (a tangential discontinuity). Use of the perfect conductivity limit thus leads to the prediction that in plasma systems space becomes divided into separate cells wholly containing the plasma and field from individual sources, and separated from each other by thin current sheets. The locations of the boundaries are determined by forces normal to the boundary, but otherwise interactions across the boundaries are weak in terms e.g. of mass and tangential momentum exchange.

This "separate cell" picture clearly often forms at least an excellent zeroth order approximation to the interaction of solar system plasma systems, as witness e.g. the existence of well-defined planetary magnetospheres. It must now be remembered, however, that the large R_m values on which use of the perfect conductivity approximation was justified were derived using the large over-all spatial scales of the systems involved. But strict application of this limit to the problem of the interaction of separate plasma populations then immediately leads to the conclusion that structures will almost inevitably be formed having small spatial scales, at least in one dimension i.e. the thin current sheets constituting the cell walls. It is certainly not guaranteed by the above arguments that the effects of diffusion can be neglected in discussing the physics of these boundaries, even though it may be totally negligible in describing the large-scale behaviour within the cells themselves. We will therefore now go on to consider the effects of diffusion in these boundaries, and will in fact show that the localized breakdown of the perfect conductivity approximation in the boundary regions which diffusion produces not only has impact on the properties of the boundary regions themselves but can also have a decisive influence on the behaviour of the large-scale plasma regimes where the perfect conductivity approximation does remain valid. In this lies the subtlety and significance of the reconnection process.

If for simplicity we first assume that the plasma in the boundary is not flowing such that (5.2) reduces to a diffusion equation, then the effect of diffusion in the boundary is to cause the current sheet to widen with time. The decreasing magnetic energy of the system is converted to plasma energy by Joule heating. Suppose, for example, that equal and opposite fields of strength B_0 occur on either side of the boundary separated by a magnetic neutral sheet in the geometry shown in Figure 5.1 (a). If at t=0 there exists a step function in the magnetic field across the neutral sheet (i.e. a current δ-function), so

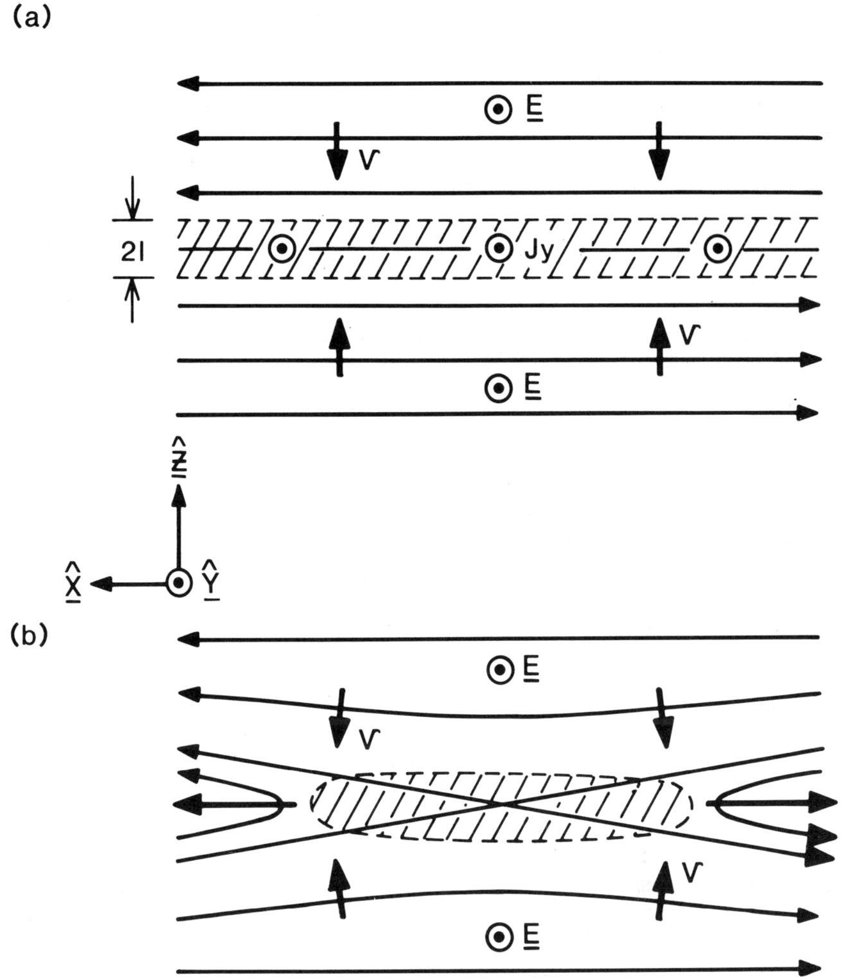

Figure 5.1 Sketch of the magnetic field (solid lines) and flow geometry (arrows) appropriate to (a) field annihilation and (b) magnetic reconnection. The diffusion regions where finite conductivity is important is shown by the hatched areas, surrounded by the unhatched convection regions. The directions of the electric field and current flow out of the plane of the diagram are shown by the circled dots.

$B_x = \pm B_0$ for $Z \gtrless 0$ then the solution of (5.2) for the variation of the field with time, with $\underline{v} = \underline{0}$ is

$$\frac{B_x}{B_0} = \frac{2}{\sqrt{\pi}}\int_0^{\zeta} e^{-u^2}\,du = \text{erf}\,(\zeta) \text{ where } \zeta = \left(\frac{\mu_0\sigma}{t}\right)^{\frac{1}{2}} Z \qquad (5.4)$$

(see e.g. Axford (1984)). The current distribution is gaussian in Z but has a width which increases in time as $\sqrt{t}$ i.e. the expansion of the current layer and the conversion of field energy to plasma energy is initially rapid but continuously decreases due to the decrease in the field gradients. The behaviour of the field is illustrated in Figure 5.2 (a) which shows B_x (Z) at various times on the left and the corresponding current distributions on the right. One may picture the antiparallel field lines diffusing down field gradients, through the plasma towards the neutral sheet on either side and there annihilating each other. However the process is self-limiting since the diffusion destroys the gradients which lead to the energy conversion, resulting in an overall energy conversion which decreases in time as $t^{-\frac{1}{2}}$.

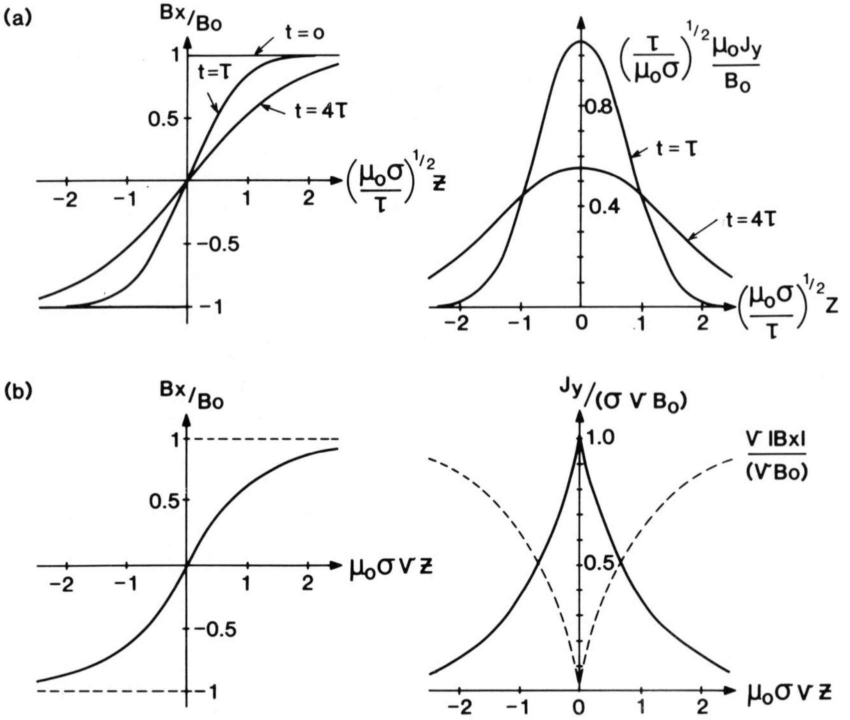

Figure 5.2 Graphs of magnetic field (left) and current density (right) distributions for the solution given by (a) equation (5.4) and (b) equation (5.6).

Rapid flux annihilation and energy conversion to the plasma can, however, be maintained if plasma flows in towards the neutral sheet from either side convecting flux into the current sheet to replace that lost by annihilation (see Figure 5.1 (a)). Inward convection causes the current sheet to thin with time while diffusion, as we have seen, causes it to expand. An equilibrium system can then be achieved in which inward convection maintains a sufficiently steep gradient that flux annihilation just balances the flux input. The faster the inflow, the steeper are the gradients required (i.e. the thinner is the current sheet), and the larger is the rate of energy liberated from the magnetic field to the plasma. The half-width ℓ of the equilibrium current sheet for a given inflow speed v may readily be estimated in several equivalent ways. First in terms of magnetic flux, if the inflow speed is v then the flux convected into the current sheet per unit time per unit length (y-direction, see Figure 5.1 (a)) is vB_0 from each side. For a sheet of half width ℓ the flux annihilated is, from (5.2), approximately given by $B_0/\mu_0\sigma\ell$. Equating these two then yields

$$\ell \simeq \frac{1}{\mu_0\sigma v} \tag{5.5}$$

Equivalently we may note that the inflow from either side and the current near the neutral sheet are associated with an electric field E_y (Figure 5.1 (a)), and that for a steady state system which is independent of y Faraday's law (curl $\underline{E}$ = $\underline{0}$) requires that this electric field be spatially uniform. Thus outside the current sheet where j_y is small and convection dominates (the exterior <u>convection region</u>) Ohm's law (5.1) is

$$E_y \simeq vB_0$$

while within the current sheet where j_y/σ is larger than or comparable with v_zB_x and diffusion is important (the <u>diffusion region</u>) Ohm's law is

$$E_y \simeq j_y/\sigma \quad .$$

Equating these two and putting $j_y \simeq B_0/\mu_0\ell$ from Ampere's law then again yields (5.5) for ℓ. It is then interesting to calculate the magnetic Reynolds number associated with the current sheet, based on width ℓ. From (5.3) we have

$$R_m^* = \mu_0\sigma v\ell \simeq 1$$

so that the sheet width is just sufficiently small that diffusion and convection are comparable in importance, as should be expected from the physics of the steady-state situation. An analytic solution of (5.2) exhibiting these properties may be obtained by imposing inexorable plasma motion at constant speed v into current sheet from either side so that $v_z=\mp v$ for $Z \gtrless 0$. Assuming $B_x = \pm B_0$ at $Z = \pm\infty$ as in the previous example the steady-state solution is

$$\frac{B_x}{B_0} = \pm\ (1 - \exp\ (\mp\ \mu_0\sigma\ vZ)) \tag{5.6}$$

where the upper and lower signs refer to Z>0 and Z<0 respectively. This solution is shown in Figure 5.2(b), the field again being shown on the left and the current density on the right. The dashed line on the right also shows vB_x for comparison with j_y/σ in order to illustrate the transition from the outer convection region ($vB_x \gg j_y/\sigma$) to the inner diffusion region ($j_y/\sigma \gtrsim vB_x$) of the solution. We should also note at this point that a related but more realistic exact solution of the steady-state annihilation problem has been found by Sonnerup and Priest (1975) based on stagnation-point (hyperbolic streamline) flow.

The discussion above is sufficient to show that boundary processes in the presence of diffusion may have considerably more interest than the inert boundary layers which emerge from a strict and inappropriate application of the perfect conductivity limit to such regions. However, the one-dimensional "annihilation" systems discussed above are clearly unrealistic in two important respects. Firstly, we have plasma inflow into the current sheet from either side, but have ignored the question of its exit. Secondly the field structure consists of infinitely long straight field lines with opposite directions on either side of the current layer which annihilate each other across the neutral sheet. A more realistic field and flow structure which would occur in a boundary region is shown in Figure 5.1(b), this being the geometry of magnetic reconnection. Instead of the infinite neutral sheet planes of the field annihilation systems previously discussed the field now contains a magnetic neutral line (where $\underline{B} = \underline{0}$) located at the centre of the X-type field configuration and extending out of the plane of the diagram. Field lines flow in towards the current sheet from either side, as before, again associated with an electric field pointing out of the plane of the diagram as shown, but have a concave shape as they approach the boundary so that initially they just touch their opposite numbers convecting in from the other side of the sheet at one point only (the neutral line). If we then follow any of these field lines as they convect with the flow through this structure then they appear to become 'cut' as they map to the neutral line and 'reconnected' to a similarly cut field line convecting in from the opposite side of the current sheet. This description of the field behaviour gives rise to the origin of the term "reconnection" (Dungey, 1958). The reconnected field lines form wedges of highly distended field loops lying downstream from the neutral line and threading across the boundary current sheet between the two inflow plasmas. The magnetic tension of these loops accelerates the inflowing plasma away from the neutral line on either side and expels it along the boundary, thus answering the question raised above concerning the exit of this plasma. As the reconnected field loops move with the plasma along the boundary they contract in length and steadily liberate their magnetic energy to the plasma (unlike the sudden "annihilation" along the full length of the tubes which occurs in the neutral sheet systems). The magnetic energy then appears in the form of heated, accelerated plasma jets flowing along the boundary away from the neutral line on either side. It should also be noted that we may generally expect the boundary layer plasma to be accelerated away from the neutral line to sufficiently large speeds that the $\underline{v} \wedge \underline{B}$ term again dominates $\underline{j}/\sigma$ in Ohm's law, even at the centre of the boundary layer at sufficient distance from the neutral line. Thus the "diffusion region" (shown hatched in Figure 5.1(b)) where finite conductivity is important will in general extend only a finite distance along the boundary from the neutral line. Thus although inclusion of finite conductivity is still essential in the description of the region immediately surrounding the neutral line where perforce $\underline{v} \wedge \underline{B} \not\to \underline{0}$ and $\underline{E} \simeq \underline{j}/\sigma$, this "diffusion region" will generally occupy only a small but central part of the system and will be wholly surrounded by a much larger "convection region" where the effects of diffusion are negligible.

In order to illustrate the importance of magnetic reconnection we will now discuss two solar system examples in which we will first describe expectations based on the frozen-field limit, and will then discuss the effects of the inclusion of finite conductivity and reconnection. In Figure 5.3 we show the structure which would result from the

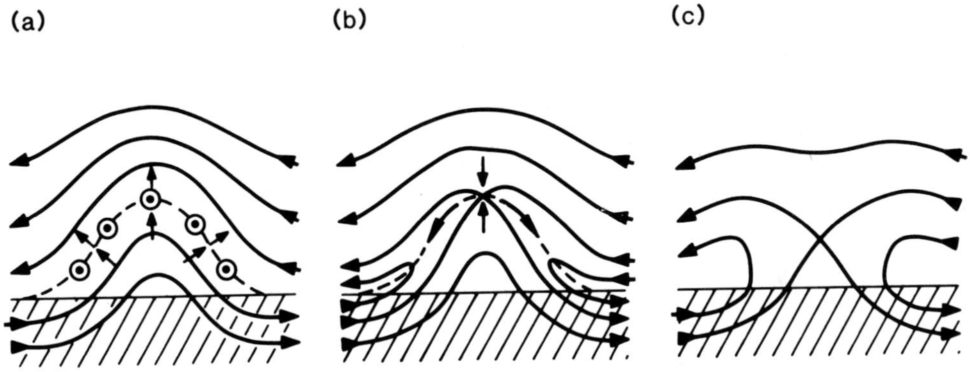

Figure 5.3 Flux emerges from the solar photosphere (hatched region) and interacts with overlying coronal plasma and field. (a) In the frozen-field limit the two plasmas remain distinct and a thin current sheet (dashed line) forms between them. (b) Reconnection results in field interconnection and plasma mixing across the boundary current sheet. (c) The end state is a potential field obeying the new photospheric boundary condition; the magnetic energy is reduced compared with (a), having been fed into the surrounding plasma.

emergence of new flux from beneath the Sun's photosphere which rises and pushes aside the pre-existing overlying coronal plasma and field. Figure 5.3 (a) shows the structure expected from the frozen-field limit with the new and pre-existing plasmas and fields separated by a thin current sheet (circled dots indicating current flow out of the plane of the diagram). Figure 5.3(b) illustrates the onset of reconnection forming field loops connecting across the current sheet which accelerate the plasma away from the neutral line as shown, as well as allowing "new" and "old" plasmas to mix along the reconnected field lines. Figure 5.3(c) shows the end state as a potential (curl-free) field in the absence of further perturbations, completely different from the equilibrium expected on the basis of the frozen-field limit. It may be noted that this reconnection picture has been suggested as a model for small solar flares by Heyvaerts et al. (1977).

The second example shown in Figure 5.4 concerns the interaction between the solar wind flow and a planetary dipole field. The frozen-field picture is shown in Figure 5.4(a) in which the planetary field is confined to an asymmetric cavity surrounding the planet constituting

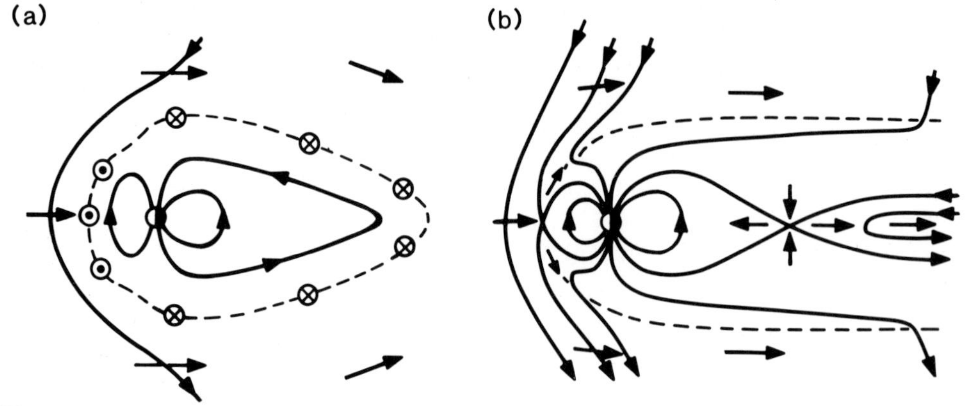

Figure 5.4 Magnetospheric models based (a) on strict application of the frozen-field limit and (b) on the occurrence of reconnection at the dayside current sheet between the planetary field and the solar wind magnetic field.

its magnetosphere, while the solar wind field is draped over and around the cavity as shown, before slipping with the flow around the sides. Planetary and solar wind plasmas are separated by a thin current sheet (the magnetopause) and plasma mass and momentum transfer across the boundary is weak. Figure 5.4(b) then shows the effect of reconnection at the dayside boundary current sheet (after Dungey (1961)). Field lines from the polar regions of the planet become directly connected into the solar wind, allowing the latter plasma ready access to the magnetospheric interior. The solar wind flow furthermore carries these field lines from the dayside to the nightside of the planet and stretches them out into a long magnetic tail. Subsequent reconnection of the oppositely-directed field lines of the two tail lobes at the tail centre then returns closed tubes back to the planet which flow through the central regions back to the dayside, where the process will repeat. Reconnection in this case thus also allows efficient coupling of solar wind momentum into the magnetosphere driving a large-scale internal cyclic convective flow. In the case of the Earth's magnetosphere this flow dominantly determines the structure and properties of the interior plasma populations.

These two examples should serve to show how a localized breakdown of the perfect conductivity approximation in the vicinity of a reconnection neutral line can lead to consequences of global significance, even though on the global scale itself the perfect conductivity approximation is quite valid. Not only does it allow reconfigurations of the magnetic field to take place towards states of lower energy, the energy appearing as accelerated and heated plasmas, but also the formation of reconnected field lines connecting across the boundaries between plasmas from differing sources allows effective exchanges of plasma mass, momentum and energy between them.

One question we have not so far considered is that of the generality of the reconnection process. In the above discussion we have effectively assumed throughout that the magnetic fields on either side of the boundary current sheet are antiparallel, whereas arbitrary relative angles will occur in general. Certainly in the fluid (MHD) picture, however, an arbitrary uniform field can always be added in the direction along the neutral line to any two-dimensional antiparallel field solution (e.g. a field out of the plane of the diagram in Figure 5.1 (b)) without disturbing the equilibrium. Thus solutions with twisted fields may readily be obtained (even though they may not be the most general twisted field case, as discussed by Cowley (1976)). Indeed in tokamak fusion machines, reconnection involves only the weak poloidal field components, and the much larger toroidal fields lying along the X-lines are not directly involved. We thus conclude that reconnection can occur for arbitrary relative orientations of fields on either side of a plasma boundary. The only case where reconnection cannot occur is the singular case where the fields on either side are exactly parallel.

Finally, before discussing reconnection models in more detail in subsequent sections we will now give some terminology and definitions. Because of the existence of the approximately uniform field parallel to the "neutral line" in the general case discussed above, it is not actually appropriate to call this line a "neutral line" (implying $\underline{B} = 0$) in this case. In this more general case the line is then often called simply a magnetic X-line or a magnetic separator line. In either case the field lines mapping to the X-line form surfaces in three dimensions which divide space into regions of field lines of different type (i.e. the field lines of the two inflow regions and the two wedges of reconnected field lines). These surfaces are called separatrices (singular separatrix). Lastly, we will briefly consider the definition of "reconnection" itself. As described above the essential feature of reconnection is the transfer of magnetic flux from the two inflow regions, across the separatrices, and into the reconnected field wedges. This process requires the existence of an electric field along the X-line. Following Dungey (1978), therefore, we define reconnection to be occurring if an electric field E exists along an X-line in a magnetic field. Faraday's law then shows directly that the magnetic flux reconnected per second per meter of X-line is just E Wb s^{-1} m^{-1} (= volts m^{-1}) so that the electric field strength also serves as a direct measure of the reconnection rate. We note, however, that alternative definitions of reconnection have also been proposed. Notably Vayliunas (1975) defines reconnection as occurring when a plasma flows across the separatrices of an X-type field configuration. Now under most reasonable circumstances the existence of such a flow, and its implied $\underline{v} \wedge \underline{B}$ electric field, would also indicate the existence of an electric field at the X-line in conformity with the definition given above. Indeed in a 2-d steady system the (uniform) electric field could be expressed as $v_\perp B_\perp$ where $v_\perp$ is the flow speed perpendicular to a separatrix and $B_\perp$ is the field strength at the separatrix perpendicular to the neutral line (assuming the perfect conductivity approximation is valid, in the convection region). In principle, however, the existence of motional electric field associated with flow across a separator some distance from an X-line does not absolutely guarantee the existence of an electric field at the X-line itself. The electric fields at the two locations could be decoupled by the existence of large field-aligned voltage drops between the two locations, as effectively assumed in the magnetospheric model discussed by Heikkila (1978, 1982). This model has an "open" magnetic structure as shown in Figure 5.4 (b), and solar wind plasma flows across its separatrices, but by hypothesis no reconnection occurs in the sense of the existence of an electric field along the X-line of the field structure. Large field-aligned voltages must then occur along the open field lines on either side of the magnetopause, although we know of no means how such voltages could possibly be sustained in a plasma, and Heikkila suggests none. Heikkila's model has thus been criticized as being most implausible on this basis by Cowley (1984a). However, for this reason it seems preferable to adopt a direct definition of reconnection in terms of an electric field along an X-line, this definition also having the advantage of being directly related to the rate of change of reconnected flux via Faraday's law alone.

Having now introduced the basic concepts involved in magnetic reconnection, and having also indicated why it is a rather subtle but important process in solar system (and other) plasmas, we will now go on to consider theoretical models of reconnection in more detail.

5.3 FLUID (MHD) MODELS OF RECONNECTION

In this section we will describe the principal features of magnetic reconnection models based upon fluid (MHD) theory. We wish to emphasize from the outset that there is no unique "correct" solution of the reconnection problem of universal applicability. Rather, "reconnection" involves a class of problems in which the detailed solutions vary from case to case depending upon the properties of the plasma and the physical situation (the boundary conditions) involved. Nevertheless, one's physical ideas about the nature of the problem are necessarily tied to concrete examples. Here we will discuss the "classic" solutions of Sweet and Parker, Petschek and Sonnerup which have formed the basis of most recent work, and hence we will essentially follow the historical development of the subject.

We will first consider the Sweet-Parker model, discussed initially by Parker (1957a) and Sweet (1958), and later in more mathematical detail by Parker (1963). The main features of the field and flow configuration envisaged is shown in Figure 5.5. The system is two-dimensional and steady state (i.e. $\frac{\partial}{\partial y} \equiv 0$ and $\frac{\partial}{\partial t} \equiv 0$), and the inflow magnetic fields are antiparallel and of equal strength B_i on either side of the current sheet. For simplicity we will consider the incompressible plasma problem where the mass density ρ is everywhere uniform; the plasma flows in towards the current sheet from both sides at speed V_i and out along the current sheet at speed V_o. From the arguments of the previous section concerning the balance between convection of flux into the current sheet and flux annihilation due to diffusion within it, the half-width of the current sheet (diffusion region) is given by equation (5.5) as

$$\ell \simeq \frac{1}{\mu_0 \sigma V_i} \tag{5.7}$$

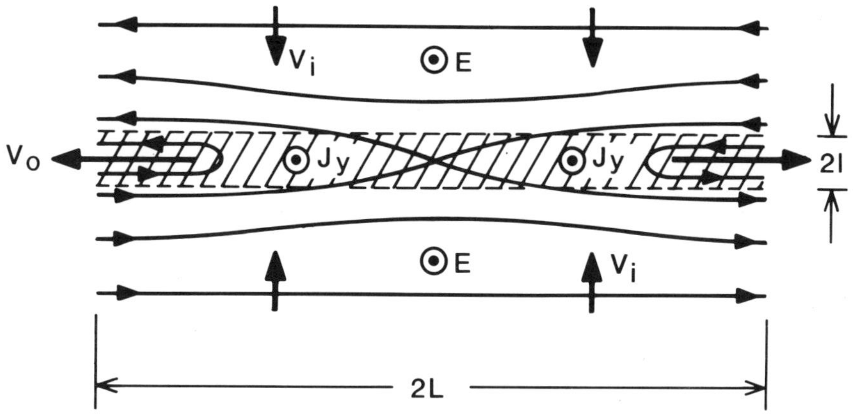

Figure 5.5 Sketch of the field and flow configuration (solid lines and arrows respectively) assumed in the Sweet-Parker solution. The current sheet (diffusion region) shown hatched, has the small half-with ℓ set by the scale of diffusion over its whole half-length L, the latter length being assumed to be equal to the scale size of the system.

It is assumed that the current sheet width remains at this value for a distance L on either side of the neutral line, and (this is the crucial bit) that the length L is known and equal to the scale length of the system. In effect it is assumed that the diffusion region extends from the neutral line to cover the whole of the boundary region between the two inflow plasmas. If we consider conservation of plasma mass we then have, if the density is constant

$$V_i L = V_o \ell \tag{5.8}$$

In this equation ℓ is given by (5.7) and L is assumed known, so that if V_o can be calculated the inflow speed V_i can be determined and hence the reconnection rate ($E = V_i\ B_i$ in the inflow region). Now under very general circumstances the outflow speed in a reconnection system can be calculated, and can be shown to be approximately equal to the Alfven speed in the inflow region, to within a numerical factor of order unity. This result is to a first approximation almost independent of the reconnection rate. All the reconnection rate actually determines is how much plasma is accelerated to the inflow Alfven speed per unit time. This result may be derived in several ways e.g. by considering the forces which accelerate the current sheet plasma away from the neutral line, or by energy considerations. Here we will adopt the latter procedure, assuming that the field energy liberated to the plasma in the current sheet appears wholly as flow kinetic energy away from the neutral line, ignoring the fact that part of this energy at least will actually appear as heat.

From an energetic point of view Poynting flux of electromagnetic energy $\underline{S} = \underline{E} \wedge \underline{H}$ is flowing into the current sheet from either side where it is converted into plasma energy ($\underline{j} . \underline{E} > 0$). Conservation of energy in the steady state is expressed by $\text{div}\, \underline{S} + \underline{j} . \underline{E} = 0$. The electromagnetic energy flowing into each side of the current sheet per unit area per unit time is given by

$$W_E = \frac{EB_i}{\mu_0} = \frac{V_i\, B_i^2}{\mu_0}$$

The mass of plasma flowing into the current sheet from one inflow region per unit area, per unit time is similarly

$$F_M = \rho V_i$$

and this plasma is accelerated to speed V_o before it exits the system. Ignoring the initial kinetic energy of the plasma associated with V_i the energy delivered to this plasma is then

$$W_p = \rho \frac{V_i V_o^2}{2}$$

Similarly ignoring the Poynting flux flowing out of the ends of the current sheet associated with the weak field component threading across the latter, conservation of energy is given by equating W_E and W_p giving the outflow speed V_o as

$$V_o \simeq \sqrt{2}\, \frac{B_i}{\sqrt{\mu_0 \rho}} \simeq \sqrt{2}\, V_{Ai} \qquad (5.9)$$

i.e. equal to the inflow Alfven speed to within a numerical factor, as stated above.

If we substitute $V_o \simeq V_{Ai}$ into (5.8) and use (5.7) to determine ℓ we can now obtain the inflow speed V_i (proportional to the reconnection rate) as

$$V_i \simeq \sqrt{\frac{V_{Ai}}{\mu_0 \sigma L}}$$

Introducing a magnetic Reynolds number based upon the inflow Alfven speed (the plasma outflow speed) and overall system scale L i.e. $R_m = \mu_0 \sigma\, V_{Ai} L$ (more properly a Lundquist number) we finally have the inflow speed

$$V_i \simeq \frac{V_{Ai}}{\sqrt{R_m}} \qquad (5.10)$$

Now we have pointed out in the previous section that the magnetic Reynolds numbers based upon the overall scales of solar system plasmas are very large indeed e.g. $R_m \sim 10^8$ to 10^{10} for the solar flare problem. Equation (5.10) then implies that the inflow speed to the current sheet is very small compared with V_{Ai} i.e. that the magnetic energy of the system is liberated to the plasma only very slowly. If we take e.g. $V_{Ai} \simeq 100$ km s^{-1} and L $\approx 10^4$ km for the solar flare problem, then the time required to convert the magnetic energy in a region of dimension L transverse to the current sheet to plasma energy by reconnection would be a few tens of days, compared with observed flare release times of order minutes to hours. In order to make the process significant in this and other problems the inflow speed (reconnection rate) must in fact be much larger than that given by (5.10), and comparable itself with the inflow Alfven speed V_{Ai}.

The physical reason for the small reconnection rate in the Sweet-Parker system is easy to understand. Plasma flows into the boundary region from either side over a large area of length L, but by assumption can only exit along a very narrow channel (the current sheet or diffusion region) whose width is set by diffusion. Since the plasma exits at the Alfven speed, continuity requiresthat the inflow speed must be much lower, by the ratio of the current sheet width to the current sheet length $\ell/L \simeq R_m^{-\frac{1}{2}}$. The conclusion that reconnection can only occur very slowly is then inescapable.

A resolution of this problem was rapidly provided, however, in a classic paper by Petschek (1964) which provided the basis for essentially all subsequent theoretical study of reconnection in the framework of MHD. He pointed out (crediting A.R. Kantrowitz with the initial suggestion) that magnetic energy could be converted to plasma energy as a result of shock waves being set up in the plasma, as well as due to diffusion. The configuration he envisaged is shown in Figure 5.6. Two waves (slow mode shocks) stand in the flow on either side of the neutral line marking the boundaries of the plasma outflow regions. A small diffusion region still surrounds the neutral line, however, constituting a miniature (in length) Sweet-Parker system. The width of this current sheet is given by (5.7) as before, but now we do not assume that its length L* is comparable to the scale size of the system L. Rather, its length may be much smaller and is determined self-consistently by the continuity condition

$$L^* \simeq \frac{V_{Ai}}{V_i}\,\ell \qquad (5.11)$$

where we have assumed incompressible flow and an outflow speed $\simeq V_{Ai}$ as previously. Thus, if the inflow speed V_i is much less than V_{Ai} the length of the diffusion region is much larger than its width ℓ as assumed by Sweet and Parker. But if we now allow $V_i \underset{\sim}{\to} V_{Ai}$ all that

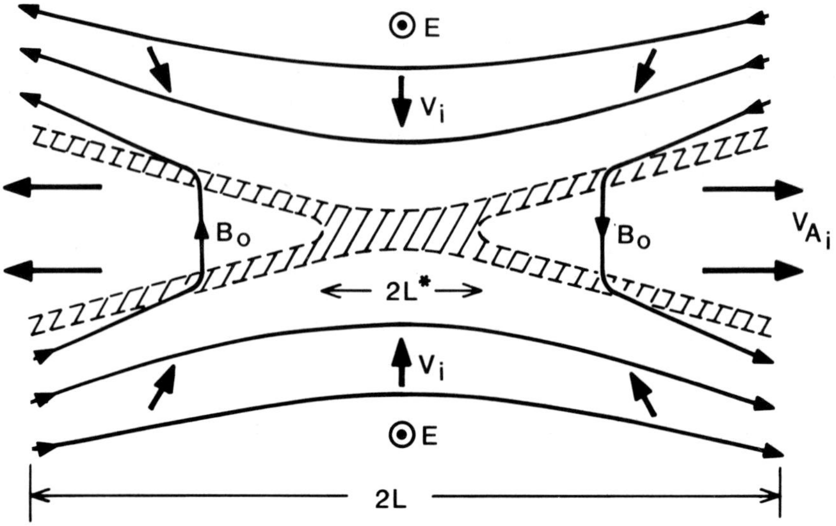

Figure 5.6 Sketch of the magnetic field and plasma flow configuration (solid lines and arrows respectively) in the Petschek reconnection solution. A small central diffusion region of length 2L* surrounding the neutral line bifurcates into two standing-wave current sheets in the downstream flow (current-carrying regions are shown hatched). Most of the inflowing plasma is accelerated to flow rapidly ($\simeq V_{Ai}$) along the boundary between the two inflow regions as it crosses these standing waves.

happens is that the diffusion region shrinks in length so that L* becomes comparable with ℓ.

For reasonably large reconnection rates, therefore, the length of the diffusion region will be much smaller than the scale size of the system L, so that most of the plasma flowing into the boundary region will flow across the standing waves rather than through the central diffusion region. The standing waves in effect produce a bifurcation of the diffusion region current into two separate sheets, and may be considered to represent an image of the sharp kink formed in the field lines as they are reconnected at the neutral line which then propagates as waves in either direction along the reconnected field lines while being convected with the flow, thus producing a standing structure. The effects of diffusion dominate the early evolution of the field however (forming the diffusion region), since as pointed out in the previous section, the effects of diffusion expand as $t^{\frac{1}{2}}$, while that of the waves should go as t. As the plasma crosses these standing waves it is accelerated along the boundary by the tension in the magnetic field ($\underline{j} \wedge \underline{B}$ force), and energy considerations are sufficient to show that the outflow speed will be comparable with the inflow Alfven speed in this case also. The waves lie at an angle to the boundary determined by the requirement that they remain stationary in the flow. In the incompressible case the required condition is that the plasma flow speed normal to the plane of the wave should be equal to the Alfven speed based on the magnetic field component also normal to the plane of the wave. We will now apply this condition to the outflow region, assuming that the outflow speed is $\simeq V_{Ai}$ parallel to the boundary, while the outflow magnetic field is B_o normal to the boundary (see Figure 5.6). If the wave lies at angle θ to the boundary the normal flow speed is $V_n \simeq V_{Ai} \sin\theta$, while the normal magnetic field is $B_n = B_o \cos\theta$. The shock angle is then given by

$$V_{Ai} \sin\theta \simeq \frac{B_o \cos\theta}{\sqrt{\mu_o \rho}} \quad \text{giving} \quad \tan\theta \simeq \frac{B_o}{V_{Ai}\sqrt{\mu_o \rho}} .$$

However, Faraday's law (E_y = constant) allows us to determine B_o in terms of the inflow speed V_i and field strength B_i as follows:

$$V_{Ai}\, B_o = V_i\, B_i \text{ giving } B_o \simeq \sqrt{\mu_o \rho}\, V_i$$

so that finally, to within a numerical factor

$$\tan\theta \simeq \frac{V_i}{V_{Ai}} \tag{5.12}$$

It can thus be seen that for small inflow speeds V_i compared with V_{Ai} the outflow is confined to a narrow wedge along the boundary, but that as the inflow speed increases the angle of the outflow wedges (the standing waves) also increases to accommodate the increased flow. The wave angle given by (5.12) is, of course, quite compatible with mass conservation. Application of a condition similar to equation (5.8) to this configuration immediately results in a required wave angle which is the same as that given by equation (5.12).

We therefore arrive at the conclusion that in the Petschek system reconnection can in fact occur at (almost) any rate, with the outflow region responding to accommodate variations in the plasma inflow. The reason for the vastly increased reconnection rates possible in this case compared with the Sweet-Parker solution is thus clear. Rather than being constrained to flow out of a narrow channel of constant width determined by diffusion scales as assumed in the Sweet-Parker solution, the outflow in the Petschek system takes place in an expanding wedge whose angle will change to accommodate (almost) any inflow rate. The need to include the word "almost" in the above sentences arises because there does in fact exist an upper limit to the reconnection rate of the Petschek system. As the inflow speed increases, the angle of the outflow wedges also increases, causing increasingly large perturbations of the inflow region (see Figure 5.6). Detailed treatment then shows that Petschek-type solutions cease to exist when the inflow speed reaches the value

$$V_{i\,max} \simeq \frac{\pi\, V_{Ai}}{8\, \ln R_m}, \tag{5.13}$$

where R_m is the Lundquist number previously introduced into equation (5.10). It should be noted that this maximum rate depends inversely on the logarithm of R_m rather than on its square root as in the Sweet-Parker system, so that $V_{i\ max}$ is much larger than the inflow speed of the latter solution. For typical R_m values, $V_{i\ max}$ lies in the range 0.01 to 0.1 V_{Ai}, compared with 10^{-4} to 10^{-5} V_{Ai} for the Sweet-Parker system. The latter values then represent the minimum inflow speeds required to self-consistently maintain an equilibrium boundary current sheet against ohmic dissipation in an incompressible plasma.

Subsequent work over the past twenty years on steady reconnection in a MHD fluid has concentrated on elaborating Petschek's wave model, first by putting Petschek's solution itself on a firm mathematical basis (Vasyliunas, 1975) and second by extending the model to include situations in which the inflow plasmas have differing properties on either side of the boundary, including arbitrarily sheared fields, and for different external boundary conditions (e.g. Cowley, 1976, Semenov et al., 1983). In the latter regard special mention should be made of a model for the convection region derived by Sonnerup (1970) which is sufficiently simple that solutions for general inflow conditions may readily be found. The basic Sonnerup solution for symmetrical inflow and outflow conditions is shown in Figure (5.7). In this figure the

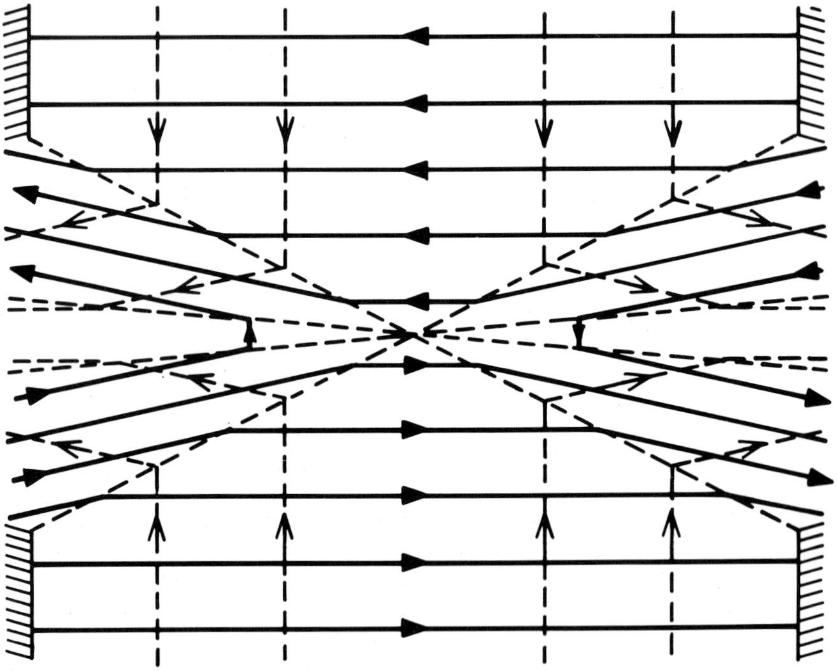

Figure 5.7 Sonnerup convection region solution for $V_i = 0.5\ V_{Ai}$. Magnetic field lines are shown by solid lines, streamlines by long-dashed lines and the standing waves by short-dashed lines. The inflow may be considered to take place between plane parallel walls (shown hatched), while the outflow takes place through gaps in the walls. The outer waves originate at the gap edges, and the width of the gaps determines the inflow (and reconnection) rate. The central diffusion region is not represented.

standing waves are shown as small-dashed lines and no attempt has been made to indicate the diffusion region. The solution, which assumes an incompressible plasma, is constructed from regions of uniform magnetic field and plasma velocity separated by the plane wave discontinuities standing in the flow. The condition for the existence of these standing waves is the same as that stated above for the Petschek solution (i.e. $v_n = \pm B_n/\sqrt{\mu_0 \rho}$, where subscript n indicates the component normal to the plane of this wave),and their angles are determined by the requirement (in the symmetrical case) that both the magnetic field and the fluid velocity are rotated through a right angle between the inflow and outflow regions. In order to accomplish this, at least two standing waves must be present in each quadrant of the solution as shown in Figure (5.7). The waves nearest the boundary, bounding the central outflow region, correspond to Petschek's waves since consideration of the angle with which these waves lie relative to the flow shows that they originate in the central diffusion region and propagate outward along the reconnected field lines. Similar considerations applied to the four outer waves, however, show that they must be generated in the outer part of the flow at the "corners" shown in the figure and then propagate in towards the diffusion region.

Exact analytic solutions for the field and flow in each region together with the wave angles may readily be found for the Sonnerup model from the jump conditions across the wave discontinuities, and solutions exist for essentially arbitrary inflow speeds (reconnection rates) relative to the inflow Alfven speed. In particular, if the plasma speed and the Alfven speed in the inflow region are V_i and V_{Ai} respectively, then the angle which the Petschek-waves and the outer (Sonnerup) waves make to the boundary (θ_p and θ_s respectively) are given by

$$\tan\theta_p = \frac{1}{(\sqrt{2}+1)^2}\left(\frac{V_i}{V_{Ai}}\right) \quad \text{and} \quad \tan\theta_s = \left(\frac{V_i}{V_{Ai}}\right), \tag{5.14}$$

while the outflow speed along the boundary and the outflow region magnetic field are given by

$$V_o = (\sqrt{2}+1)\,V_{Ai} \qquad B_o = (\sqrt{2}-1)\sqrt{\mu_0\rho}\,V_i\ , \tag{5.15}$$

values which are in conformity with the general discussion given above. It should be noted that when V_i is small compared with V_{Ai} both waves lie close to the boundary so that the outflow is confined to a narrow wedge. As V_i increases the wave angles and outflow wedge expand to accommodate the increased inflow in the same way as previously discussed for the Petschek model. When V_i reaches the value $(\sqrt{2}+1)\,V_{Ai}$ the inflow and outflow regions become identical and there is then no net energy exchange between the plasma and the field, while if V_i exceeds this value the outflow speed becomes less than the inflow speed and energy is converted from the flow to the field rather than vice versa. This situation seems physically rather unrealistic (e.g. matching a diffusion region solution would be rather problematic), so that it is usually stated that an upper limit to the inflow speed in the Sonnerup model is $(\sqrt{2}+1)\,V_{Ai}$ (considerably larger, it may be noted,than the upper limit in the Petschek solution given by equation (5.13)),although formally convection region solutions do exist for arbitrarily large V_i.

If solutions in the wave model of reconnection (e.g. either Petschek's or Sonnerup's) exist over a wide range of inflow speeds (reconnection rates) the question then arises as to what actually determines that rate in any given situation. The most reasonable answer to this question is that the overall reconnection rate is ultimately determined by the external boundary conditions of the physical set-up, although this does not preclude the possibility that more local conditions could influence the system over shorter time scales. The influence of the external region is indeed explicit in Sonnerup's solution (as recently discussed by Axford (1984)) via the presence of the standing waves propagating toward the diffusion region from the outside. We may regard these waves as emanating from the corners of plane parallel walls, shown hatched in Figure 5.7, between which the inflow plasma is moving. Since these waves must meet in the vicinity of the diffusion region, their angle relative to the outflow direction is determined by the size of the gap in the walls where the plasma flows out relative to the distance between the walls across the inflow region. For narrow outflow gaps the wave angle to the outflow must be small, corresponding to a small inflow speed relative to V_{Ai}. If the width of the outflow gaps is then increased, the wave angle must also increase, leading to faster inflow.

5.4 THE SINGLE-PARTICLE APPROACH IN A COLLISION-FREE PLASMA

The reconnection theory described in the previous section is directly applicable to collisional plasmas (e.g. the lower solar atmosphere) in which the current carried by the plasma is simply related to the rest frame electric field by a scalar conductivity.However, most solar system plasmas (e.g. the solar wind and planetary magnetospheric plasmas) are in fact collision-free i.e. the mean free path for particle collisions is comparable to or larger than the size of the system, and it is then not clear that such a description is appropriate. Over the past thirty years, therefore, a separate branch of reconnection theory has developed in parallel with the fluid theories of section (5.3), intended for application to the collision-free situation. Since by hypothesis collisions are unimportant, a simple and direct approach to the problem is then possible by considering the motion of individual particles in model electromagnetic fields appropriate to reconnection, and then applying self-consistency conditions. When this is done it is found that the resulting equilibria closely resemble their fluid counterparts discussed in the previous section.

It is a little-appreciated fact that the first description of single-particle particle motion in an X-type field geometry in the presence of reconnection (an electric field along the X-line) is contained in the classic paper by Dungey (1953), in which the significance of magnetic neutral line regions and the 'breaking'and 'rejoining' of magnetic field lines which can occur there was also first properly elucidated. However, since it took the remainder of the field another

∿ 15 years to catch up with these ideas, via the consideration of a succession of simpler problems, we will here follow the latter route for pedagogical reasons.

The simplest system of interest consists of a neutral sheet magnetic field configuration with no electric field as shown in Figure (5.8). The diagram on the right shows the assumed variation of the

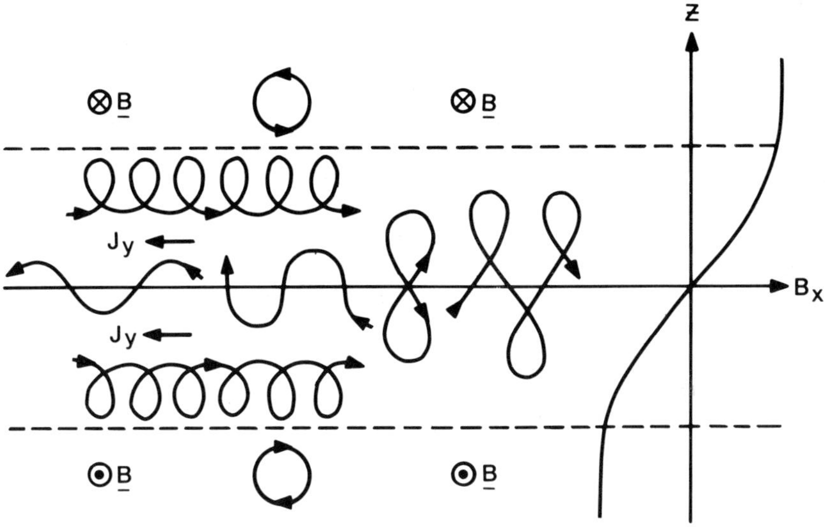

Figure 5.8 Sketch of ion trajectories in a magnetic neutral sheet system with zero electric field. The variation of the field strength is shown on the right and the dashed lines indicate the boundaries of the current layer. The horizontal solid line is the neutral sheet.

field $B_x(z)$. In the vicinity of the neutral sheet the field varies approximately linearly with z due to a uniform current, before becoming constant with distance (zero current) further away. The sketch on the left then shows how ions move in this field, it being noted that the plane of this diagram is rotated through a right angle compared with related diagrams shown previously (the current flows from right to left rather than out of the page). The motion of electrons is obtained simply by reversing the directions of the arrows on the ion orbits (or the direction of the field). Outside the current sheet the particles simply gyrate in circular orbits about the magnetic field lines, the sense of gyration being opposite on opposite sides of the neutral sheet due to the reversal in the field direction. Motion within the current sheet (specifically for the case of a constant field gradient) was first studied quantitatively by Parker (1957b) and in more detail by Seymour (1959). The motion was found to be of two types, depending on whether or not the trajectories cross the neutral sheet. If they do not, then the ions drift in the direction opposite to the current as shown, the motion essentially being an adiabatic $\underline{\nabla} B$ drift if the particle is moving several Larmour radii from the neutral sheet such that the fractional change in $\underline{B}$ over the orbit is not large. Within one or two Larmour radii of the neutral sheet, however, the latter condition clearly cannot be met such that particle motion in this vicinity is clearly non- adiabatic. For particles which actually cross the neutral sheet a variety of motions are possible, depending on the angle of the particles velocity vector at the neutral sheet plane. Ions may drift in either direction along the sheet, or, in a singular case, perform a figure-of-eight trajectory with zero net drift. The preferred motion, however (e.g. for a uniform distribution of velocity vector angles at the neutral sheet), is the snake-like orbits about the neutral sheet arising from the reversal in the sign of the field, in which the ions' net motion is in the direction of the current. It should be noted that for electrons the preferred motion will then be in the opposite direction, thus also contributing to the current due to the opposite sign of the electron charge.

It may thus be concluded that equilibrium neutral sheet solutions may be constructed if an appropriate population of 'snake-orbiting' particles is set up to carry the current in the field reversal region. Exact equilibria of this nature were first derived by Harris (1962), based upon assumed drifting maxwellian distributions of ions and electrons at the neutral plane which are mapped into the surrounding region using Liouville's (Jeans') theorem and conservation of energy, and canonical momentum in the plane of the neutral sheet. The essential features, are, however, easy to derive in an approximate way from the ion and electron equations of motion. Assuming for simplicity a linear variation of the field away from the neutral sheet $B_x(z) = \left(\frac{Bz}{h}\right)$ where B is the field strength outside the current sheet and h is the half-thickness, the z equation of ion motion is

$$\frac{d^2z}{dt^2} = -\frac{e}{m_i}\left(\frac{V_yB}{h}\right)Z \tag{5.16}$$

with a similar expression for electrons. Now if the particle speed V_z associated with the oscillation about the current sheet is small compared with V_y, such that the particles to a first approximation form a beam moving parallel to the current, then V_y will be approximately constant on the trajectory and equation (5.16) reduced to SHM about the neutral sheet. (Note that V_x is separately constant on the trajectory). The angular frequency of the z-oscillation ω_{z_i} is given approximately by

$$\omega_{z_i}^2 = \frac{e}{m_i}\left(\frac{V_{yi}\,B}{h}\right) \tag{5.17}$$

with a similar expression for electrons. Now if the typical ion speed normal to the neutral sheet is V_{z_i} the typical oscillation amplitude is

$$a_i = V_{z_i}/\omega_{z_i}$$

If the ion beam temperature is then T_i such that $m_i V_{z_i}{}^2 \simeq 2kT_i$ then the ion oscillation amplitude is given by

$$a_i^2 \simeq \frac{2kT_ih}{eV_{yi}B} \tag{5.18}$$

Now in the self-consistent situation the particle oscillation amplitudes must satisfy two criteria. First, charge neutrality demands that ions and electrons have the same amplitude (if no other charge-neutralizing populations are present). From (5.18) this requires that $T_i/V_{yi} = -T_e/V_{ye}$ (the minus sign since electrons travel in the opposite direction to ions), which amounts to a particular choice of frame of reference in the y direction. Second, since it is just these oscillating particles which carry the sheet current, their oscillation amplitude must match the thickness of the current sheet i.e. we must have $a_i \simeq a_e \simeq h$. Making this substitution into (5.18) then yields

$$h \simeq \frac{2kT_i}{eV_{yi}B} \simeq \frac{2kT_e}{e\,|V_{ye}|\,B} \tag{5.19}$$

which is the equilibrium thickness of the Harris neutral sheet as confirmed by detailed computation. The number density n of the current carrying ions or electrons within the sheet is then determined from Amperes law

$$\frac{B}{\mu_0 h} \simeq neV_{yi}\left(1 + \frac{Te}{T_i}\right) \tag{5.20}$$

where use has been made of the above relation between V_{yi} and V_{ye}. It may then be noted that the particle pressure at the neutral sheet is

$$P = nk\,(T_i + T_e) = B^2/2\mu_0$$

i.e. equal to the field pressure outside the sheet, as might have been expected. In the exact Harris solution it is shown that $P + nk(T_i+T_e)$ remains exactly constant at all distances z from the neutral sheet.

It should be noted that in the above system the neutral sheet current is carried by the snake-orbiting particles in the absence of any electric fields. Once this particle population is set up it will continue to carry the sheet current indefinitely, provided, of course, that the sheet is infinite in spatial extent. There is no field annihilation taking place and no conversion of energy between field and plasma. We must then enquire what will happen if losses of these particles take place. In particular, suppose that the neutral sheet is of finite extent in the direction of the current and bounded by conducting walls which do not emit current-carriers, but instead, absorb the neutral sheet particles which hit them. The current-carrying particles then move along the neutral sheet to one or other of the walls and are there 'lost' (but provide a source for the 'return' current on these walls, as shown in Figure 5.10 (a)). These particles must then be replaced to provide the current required by curl $\underline{B}$, and since by the above assumption they are not provided by the boundary walls they must instead be provided by plasma inflow from the uniform field region surrounding the current sheet. This flow must be associated with an electric field E_y parallel to the current, giving $\underline{E} \wedge \underline{B}$ drift into the sheet from both sides.

We must then enquire how particle motion in a neutral sheet is affected by the addition e.g. of a uniform E_y to the magnetic field of Figure 5.8. This is shown in Figure 5.9 for both ions (left) and electrons (right).

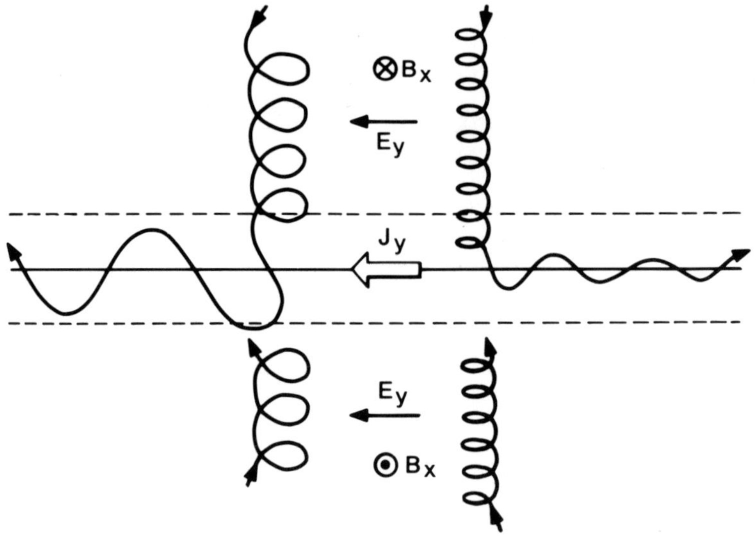

Figure 5.9 Ion (left)and electron (right) motion in the neutral sheet field of Figure 5.8 when a uniform electric field E_y is added in the direction of the current.

Particles of both signs first $\underline{E} \wedge \underline{B}$ drift towards the current sheet from either side, but on reaching the neutral sheet they start to oscillate about it due to the field reversal and are then accelerated along it by the electric field. Ions and electrons are accelerated in opposite directions, and both species therefore contribute to the sheet current (Speiser, 1965). Thus on entering the current sheet and crossing the neutral sheet there occurs a transition in the motion from an adiabatic drift toward the centre to a non-adiabatic current-carrying motion along the neutral sheet. After a few oscillations the latter motion may be approximated as a field-free acceleration in the cross-system electric field so that $V_y \simeq \frac{q}{m} E_y t$. Substitution of this form into the z-equation of motion (5.16) then yields Airy's differential equation, approximate solutions of which may be obtained by the WKB method. The amplitude of the oscillations about the neutral sheet is found to slowly decrease with time as the effective 'spring constant' of the motion increases with V_y. Again, the equilibrium thickness of the current sheet must be matched to the typical oscillation amplitude of the current-carrying particles.

When E_y is included, magnetic field annihilation takes place in the system (i.e. there is a Poynting flux of electromagnetic energy into the current sheet from either side) and the above description then shows how this energy is fed into plasma particles accelerating along the neutral sheet. Self-consistency (energy conservation) must then be imposed, equivalent to requiring that the inflowing plasma provide the correct total current to satisfy Ampere's law. The required condition is simple to calculate with the aid of Figure 5.10 (a) which shows the neutral

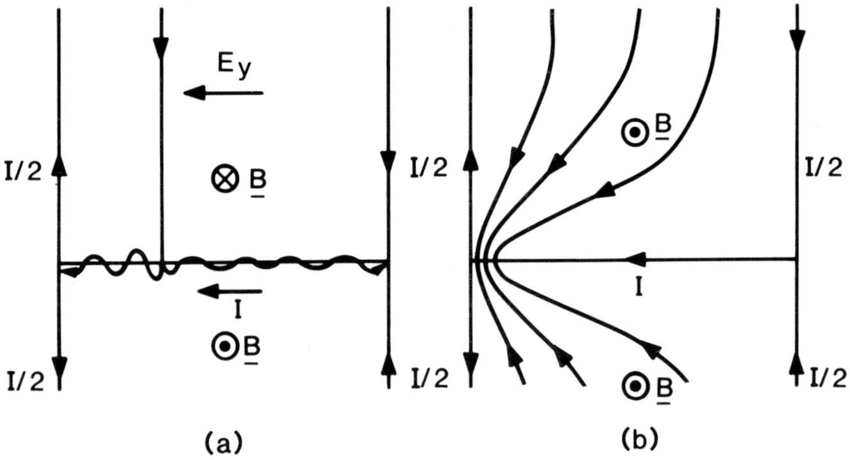

Figure 5.10 (a) Sketch of the Alfven neutral sheet system for assumed uniform cross-system electric field E_y. A charge-neutral stream of plasma is indicated which $\underline{E} \wedge \underline{B}$ drifts into the neutral sheet and there charge-separates to produce an ion stream moving to the left-hand boundary and an electron stream moving to the right-hand boundary. The electric field produced by this charge-separation then modifies the equipotentials (streamlines) in the inflow region as shown in (b), after Cowley (1973).

sheet system bounded by plane conducting walls separated by distance L. Each charge-neutral stream of plasma flowing into the current sheet charge-separates at the sheet to produce a beam of ions moving in the direction of the current towards one wall, and an equal flux of electrons moving towards the other wall. It is thus seen that the current (charge flux) flowing in the sheet due to this stream is spatially uniform. Thus the total current must also be spatially uniform, though carried wholly by ions at one boundary of the system and wholly by electrons at the other (and half and half in the middle). In this case, therefore, the total current can be simply calculated from the total ion flux into the current sheet (which also equals the total electron flux into the sheet). The current per unit length (x direction) of system, I, is thus given by

$$I = \frac{2neEL}{B}$$

where B is the field strength outside the current sheet such that the inflow speed is E/B, and n is the plasma density in this region. Using Ampere's law $\mu_0 I = 2B$ and noting that $EL = \Phi_A$, the total cross-system potential, then yields the consistency condition

$$\Phi_A = \frac{B^2}{\mu_0 ne} \qquad 95.21)$$

This result was first derived by Alfven (1968), and Φ_A is therefore called the Alfven potential. It should be noted that self consistency in this case determines the total voltage across the system for given inflow B and n, rather than the electric field itself i.e. it is the total ion/electron flux and the total energy input to the current sheet ($I\Phi_A$) per unit x-length of the system which is determined, independent of the width L.

In the above discussion it was effectively assumed that the plasma flow into the neutral sheet is uniform along its width. It should be mentioned, however, that even if this is true at large distances from the neutral sheet, the flow will become very non-uniform near to the current sheet due to electric fields which result from the ion and electron charge separation at the neutral sheet, as pointed out by Cowley (1973). The major effect is that the current sheet tends to become positively charged due to the fact that the accelerating ions tend to remain within it for a much longer time than the accelerating electrons as a result of their much larger mass. Consideration of how a positively charged current sheet would perturb the equipotentials (streamlines) of the external flow then immediately shows that the flow into the sheet will tend to concentrate near the boundary where the ions reach the wall, thus reducing the time the ions spend in the sheet and hence their charge. A sketch of the flow in the Cowley (1973) solution is shown in Figure 5.10 (b). Most of the flow enters the current sheet over a distance λ_i (the ion inertial length, equal to the speed of light divided by the ion plasma frequency) at the left-hand boundary, so that over most of the system the sheet current is carried by an accelerated electron beam in a region where $E_y \simeq 0$. Note, however, that the Alfven potential self-consistency condition is in no way affected by these developments.

In comparing these results with those of the previous section, it is first interesting to note that just as in the fluid picture, the ions in the Alfven-Cowley neutral sheet model exit the system at a speed comparable with the inflow Alfven speed, though now in a direction along the current rather than orthogonal to it. This result follows from the same energy argument given previously, or more directly from equation (5.21). The average ion falls through a voltage $\Phi_A/2$ between entry and exit, so that neglecting the inflow speed, the exit speed of the average ion is

$$V_{yi} = \left(\frac{e\Phi_A}{m_i}\right)^{\frac{1}{2}} = \frac{B}{\sqrt{\mu_0 n m_i}} = V_A$$

as stated. In this case, however, the electrons also take up an equal amount of electromagnetic energy and exit from the opposite boundary at a speed which is a factor $(m_i/m_e)^{\frac{1}{2}}$ higher.

If we now consider the annihilation rate which can occur in the Alfven-Cowley system we first note that the (uniform) inflow speed towards the sheet at large distances is given from (5.21) as

$$U = \frac{\Phi_A}{BL} = V_A \left(\frac{\lambda i}{L} \right) \qquad (5.22)$$

where λ_i is the ion inertial length as before, and V_A is the inflow Alfven speed. Now for a proton plasma λ_i is given numerically by $227/\sqrt{n(cm^{-3})}$ km i.e. λ_i is generally much smaller than the typical scale sizes of solar system plasma regimes L. Thus U will be much smaller than V_A, and as discussed above in relation to the Sweet-Parker system, this then means that the annihilation rate in the Alfven-Cowley system is not very rapid, and for somewhat similar reasons. All the particles entering the current sheet in this system are constrained to flow along it to the outer boundaries, separated by the large system scale L. The plasma influx required to maintain the current is then small. It will now be shown, however, that much larger inflow speeds can occur, up to values comparable with the inflow Alfven speed, if X-type magnetic systems are considered. As the inflow speed increases in this case the inflowing particles traverse a decreasingly small fraction of the total system width in their interaction with the current sheet, and instead exit on reconnected field lines downstream from the neutral line just as in the fluid models, in outflow wedges which expand to accommodate the inflow. The Alfven-Cowley neutral sheet system thus represents a limiting case of the X-type systems discussed below which is entirely amalogous to the relationship between the Sweet-Parker limiting case and the standing wave MHD models discussed previously.

We thus now turn to an X-type magnetic field configuration in the presence of E_y and again begin by considering the motion of particles. Near to the neutral line itself, of course, the motion will be essentially the same as in the case of the neutral sheet discussed above. We therefore need to consider how the motion is modified for particles drifting into the current sheet away from the neutral line, where the magnetic field description is modified by the existence of a B_z component threading across the current sheet. This is shown in Figure 5.11 (a) which depicts the ion motion (in the presence of an appreciable B_z the electron motion will generally be adiabatic). Ions (and electrons) $\underline{E} \wedge \underline{B}$ drift towards the current sheet as before, and on reaching it the ions start to oscillate about the field reversal plane and accelerate along it due to E_y in the direction of the current. A Lorentz force in the x-direction $eV_y B_z$ now acts on the particles, however, and turns their motion around in the plane of the current sheet away from the neutral line (on both sides of the latter). The ions continue to turn in the plane of the current sheet until they are moving principally in the $\hat{x}$-direction, along the direction of the field outside the current sheet. At this point the particles are no longer trapped by magnetic forces within the current sheet (the Z equation of motion changes character from SHM to exponential as V_y changes sign) and they then exit as a (nearly) field-aligned beam moving (almost) along the reconnected field lines away from the neutral line ('nearly' and 'almost' because the particles are still subject to the $\underline{E} \wedge \underline{B}$ drift outside the current sheet which deflects their net motion away from the direction of $\underline{B}$ towards the

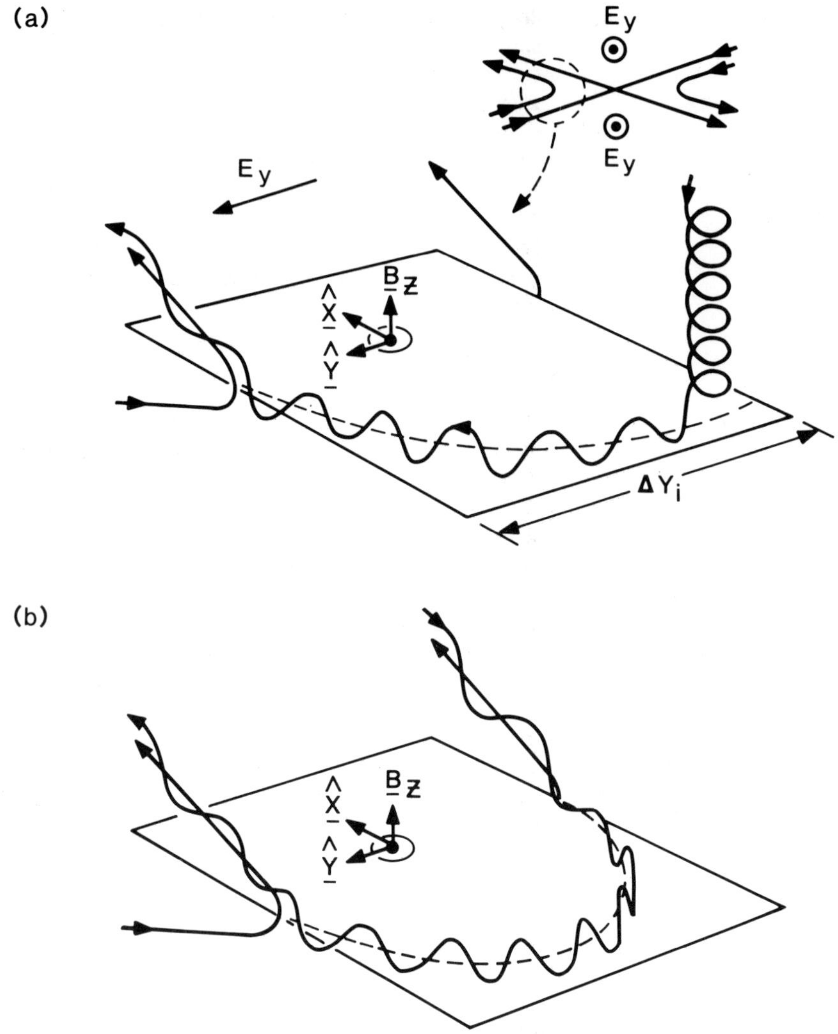

Figure 5.11 Sketch of non-adiabatic ion motion in a current sheet ("Speiser motion") downstream from a neutral line where a field component B_z threads across the centre plane, for (a) the neutral line rest frame ($E_y \neq 0$), and (b) the field line rest frame ($E_y = 0$). The latter moves with a speed $V_F = E_y/B_z$ away from the neutral line.

current sheet). In this interaction with the current sheet the ions can, of course, gain a significant amount of energy, equal to $eE_y\,\Delta y_i$, where Δy_i is the total cross-system displacement. This basic behaviour was first described in detail by Speiser (1965) and for this reason is often called 'Speiser-motion', although we note that the basic features were briefly described earlier by Dungey (1953) as stated above.

We now turn to more detailed considerations, and begin by noting that considerable simplification is achieved by taking the B_z field to be (locally) uniform i.e. we assume that B_z does not vary significantly over the scale of the ion orbits within the current sheet, as seems plausible at sufficient distance from the neutral line. We thus consider a one-dimensional planar current sheet in which the fields are given by $\underline{B} = (B_x(z), 0, B_z)$ with B_z a non-zero constant, and E_y = constant. A major simplification may then be introduced by noting that electric field E_y may be transformed away everywhere by moving to a frame of reference travelling in the $\hat{\underline{x}}$ direction away from the neutral line with the "field line" speed $V_F = E_y/B_z$ (the "de Hoffman-Teller" frame). In this frame the electric field is

$$\underline{E}' = \underline{E}_y + \underline{V}_F \wedge \underline{B} = (E_y - V_F B_z)\,\hat{\underline{y}} = \underline{0}$$

i.e. zero for all Z. The particle motion is then easier to comprehend because without $\underline{E}$ there are no $\underline{E} \wedge \underline{B}$ drifts, and also the speed of the particles is preserved in their interaction with the current sheet. Non-adiabatic 'Speiser' motion in this frame is shown in Figure 5.11(b). Ions spiral into the current sheet along magnetic field lines (with speed $\simeq V_F$ if there was little $\hat{\underline{x}}$-directed flow of the inflow plasma in the untransformed frame) and then perform a half-turn in the plane of the current sheet before again exiting at the same speed along the field lines. While it is moving in the current sheet the ion motion in the plane of the sheet is approximately decoupled from the rapid oscillations about it (Speiser, 1968), and is given approximately by a half-circle whose radius is the Larmour radius of an ion in the field B_z i.e. if the speed of the ion in the plane of the sheet as it enters the sheet is V_x then the circle radius will be V_x/Ω_z such that the total cross-system displacement will be $\Delta y_i \simeq 2V_x/\Omega_z$ where $\Omega_z = eB_z/m_i$. This may readily be verified by direct integration of the $\underline{x}$-equation of motion which is

$$\frac{dv_x}{dt} = \frac{e}{m_i} V_y B_z = \Omega_z \frac{dy}{dt}$$

A first integral may be obtained immediately giving $(v_x - \Omega_z y)$ = constant, so that if on entry and exit $v_x = -V_x$ and $v_x = +V_x$ respectively then the cross-system displacement is $\Delta y = 2V_x/\Omega_z$ as stated above. It can be seen that this result does not actually depend on whether the ion motion in the current sheet is or is not adiabatic (it may be adiabatic if the current sheet is sufficiently wide for finite B_z), and also that it applies equally to both ions and electrons. It can thus be seen that unlike the Alfven-Cowley neutral sheet system the cross-system electron displacements in the current sheet interaction in this case will generally be much less than that of the ions, and since the current carried by the species is proportional to this displacement (for equal fluxes into the current sheet), the electrons will thus generally carry a negligible fraction of the sheet current.

Turning now to consider the energization of particles in the current sheet interaction in the neutral line rest frame (which has uniform E_y present), we note that if in the field line rest ($E_y = 0$) frame a particle enters and exits the current sheet at speed V^*, then in the neutral line rest frame it enters the sheet with speed $(V^* - V_F)$ along $-\hat{x}$, and exits with the speed $(V^* + V_F)$ along $+\hat{x}$, thus gaining energy (we are assuming here that $B \gg B_z$ outside the current sheet). The gain in energy is then $2m\, V_F\, V^*$, compatible with the cross-system displacement in the electric field E_y, since

$$eE_y\, \Delta y = \frac{2eE_y\, V^*}{\Omega_z} = 2mV_F\, V^*$$

as required. It should be noted that if the inflow plasma bulk speed in the $\hat{x}$-direction is small so that $V^* \simeq V_F$, then the ion exit speed will be just $2V_F$ i.e. the ions pick up a field-aligned speed equal to twice the field-line speed V_F in their interaction with the current sheet. The same argument applies also to inflow electrons whose initial speed is small compared with V_F (such that they gain a negligible amount of energy compared with the ions), but usually inflow electron thermal speeds V_{Th} will be much larger than V_F so that the fractional increase of electron energy in one interaction with the current sheet $\Delta W/W \simeq 4\, V_F/V_{Th}$ is very small. In general, therefore, the current sheet acts as an effective ion accelerator, but its effect on electrons will be small unless energy can be coupled from ions to electrons by another process.

We now turn to consider current sheet equilibrium, and note that this may be approached in a number of equivalent ways, either through considerations of energy conservation, or using Amperes law, or momentum balance. The first two methods have been applied above to previous problems, so we will here use the third, which is also the most satisfactory in this case. In a one-dimensional current sheet the field lines typically constitute sharply-bent hairpins which exert a force on the plasma in the $\hat{x}$ direction along the current sheet. In the field line rest frame these field lines must be maintained in equilibrium by the dynamic pressure exerted by the plasma as it streams into the current sheet at speed V^* and is reversed in direction. By balancing the field force with the rate of change of plasma momentum the equilibrium condition may be derived (see Rich et al. (1972) and Eastwood (1974) for more sophisticated treatments). The $\hat{x}$-force per unit area of the current sheet exerted by the field on the plasma is

$$F_x = \int_{-\infty}^{\infty} j_y\, B_z\, dz = \frac{2\, B_x B_z}{\mu_0}$$

where B_x is the uniform field value outside the current sheet. Neglecting electrons, the change in $\hat{x}$ momentum of each ion which interacts with the current sheet after flowing in with field-aligned speed V^* is $2m_i\, V_x = 2m_i\, V^*\, B_x/B$, and since the total ion influx per area of sheet from both sides is $2n\, V_z = 2n\, V^*\, B_z/B$, where n is the number density of the inflow plasma, the change in ion momentum per area of sheet per time is

$$\dot{P}_x = 4nm_i\, V^{*2}\, \frac{B_x B_z}{B^2}$$

Equating F_x and $\dot{P}_x$ then yields the equilibrium condition

$$V^* = \frac{B}{\sqrt{2\mu_0 n m_i}} = \frac{V_A}{\sqrt{2}} \qquad (5.23)$$

i.e. the equilibrium field-aligned inflow speed of the plasma in the field line rest frame must be equal to the inflow Alfven speed to within a numerical factor. If as before we then assume that the inflow plasma speed ($\hat{x}$-direction) is small in the neutral line rest frame we then must have $V_F \simeq V^* = V_A/\sqrt{2}$ (assuming $|B_x|>>B_z$ outside the current sheet so that $|V_x^*| \simeq V^*$ in the field line rest frame), and hence the outflow speed of the plasma, away from the neutral line is

$$V_o = 2V_F \simeq \sqrt{2}\; V_A \qquad (5.24)$$

Thus, in common with all the other systems considered in this chapter, the plasma outflow speed is equal to the inflow Alfven speed to within a numerical factor of order unity, approximately independent of the reconnection rate.

It is now important to note that the above equilibrium condition determines only the field line speed $V_F = E_y/B_z$, and not a unique reconnection rate i.e. E_y value. For any given E_y the field B_z is then determined from

$$B_z = \frac{E_y}{V_F} = \frac{\sqrt{2}\, E_y}{V_A}$$

Thus as E_y increases (the reconnection rate increases), B_z also increases in proportion. This means first of all that the cross-system ion displacement in the current sheet decreases as E_y increases (as previously stated), since from above

$$\Delta y \simeq \frac{2V_F}{\Omega_z} = \frac{mV_A{}^2}{eE_y}$$

This result also follows from the fact that the ion outflow energy is approximately independent of the reconnection rate (i.e. E_y) as E_y increases. More importantly, however, as E_y increases and B_z increases the reconnected field wedge lying downstream from the neutral line expands in size. The angle of the field wedge to the boundary between the inflow regions (the plane of the current sheet) is given by

$$\sin\Theta_F \simeq \frac{B_z}{B} = \frac{\sqrt{2}\, E_y}{V_A B} = \frac{\sqrt{2}\, V_i}{V_A} \qquad (5.25)$$

The angle of the wedge is therefore small when V_i is small compared with V_A, and increases to values such that significant perturbations of the inflow region occur when V_i approaches V_A (just as in the Petschek solution). Due to the $\underline{E} \wedge \underline{B}$ drift the outflow plasma streamlines are, however, inclined at a smaller angle to the current sheet than the field angle Θ_F, and it is easy to show that in the present case (zero $\hat{\underline{x}}$ velocity in the inflow) their inclination is just given by

$$\sin \Theta_p \simeq \tfrac{1}{2} \sin \Theta_F \simeq \frac{V_i}{\sqrt{2}\, V_A} . \qquad (5.26)$$

The overall field and flow configuration in this case is therefore as shown in Figure 5.12, specifically for the case $V_i = V_A/4$. The central

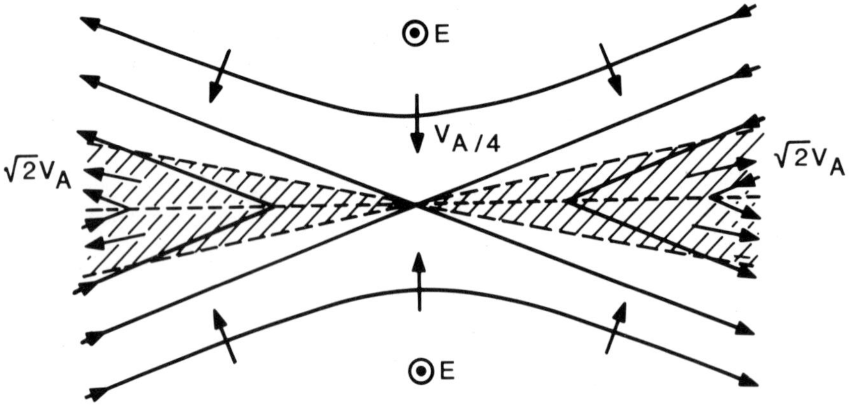

Figure 5.12 Field and flow configuration (solid lines and arrows respectively) for reconnection based on the collision-free cold ion beam picture for the case $V_i = V_A/4$. The central current sheet along the boundary is shown as a short-dashed line, while the region occupied by the accelerated ion beams is shown by the hatching bounded by the long-dashed lines. When finite outflow plasma pressure is included the latter lines will also become current sheets with the field strength reduced inside, corresponding to slow mode waves expanding along the reconnected field lines, akin to Petschek's shocks.

current sheet is indicated by the horizontal dashed line, while the region occupied by the accelerated outflow plasma is shown by the hatched region bounded by the long-dashed lines. By virtue of (5.25) and (5.26) the accelerated plasma occupies just the central half of the wedge of reconnected field lines (see e.g. Hill (1975) and Cowley (1984b), where discussion is also given concerning the effect on these results of $\hat{\underline{x}}$-directed flow in the inflow region in the neutral line rest frame). Comparison of Figure 5.12 with Figures 5.6 and 5.7, equation (5.24) for the outflow speed with equation (5.15) of the Sonnerup model, and equation (5.26) for the angular width of the outflow wedge with equations (5.12) and (5.14) for the wave models then demonstrates the close similarity in results which one ultimately obtains from the particle approach and from the MHD approach. In fact the correspondence will be rather greater than indicated in the discussion so far, since we have assumed that the outflow plasma consists to a first approximation of a cold beam moving (in the field line rest frame) along the reconnected magnetic field lines. In fact, however, part of the energy liberated to the ion plasma in the current sheet will appear as increased temperature, in addition to the increased bulk speed already discussed. The diamagnetic effect of this hot plasma will then result in the outer surface of the outflow region (the long-dashed line in Figure 5.12) also constituting a current sheet, with the B_x field reduced inside it to preserve pressure balance. In effect, the production of hot plasma in the current sheet will result in a slow mode expansion (ΔB and Δp in antiphase) of the plasma along the reconnected field lines, which, combined with the flow, will result in a current sheet being formed at an angle to the boundary, marking the surface of the hot plasma. These current sheets correspond to Petschek's standing slow mode shocks in the MHD picture.

Finally we should make some brief remarks about the structure of the central current layer in the collision-free picture. As described above for the Harris neutral sheet, the width of the current layer is determined by the oscillation amplitude of the current-carrying ions, which in turn depends on the speed V_z associated with the ion oscillation about the current sheet centre. For a cold inflow ion beam in the field line rest frame V_z depends on the exterior field angle to the current sheet, so that the sheet thickness is similarly dependent in this case. On the other hand, when B_z is small and the inflow beam warm, V_z and the sheet thickness may depend instead on the ion temperature. The first situation corresponds to the numerical solutions of Eastwood (1972) while the second corresponds to the situation investigated theoretically by Francfort and Pellat (1976). In either case a charge neutrality problem exists because the non-adiabatic ions will in general spend a much longer time within the current sheet ($\tau \simeq \pi/\Omega_{zi}$) than electrons, so that the sheet will tend to become positively charged. Eastwood (1972) overcame this difficulty by postulating the existence of an isotropic electron population which is electrostatically trapped within the current sheet. If the temperature of this electron population is T_e then the voltage between the centre and the outside of current sheet required to trap these electrons is $\Phi_z \simeq kT_e/e$. When T_e and Φ_z are small, the ion motion is not much affected by the associated electric field E_z so that the above considerations apply, but when $e\Phi_z \gtrsim m_i V_z^2/2$, where V_z is the z-component of the inflow ion beam velocity, the ion motion is affected and the ions are initially repelled from the current sheet. The numerical solutions derived by Eastwood (1974) then show that energy is coupled into the ion oscillation to enable it to overcome the potential barrier, such that the energy in the oscillation reaches $m_i V_z^2 \simeq kT_e$. In this case, therefore, the sheet thickness is determined by the trapped electron temperature, even though the current itself is carried as before by the inflow beam ions. In all these cases, however, the equilibrium sheet thickness is small, of order or less than the ion Larmour radius based on the field strength outside the current sheet and the incoming ion speed in the field rest frame.

REFERENCES

Alfven, H., Some properties of magnetospheric neutral surfaces, J. Geophys. Res., 73, 4379, 1968.

Axford, W.I., Magnetic field reconnection, in "Magnetic reconnection in space and laboratory plasmas", Geophysical Monograph 30, AGU Publishers, Washington, D.C., U.S.A., p.1, 1984

Butler, D.M., and K. Papadopoulos (Editors), Solar terrestrial physics: Present and future, NASA Ref. Publ. 1120, NASA, Washington D.C. 20546, U.S.A., 1984.

Cowley, S.W.H., A self-consistent model of a simple magnetic neutral sheet system surrounded by a cold, collisionless plasma, Cosmic Electrodyn., 3, 448, 1973.

Cowley, S.W.H., Comments on the merging of non anti-parallel fields, J. Geophys. Res., 81, 3455, 1976.

Cowley, S.W.H., Solar wind control of magnetospheric convection, in Proc. Conf. Achievements of the IMS, ESA SP-217, p.483, 1984a.

Cowley, S.W.H., The distant geomagnetic tail in theory and observation, in "Magnetic reconnection in space and laboratory plasmas", Geophysical Monograph 30, AGU Publishers, Washington, D.C., U.S.A., p.228, 1984b.

Dungey, J.W., Conditions for the occurrence of electrical discharge in astrophysical systems, Phil. Mag., 44, 725, 1953.

Dungey, J.W., Cosmic Electrodynamics, Cambridge University Press, 1958.

Dungey, J.W., Interplanetary magnetic field and the auroral zones, Phys. Rev. Lett., 6, 47, 1961.

Dungey, J.W., The history of the magnetopause regions, J. Atmos. Terr. Phys., 40, 231, 1978.

Eastwood, J.W., Consistency of fields and particle motion in the 'Speiser model of the current sheet, Planet. Space Sci., 20, 1555, 1972.

Eastwood, J.W., The warm current sheet model and its implications on the temporal behaviour of the geomagnetic tail, Planet. Space Sci., 22, 1641, 1974.

Francfort, P., and R. Pellat, Magnetic merging in collisionless plasmas, Geophys. Res. Lett., 3, 433, 1976.

Harris, E.G., On a plasma sheath separating regions of oppositely directed magnetic field, Nuovo Cim., 23, 115, 1962.

Heikkila, W.J., Electric field topology near the dayside magnetopause, J. Geophys. Res., 83, 1071, 1978.

Heikkila, W.J., Impulsive plasma transport through the magnetopause, Geophys. Res. Lett., 9, 159, 1982.

Heyvaerts, J., E.R. Priest, and D.M. Rust, An emerging flux model of the solar flare phenomenon, Astrophys. J., 216, 123, 1977.

Hill, T.W., Magnetic merging in a collisionless plasma, J. Geophys. Res., 80, 4689, 1975.

Hones, E.W., Jr. (Editor), Magnetic reconnection in space and laboratory plasmas, Geophysical Monograph 30, AGU Publishers, Washington, D.C., U.S.A., 1984.

Parker, E.N., Sweet's mechanism for merging magnetic fields in conducting fluids, J. Geophys. Res., 62, 509, 1957a.

Parker, E.N., Newtonian development of the dynamical properties of ion-

ized gases of low density, Phys. Rev., 107, 924, 1957b.
Parker, E.N., The solar flare phenomenon and the theory of reconnection and annihilation of magnetic fields, Astrophys. J., suppl. Ser., 8, 177, 1963.
Petschek, H.E., Magnetic field annihilation, in AAS-NASA Symposium on the Physics of Solar Flares, NASA Spec. Publ. SP-50, 425, 1964.
Rich, F.J., V.M. Vasyliunas, and R.A. Wolf, On the balance of stresses in the plasma sheet, J. Geophys. Res., 77, 4670, 1972.
Semenov, V.S., I.V. Kubyshkin, M.F. Heyn, and H.K. Biernat, Field-line reconnexion in the two-dimensional asymmetric case, J. Plasma Phys., 30, 321, 1983.
Seymour, P.W., Drift of a charged particle in a magnetic field of constant gradient, Aust. J. Phys., 12, 309, 1959.
Sonnerup, B.U.Ö., Magnetic-field re-connection in a highly conducting incompressible fluid, J. Plasma Phys., 4, 161, 1970.
Sonnerup, B.U.Ö., and E.R. Priest, Resistive MHD stagnation-point flows at a current sheet, J. Plasma Phys., 14, 283, 1975.
Speiser, T.W., Particle trajectories in model current sheets 1: Analytical solutions, J. Geophys. Res., 70, 4219, 1965.
Speiser, T.W., On the uncoupling of parallel and perpendicular motion in a neutral sheet, J. Geophys. Res., 73, 1113, 1968.
Sweet, P.A., The neutral point theory of solar flares, in Electromagnetic Phenomena in Cosmical Physics, edited by B. Lehnert, Cambridge University Press, London, 1958.
Vasyliunas, V.M., Magnetic field line merging, Rev. Geophys. Space Phys., 13, 303, 1975.

CHAPTER 6

MAGNETOCONVECTION

N.O. Weiss
Department of Applied Mathematics and Theoretical Physics
University of Cambridge
England CB3 9EW

From the observational point of view, the various manifestations of solar activity occur in the photosphere or above, where the optical depth is small. Yet this region contains only 10^{-10} of the total solar mass and the magnetic fields that are responsible for solar activity are generated in the subphotospheric convective zone. In order to understand the origins of solar (or stellar) activity we therefore need to study the interaction between magnetic fields and convection in a turbulent plasma. However, since the relevant parameter values cannot be attained in the laboratory, any treatment of magnetoconvection must rely on a mixture of mathematical theories with numerical experiments. In what follows I shall focus on three topics. These are, first, the formation of small flux tubes (some of whose properties have been described in Chapter 3); second, the nature of convection in a strong magnetic field; and, third, the structure of the large-scale magnetic field in the sun (which leads on to the discussion of dynamo theory in Chapter 7). Various aspects of magnetoconvection have been considered in a number of recent reviews (Priest, 1982; Proctor and Weiss, 1982; Galloway, 1984; Nordlund, 1984; Weiss, 1985).

6.1 SMALL FLUX TUBES

We may recall from Chapter 1 that observations show that solar magnetic fields have a highly intermittent structure, closely related to the pattern of convection. On a relatively large scale, the supergranules (with diameters of 30,000 km and lifetimes of a day) are surrounded by the photospheric network, containing most of the magnetic flux that is not in active regions. On a much smaller scale, photospheric convection produces a cellular pattern: hot rising elements are enclosed by a network of cooler sinking material. With poor resolution the bright regions look like grains of rice and therefore were described as granules (Bray, Loughhead and Durrant, 1984). A typical cell has a diameter of 1000 km and a lifetime of 10 minutes, and most of the magnetic flux is contained in slender tubes with intense fields of around 1500 gauss. These tubes are located in the intergranular network and especially at corners; they can be observed indirectly by studying the filigree (Dunn and Zirker, 1973) or bright points (Muller, 1983). Title, Tarbell and Ramsey (1985) have measured the fields directly and find that individual magnetic elements survive for more than 40 minutes, while tne granulation pattern changes.

From the theoretical point of view we are therefore led to study the properties of isolated flux tubes (Parker, 1979; Spruit and Roberts, 1983). These tubes are embedded in a superadiabatically stratified atmosphere (which is vigorously convecting) and this stratification produces an instability within the flux tube, which was discussed in Chapter 3. Downdrafts develop, leading to adiabatic evacuation, unless the field exceeds a critical value of about 1200 gauss. So we can expect the flux tubes to collapse until the field approaches the limit, $B_p = (2\mu_0 p)^{1/2}$, set by pressure balancing. At the photosphere, $B_p \approx 1600$ gauss. Within the flux tube there will be small-scale motion, giving rise to turbulent diffusion. Crude estimates yield a turbulent diffusivity $\eta \approx 10^6\ m^2 s^{-1}$, with a corresponding timescale $\tau_\eta \approx 25$ minutes. Since this is comparable with the lifetime of the flux tubes we need to include diffusion and transverse flow in our description of the photospheric fields. In addition, we should relate flux tubes to the pattern of convection.

Consider first the kinematic problem (Proctor and Weiss, 1982). When the velocity $\underset{\sim}{v}$, is prescribed, the magnetic field satisfies the induction equation

$$\frac{\partial \underset{\sim}{B}}{\partial t} = \underset{\sim}{\nabla} \times (\underset{\sim}{v} \times \underset{\sim}{B}) + \eta \nabla^2 \underset{\sim}{B} . \tag{6.1}$$

The ratio between the advective and diffusive terms in (6.1) is given by the magnetic Reynolds number, $R_m = UL/\eta$, where U and L are a typical velocity and lengthscale. For granules, $R_m \approx 10^6$ with molecular diffusion and $R_m \approx 10^3$ with turbulent diffusion. The formation of an axisymmetric flux tube can be studied by adopting cylindrical co-ordinates (r, ϕ, z) and taking the solenoidal velocity $\underset{\sim}{v} = (U/L^2)\underset{\sim}{\nabla} \times (rz\underset{\sim}{\phi})$, together with a field $\underset{\sim}{B} = B(r)\underset{\sim}{z}$. Then (6.1) has a steady solution

$$B = \tfrac{1}{2} R_m B_o \exp(-\tfrac{1}{2} R_m r^2/L^2) , \qquad (6.2)$$

with a Gaussian profile, and the total flux $F = \pi B_o L^2$. The field strength is enhanced by a factor R_m while the radius is compressed by a factor $R_m^{1/2}$. So intense fields can be formed in regions of converging flow.

Consider now a pattern of two-dimensional convection in rolls parallel to the y-axis of cartesian co-ordinates (x,y,z), with a velocity $\underset{\sim}{v} = (U/\pi)\underset{\sim}{\nabla} \times [\sin(\pi x/L)\sin(\pi z/L)\hat{\underset{\sim}{y}}]$. For $R_m \gg 1$ it can be shown that nearly all the flux is eventually expelled from regions with closed streamlines and concentrated into sheets between the cells. How long does this process take? Suppose that the field is initially uniform and that the eddy turnover time $\tau_o \equiv L/U$. Flux sheets are formed on a timescale given by τ_o, while the field is wound up within the cells. Reconnection starts after a time $R_m^{1/3}\tau_o$ (Parker, 1963; Weiss, 1966; Moffatt and Kamkar, 1982) and closed field lines, parallel to the streamlines, are left. This relatively weak field decays on the resistive timescale, $\tau_\eta = R_m \tau_o$ (Rhines and Young, 1983) but reconnection still continues and only ceases after a time $R_m^{3/2}\tau_o = R_m^{1/2}\tau_\eta$, (R.L. Parker, 1966). From an astrophysical point of view the relevant timescale is given by $R_m^{1/3}\tau_o$ and for reasonable estimates of the eddy diffusivity reconnection occurs after a few turnover times (Galloway and Weiss, 1981).

These arguments imply that diffuse fields cannot persist. In fact magnetic flux appears in isolated tubes, which are preserved by the mechanism described above. In a simple two-dimensional model there is a unique, symmetrical steady solution; for an axisymmetric flow most of the flux is concentrated near the centre of the cell, with a resulting loss of symmetry. In a hexagonal cell with upward motion at the centre flux is swept aside to the corners but in the final steady state much of the flux remains in the neighbourhood of the central axis, although strong fields appear only at the corners (Galloway and Proctor, 1983).

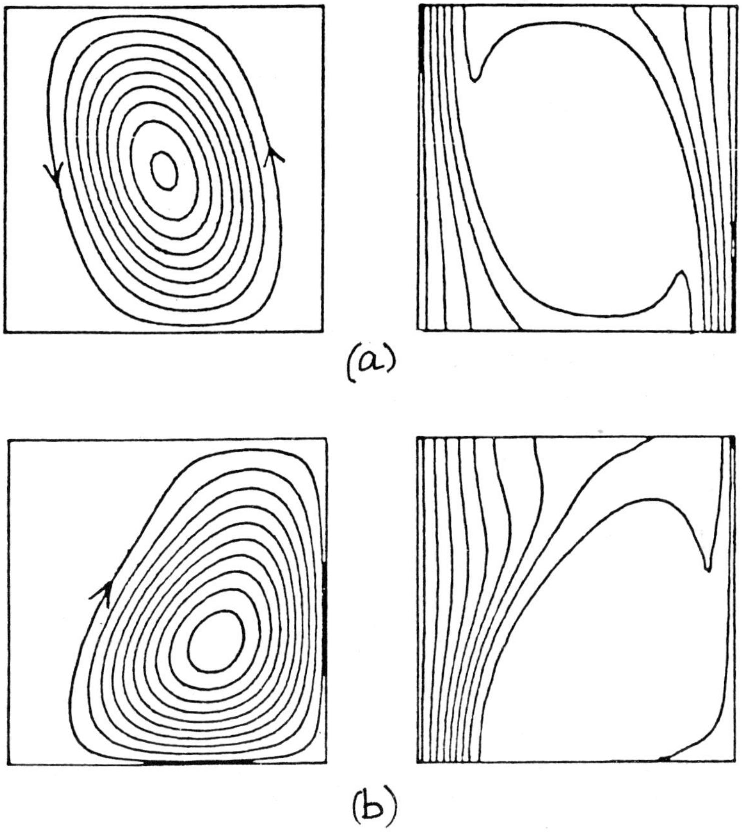

Figure 6.1 Flux expulsion and stagnant flux sheets. Streamlines and lines of force for steady rolls with (a) symmetrical solution, (b) asymmetrical solution.

If intense magnetic fields are produced, the Lorentz force can no longer be ignored. Nonlinear magnetoconvection has been extensively studied in the Boussinesq approximation (when the density, ρ, is a function of temperature only) valid if the layer depth d is much less than a scale-height and U is small compared with the sound speed. The field in an isolated tube (Galloway and Moore, 1979) or sheet (Weiss, 1981) exerts forces which exclude the motion from regions where the flux is concentrated, just as flux is expelled from regions where motion takes place. Figure 6.1 illustrates this segregation of the motion from the field for a two-dimensional flow. The symmetrical solution is unstable to perturbations that break the symmetry, leading to a stable solution with most of the flux concentrated on one side of the cell. Flux concentration is limited by Lorentz torques, owing to curvature and tension along the field lines and the peak field $B_m \approx (\nu/\eta)^{1/2} B_e$, where the equipartition field $B_e = (\mu_0 \rho)^{1/2} U$, and ν is the viscous diffusivity. For turbulent flows we expect ν and η to be comparable, so that $B_m \approx B_e$: at the photosphere this corresponds to a field of about 600 gauss, significantly less than what is found.

The best observations confirm that isolated flux tubes are stagnant (Title et al., 1985), though they are surrounded by sinking gas. Any quantitative treatment must be fully compressible, with a density that depends on pressure as well as temperature. Preliminary computations by Cattaneo (1984a) and Hurlburt and Toomre (1985) confirm that the increasing magnetic pressure leads to evacuation of the flux but systematic studies of compressible magnetoconvection have only just begun.

6.2 CONVECTION IN A STRONG MAGNETIC FIELD

So far we have dealt with situations where the overall field is relatively weak, so that magnetic structures are dominated by convection. If the imposed field is strong, the pattern of convection is altered and motion may even be suppressed. Sunspots may have radii of 20,000 km and fields of up to 3000 gauss. Normal granular convection is replaced by motions that are less efficient at carrying energy, and the spot is therefore cooler and darker than the normal photosphere. Even at sunspot maximum only 10^{-3} of the solar area is covered by spots, but more active stars show variations in luminosity suggesting that starspots may cover 20-50% of their surfaces.

In the outer part of a sunspot the field is almost horizontal. As a result, convection occurs in rolls with their axes parallel to the field, giving rise to the radial filaments characteristic of the penumbra. Towards the centre of the spot, in the umbra, the field is nearly vertical and strong enough for normal overturning convection to be inhibited. The motion that does take place, giving rise to umbral dots (e.g. Knobloch and Weiss, 1984), is inevitably more complicated. To understand what happens, let us consider the simplified problem of Boussinesq magnetoconvection (Proctor and Weiss, 1982).

Suppose that we have a plane layer of depth d, heated uniformly from below, with a uniform field $\underset{\sim}{B}_0$ in the absence of any motion. We assume periodic boundary conditions (in order to obtain convenient eigenfunctions) and seek perturbations varying as $f(x,y)\sin(\pi z/d)\exp(i\Omega t)$ etc., where $\nabla^2 f = -(a/d)^2 f$. For a non-dissipative (adiabatic) system,

$$\Omega^2 = n^2 v_A^2/d^2 - a^2|N^2|/(a^2+\pi^2) , \tag{6.3}$$

where the Alfven speed v_A and the (imaginary) buoyancy frequency N are given by

$$v_A^2 = B_0^2/\mu_0\rho_0 \qquad \text{and} \qquad N^2 = -g\alpha\Delta T/d . \tag{6.4}$$

Here α is the coefficient of thermal expansion and ΔT is the difference in potential temperature across the layer. From (6.3) there is a transition from neutral oscillations ($\Omega^2 > 0$) to exponential growth ($\Omega^2 < 0$) when

$$|N|^2 = \pi^2(a^2+\pi^2)v_A^2/a^2d^2 . \tag{6.5}$$

When dissipative effects are included behaviour is more complicated. Indeed, magnetoconvection is the most thoroughly studied example of nonlinear double diffusive convection. The nature of the solution depends on the ratio, $\zeta = \eta/\kappa$, of the magnetic to the thermal diffusivity; for $\zeta > 1$ instability leads to steady convection while for $\zeta < 1$ bifurcations to oscillatory and steady convection can occur. In laboratory experiments $\zeta > 1$ but in stars, where κ is a radiative diffusivity, $\zeta \ll 1$ (except in regions where ionization is taking place).

The simplest model problem to consider is that of two-dimensional convection in the region $\{0 < x < \lambda d;\ 0 < z < d\}$. Then the velocity and the magnetic field can be described by a stream function and a flux function such that

$$\underset{\sim}{v} = (-\partial\psi/\partial z,\ 0,\ \partial\psi/\partial x), \qquad \underset{\sim}{B} = (-\partial A/\partial x,\ 0,\ \partial A/\partial x), \qquad (6.6)$$

and the governing equations can be written in dimensionless form as

$$\frac{\partial}{\partial t}\nabla^2\psi + \frac{\partial(\psi,\nabla^2\psi)}{\partial(x,z)} = \sigma\left[\nabla^4\psi + R\frac{\partial T}{\partial x} + \zeta q\frac{\partial(A,\nabla^2 A)}{\partial(x,z)}\right], \qquad (6.7)$$

$$\frac{\partial T}{\partial t} + \frac{\partial(\psi,T)}{\partial(x,z)} = \nabla^2 T, \qquad (6.8)$$

$$\frac{\partial A}{\partial t} + \frac{\partial(\psi,A)}{\partial(x,z)} = \zeta\nabla^2 A. \qquad (6.9)$$

Here the Prandtl number $\sigma = \nu/\kappa$, the Rayleigh number $R = g\alpha\Delta T d^3/\kappa\nu$ and the Chandrasekhar number $Q = B_0^2 d^2/\mu_0\rho\nu\eta$.

Linear stability theory was discussed by Chandrasekhar (1961). In the absence of a magnetic field, the static, conducting solution with $\underset{\sim}{v} = 0$ undergoes a simple bifurcation (leading to an exponentially growing solution) at $R = R_0$, where $R_0 = \lambda^{-4}(1+d^2)^3\pi^4$. It is convenient to introduce the dimensionless quantities $r = R/R_0$ and $q = \lambda^4 Q/\pi^2(1+\lambda^2)^2$ and to consider perturbations varying as $\exp(st)$. Then the growth rate s satisfies the cubic characteristic equation

$$s^3 + (1+\sigma+\zeta)s^2 + \left[\sigma(1-r+\zeta q) + \zeta(1+\sigma)\right]s + \sigma\zeta(1-r+q) = 0. \qquad (6.10)$$

Thus there is a simple bifurcation (sometimes referred to as an exchange of stabilities) when $s = 0$, at $r = r^{(e)} = 1 + q$. In addition, there is the possibility of a Hopf bifurcation (ofter referred to as the onset of overstability) when s is complex and $\mathrm{Re}\, s = 0$, at

$$r = r^{(o)} = (\sigma+\zeta)\left[\frac{1+\zeta}{\sigma} + \frac{\zeta}{1+\sigma}q\right]. \qquad (6.11)$$

The corresponding frequency $\omega_0 = \mathrm{Im}\, s$ is real provided that

$$\omega_0^2 = -\zeta^2 + \sigma\zeta q(1-\zeta)/(1+\sigma) > 0. \qquad (6.12)$$

Thus overstability occurs only if $\zeta < 1$ and $q > q_0 = \zeta(1+\sigma)/\sigma(1-\zeta)$; in that case it can be shown that $r^{(e)} > r^{(o)}$. Consider what happens as r is increased for the astrophysically relevant case, with $q > q_0 > 0$. Of the three roots of (6.10) one is always real and negative; the other pair are initially complex conjugates with negative real parts. At $r = r^{(o)}$ there is a Hopf bifurcation: the two complex roots cross into the right-hand half of the complex s-plane. As r is further increased, these roots approach the real axis and merge when $r = r^{(i)}$. For $r > r^{(i)}$ all three roots are real. As r increases, one positive root increases monotonically, while the other decreases and eventually crosses back into the left-hand half plane when $r = r^{(e)}$. For $r > r^{(e)}$ there is only one unstable root.

To proceed further, we investigate mildly nonlinear behaviour in the neighbourhood of the bifurcations, which is amenable to analysis (Proctor and Weiss, 1982). In the neighbourhood of the simple bifurcation at $r^{(e)}$ there is a branch of steady solutions with

$$\psi = \epsilon\psi_1 + O(\epsilon^3), \qquad r = r^{(e)} + r_2^{(e)}\epsilon^2 + O(\epsilon^3), \tag{6.13}$$

where ϵ is a small parameter and

$$r_2^{(e)} = 1 + q + 2\lambda^2(1-\lambda^2)q/\zeta^2(1+\lambda^2) . \tag{6.14}$$

For $r_2^{(e)} < 0$ subcritical steady convection is possible, with $r < r^{(e)}$, but for $r_2^{(e)} > 0$ only supercritical convection can occur; in both cases, the finite amplitude solutions are unstable near the bifurcation.

The qualitative behaviour of nonlinear oscillatory solutions can be understood by studying the situation when $r^{(o)}$ is close to $r^{(e)}$ (or "unfolding a bifurcation of codimension two") (Guckenheimer and Holmes, 1983). Suppose that $q = q_0 + \epsilon^2$, where ϵ is small, so that

$$r^{(e)} = \frac{\sigma+\zeta}{\sigma(1-\zeta)} + \epsilon^2 , \qquad r^{(o)} = \frac{\sigma+\zeta}{\sigma(1-\zeta)} + \frac{\zeta(\sigma+\zeta)}{1+\sigma}\epsilon^2 \tag{6.15}$$

and $r = (\sigma+\zeta)/\sigma(1-\zeta) + \mu\epsilon^2$. Then the amplitude a of the solution satisfies a nonlinear ordinary differential equation of the form

$$\ddot{a} - Ma^3 + MNa \approx \epsilon F(a) , \tag{6.16}$$

where M,N depend on the parameters and F is a nonlinear function of a and its time derivatives (Knobloch and Proctor, 1981). If $\epsilon = 0$, (6.15) reduces to Duffing's equation, which can be solved in terms of elliptic functions; the amplitude, $\bar{a}$, of the solution is then given by requiring that $\langle \dot{a}F(a)\rangle = 0$ when averaged over a cycle. Thus we obtain $\bar{a}$ as a function of μ along the two branches of oscillatory and steady solutions. This behaviour is best described by bifurcation diagrams and two cases can now be distinguished.

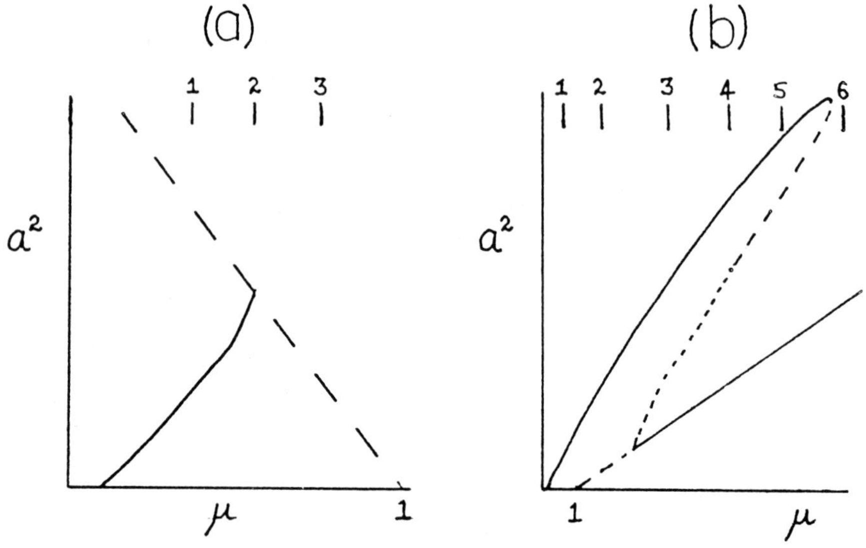

Figure 6.2 Bifurcation diagrams for (a) $r_2^{(e)} < 0$ and (b) $r_2^{(e)} > 0$. Unstable solutions are shown by broken lines.

When $r_2^{(e)} < 0$ the unstable steady branch bifurcates in the direction of decreasing μ, i.e. towards the left in Fig. 6.2(a), while the stable oscillatory branch bifurcates towards the right. The period of the oscillations increases monotonically along the oscillatory branch, becoming infinite where it terminates on the steady branch. The pattern of behaviour for fixed values of μ can be represented by sketching phase portraits in the $a\dot{a}$-phase plane, as shown in Fig. 6.3(a). The static solution loses stability at $r^{(0)}$ and sheds a symmetrical limit cycle which expands until it is destroyed in a heteroclinic bifurcation. Thereafter all trajectories escape from the region where the truncated equations are valid.

When $r_2^{(e)} > 0$ the steady branch bifurcates in the direction of increasing μ, as shown in Fig. 6.2(b). Near the bifurcation at $r^{(e)}$ steady solutions are unstable but they gain stability after a Hopf bifurcation at some $\mu > 1$. The unstable limit cycles shed here correspond to vacillations about the steady state; these limit cycles expand until they eventually merge at the origin, forming large-scale oscillations like those on the branch that bifurcates from $r^{(0)}$. Once again, details of this behaviour can be understood by studying the phase portraits in Fig. 6.4, which were established, in a slightly different context, by Arnold (1982).

(a)

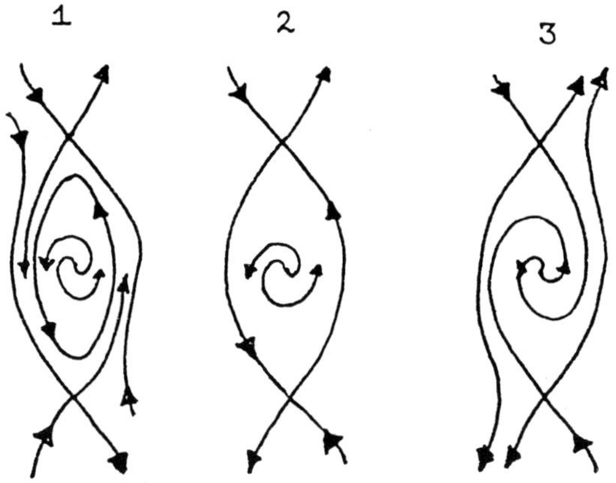

(b)

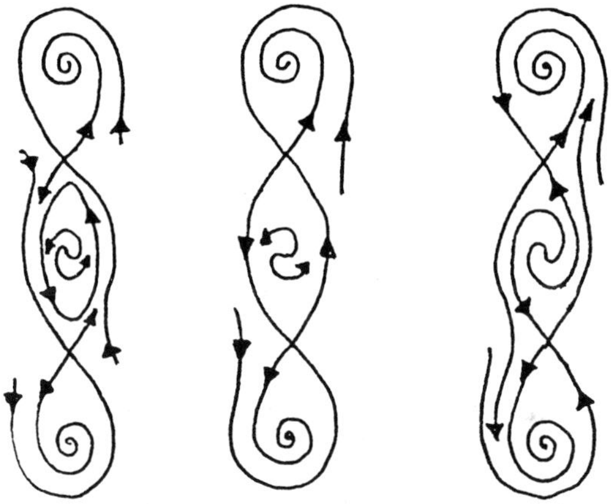

Figure 6.3 Phase portraits for $r_2^{(e)} < 0$ (a) for solutions of (6.16) and (b) for solutions of the full equatins. The numbers correspond to the values of μ indicated in Fig. 6.2(a).

In order to explore fully nonlinear behaviour we must turn to numerical experiments (Weiss, 1981). Fig. 6.5 shows two sets of results, for cases corresponding to those in Fig. 6.2. When $r_2^{(e)} > 0$, the qualitative behaviour is identical to that in Fig. 6.2(b) and Fig. 6.4. When $r_2^{(e)} < 0$, we see that the steady branch turns back and recovers stability (as it must if trajectories spiral inwards from infinity and the static solution is globally stable for $r < 1$). The appearance of stable steady solutions makes it possible to complete the phase

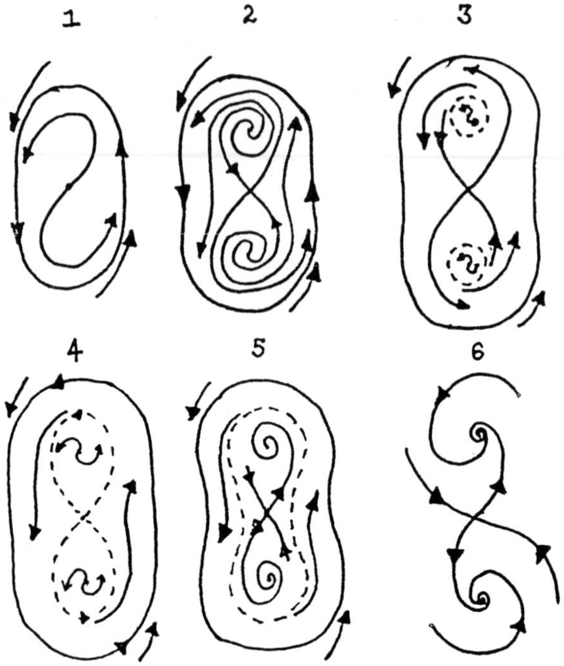

Figure 6.4 Phase portraits for $r_2^{(\theta)} > 0$. The numbers correspond to the values of μ indicated in Fig. 6.2(b).

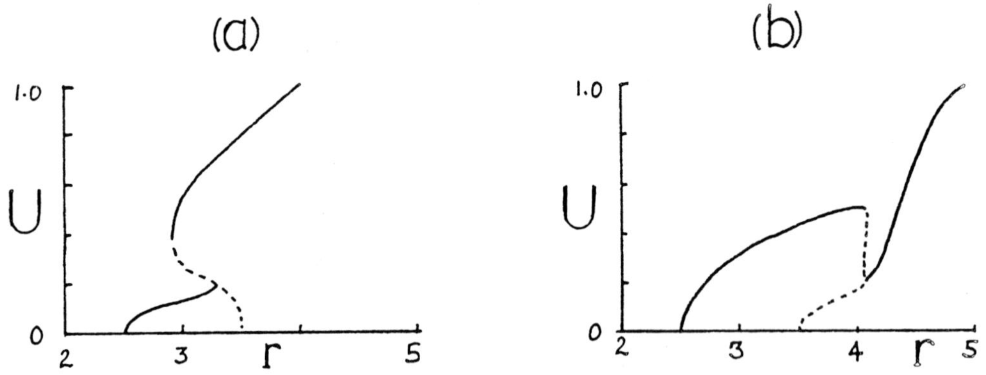

Figure 6.5 Numerical solutions of the partial differential equations for $q = 2.5$, $\sigma = \zeta = 0.2$ (a) $\lambda = 1\,(r_2^{(\theta)} < 0)$ and (b) $\lambda = \tfrac{1}{2}\,(r_2^{(\theta)} > 0)$. The ordinate is the rms velocity U and broken lines show conjectured unstable segments of solution branches.

portraits, as shown in Fig. 6.3(b). It is of interest to establish the lowest value of the Rayleigh number, R_{min}, for which steady convection can occur. Fig. 6.6 shows the stable parts of the steady solution branches for $\zeta Q = 100$, $\zeta = 0.2$ and 0.1 (in physical terms, this corresponds to fixing B_0 and varying η). When ζQ is sufficiently large and ζ sufficiently small, R_{min} is independent of ζ. Simple physical arguments suggest that $R_{min} \propto \zeta Q$.

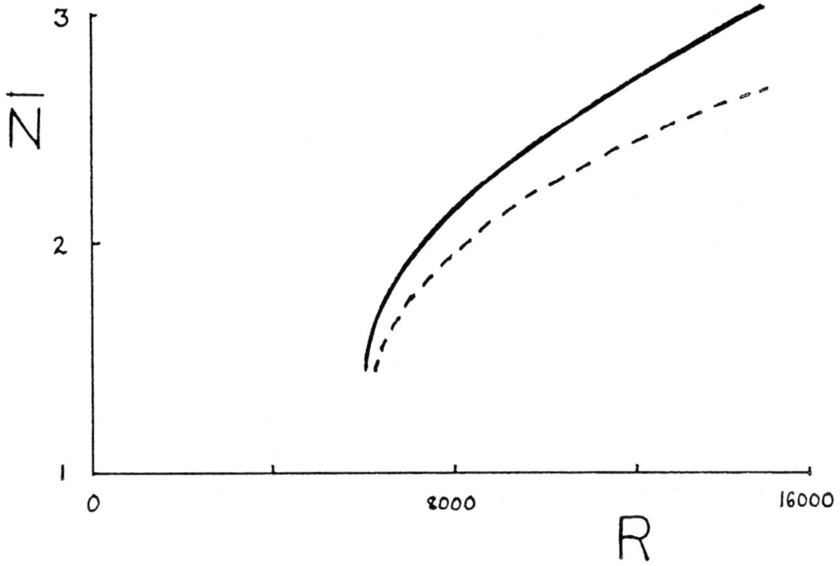

Figure 6.6 Steady solutions for $\zeta Q = 100$, $\sigma = 1$ and $\zeta = 0.2$ (full line), $\zeta = 0.1$ (broken line).

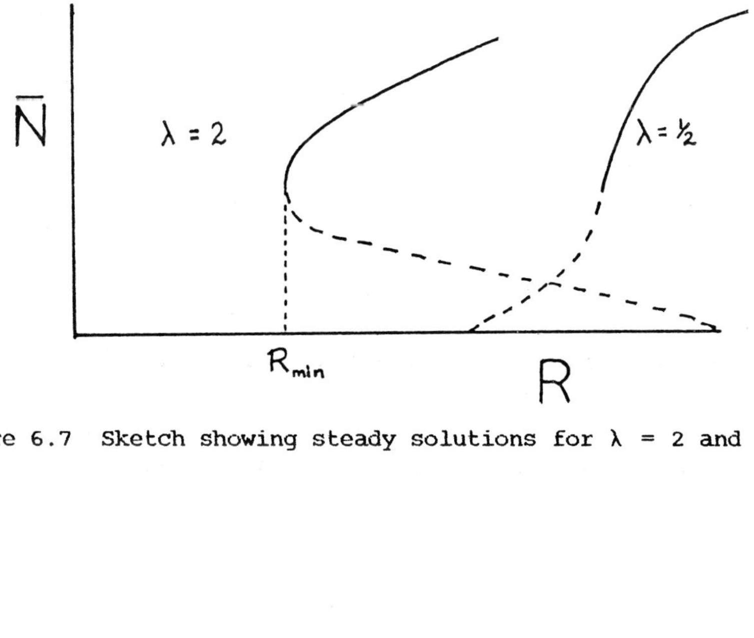

Figure 6.7 Sketch showing steady solutions for $\lambda = 2$ and $\lambda = \frac{1}{2}$.

It is instructive to compare predictions based on linear theory with results derived from nonlinear computations. Let us assume that $\zeta \ll 1 \ll \zeta Q$. Then according to linear theory convection first appears in narrow cells, with $\lambda \ll 1$, and the Hopf bifurcation occurs at $R^{(o)} \approx \pi^2(1 + \lambda^2)\zeta Q\sigma/(\sigma + 1)$. The complex eigenvalues merge at $R^{(i)} \approx \pi^2(1 + \lambda^2)\zeta Q$ and the simple bifurcation follows at $R^{(e)} \approx \pi^2(1 + \lambda^2)Q$; so $R^{(o)} < R^{(i)} \ll R^{(e)}$. From the nonlinear results we find that oscillatory convection is inefficient at transporting heat. Stable overturning convection appears when $R = R_{min}$ but when $\lambda \ll 1$ only supercritical convection is possible and $R_{min} > R^{(e)}$. This situation is sketched in Fig. 6.7. In fact, cells with $\lambda \approx 2$ are preferred, contrary to what might be expected on the basis of extrapolation from linear theory. On the other hand, efficient convection occurs for $R \geqslant R_{min} \approx R^{(i)}$ $\pi^2\zeta Q$, although the appearance of real eigenvalues is irrelevant in the nonlinear regime. Indeed, the condition $R = \pi^2\zeta Q$ reduces to $|N^2| = \pi^2 v_A^2/d^2$, which is similar to the criterion for adiabatic instability in (6.5). This is, of course, the only simple dimensionless result that is independent of the diffusivities. In mixing length theory it is often assumed that the kinetic energy gained by a blob of fluid falling through the layer is comparable with the potential energy lost, so that $U^2 \approx g\alpha\Delta T d$: then it follows that overturning convection with a typical speed U can be suppressed by a magnetic field B_o if B_o is greater than the equipartition field B_e. This seems a very plausible result. In sunspot umbrae ($B_o \approx 3000$ G, $B_e \approx 600$ G) only oscillatory convection, in vertically elongated cells, is likely to occur.

Boussinesq theory can only provide a rough guide to behaviour in the solar photosphere and compressible magnetoconvection is inevitably more complicated still. The overstable modes correspond to magnetoacoustic (rather than hydromagnetic) oscillations that are destabilized by the thermal stratification. When $B \approx B_p$ anomalous behaviour can be found,with overstability even when the atmosphere is subadiabatically stratified (Cattaneo, 1984a,b). Hurlburt (1985) has studied two-dimensional behaviour in the nonlinear regime and Nordlund (1983, 1984) has carried out three-dimensional simulations of the solar granulation, using the anelastic approximation, with $\nabla . (\rho \underset{\sim}{v}) = 0$.

6.3 STRUCTURE OF THE LARGE-SCALE MAGNETIC FIELD

Our understanding of magnetic fields within the sun is based on theoretical studies of the kind described above, coupled with observations of magnetic activity in the sun and other stars with deep convective zones. These considerations suggest that magnetic fields within a turbulent region are bound to be highly intermittent (Galloway and Weiss, 1981). Most of the flux will be confined to isolated (though perhaps ephemeral) tubes with fields that are relatively intense. It is only near the surface that field strengths are likely to approach B_p; deeper down the limit is probably set by B_e, which rises to 10^4 gauss at the base of the convective zone (where $B_p/B_e \approx 3000$).

In considering the structure of the large-scale field other instabilities must be borne in mind. Among these the most prominent are interchange (or flute) instabilities, such as the Rayleigh-Taylor instability discussed in Chapter 4. In a plane layer a plasma that is partially supported by a horizontal magnetic field is unstable. In an axisymmetric sunspot the magnetic field at the boundary is concave towards the external plasma. Such a configuration is prone to instability but here the flux tube fans out, owing to the stratification, so that less dense magnetised material is supported by denser field-free plasmas, and the spot is stable if the radial field decreases upwards on the boundary (Meyer, Schmidt and Weiss, 1977). Other instabilities are driven by magnetic buoyancy. Consider a slowly varying field $\mathbf{B} = B(z)\hat{\mathbf{y}}$: this is unstable to adiabatic perturbations if

$$\frac{d}{dz}\left[\frac{2B^2}{B_p^2}\,\ln\left(\frac{B}{\rho}\right) + \ln\left(\frac{P}{\rho^\gamma}\right)\right] < 0 , \qquad (6.17)$$

i.e. if the field decreases upwards sufficiently rapidly. With diffusion, it turns out that the field is unstable if its strength increases upwards as well (Hughes, 1984,1985).

One important question to be settled is the location of the dynamo responsible for the sun's magnetic cycle. A number of arguments suggest that the dynamo operates near the base of the convective zone and generates a predominantly toroidal field. This field fills a magnetic layer which is liable to instabilities driven by magnetic buoyancy. As a result, stitches of flux escape and float upwards to emerge in active regions and in sunspots. These flux tubes are, however, only the largest examples of the intermittent fields that exist in the convective zone.

REFERENCES

Arnold, V.I.: 1982, Geometrical Methods in the Theory of Ordinary Differential Equations, Springer, Berlin.

Bray, R.J., Loughhead, R.E. and Durrant, C.J.: 1984, The Solar Granulation, Cambridge University Press.

Cattaneo, F.: 1984a, Ph.D. dissertation, University of Cambridge.

Cattaneo, F.: 1984b, in The Hydromagnetics of the Sun, ed. T.D. Guyenne, p.47, ESA SP-220.

Chandrasekhar, S.: 1961, Hydrodynamic and Hydromagnetic Stability, Clarendon Press, Oxford.

Dunn, R.B. and Zirker, J.B.: 1973, Solar Phys. 33, 281.

Galloway, D.J.: 1984, preprint.

Galloway, D.J. and Moore, D.R.: 1979, Geophys. Astrophys. Fluid Dyn **12**, 73.

Galloway, D.J. and Proctor M.R.E.: 1983, Geophys. Astrophys. Fluid Dyn. **24**, 109.

Galloway, D.J. and Weiss, N.O.: 1981, Astrophys. J. **243**, 945.

Guckenheimer, J. and Holmes, P.: 1983, Nonlinear Oscillations, Dynamical Systems and Bifurcations of Vector Fields, Springer, New York.

Hughes, D.W.: 1984, in The Hydromagnetics of the Sun, ed. T.D. Guyenne, p.55, ESA SP-220.

Hughes, D.W.: 1985, Geophys. Astrophys. Fluid Dyn. (in press).

Hurlburt, N.E. and Toomre, J.: 1985, Astrophys. J. (submitted).

Knobloch, E. and Proctor, M.R.E.: 1981, J. Fluid Mech. **108**, 291.

Knobloch, E. and Weiss, N.O.: 1984, Mon. Not. Roy. Astr. Soc. **207**, 203.

Meyer, F., Schmidt, H.U. and Weiss, N.O.: 1977, Mon. Not. Roy. Astr. Soc. **179**, 741.

Moffatt, H.K. and Kamkar, S.: 1982, in Stellar and Planetary Magnetism, ed. A.M. Soward, p.91, Gordon and Breach, New York.

Muller R.: 1983, Solar Phys. **85**, 113.

Nordlund, A.: 1983, in Solar and Stellar Magnetic Fields, ed. J.O. Stenflo, p.79, Reidel, Dordrecht.

Nordlund, A.: 1984, in The Hydromagnetics of the Sun, ed. T.D. Guyenne, p.37, ESA SP-220.

Parker, E.N.: 1963, Astrophys. J. **138**, 552.

Parker, E.N.: 1979, Cosmical Magnetic Fields, Clarendon Press, Oxford.

Parker, R.L.: 1966, Proc. Roy. Soc. A **291**, 60.

Priest, E.R.: 1982, Solar Magnetohydrodynamics, Reidel, Dordrecht.

Proctor, M.R.E. and Weiss, N.O.: 1982, Rep. Prog. Phys. **45**, 1317.

Rhines, P.B. and Young, W.R.: 1983, J. Fluid Mech. **133**, 133.

Spruit, H.C. and Roberts, B.: 1983, Nature **304**, 401.

Title, A.M., Tarbell, T.D. and Ramsey, H.E.: 1985, preprint.

Weiss, N.O.: 1966, Proc. Roy. Soc. A **293**, 310.

Weiss, N.O,: 1981, J. Fluid Mech. **108**, 247.

Weiss, N.O.: 1985, in High Spatial Resolution in Solar Physics, ed. R. Muller, Springer, Berlin.

CHAPTER 7

ASPECTS OF DYNAMO THEORY

H.K. Moffatt
Department of Applied Mathematics and Theoretical Physics
University of Cambridge
England CB3 9EW

The elements of dynamo theory are discussed, with particular attention to the particular problems that arise when, as in the solar context, the magnetic diffusivity is very small. The growth of the dipole moment of a localised current system is essentially diffusive in character; in the limit of vanishing diffusivity, the spatial structure of any dynamo must become increasingly complex; this is the 'fast dynamo' limit.

When convective eddies are persistent, the phenomena of flux expulsion and topological pumping play an important part in the dynamo process. These effects appear in the 'mean-field' theory of the turbulent dynamo via an 'effective velocity' of transport of the mean magnetic field relative to the fluid.

These effects are all discussed in the context of the solar dynamo, regarded as a dynamo of $\alpha\omega$-type, with magnetic buoyancy providing an equilibration mechanism.

7.1 THE HOMOPOLAR DISC DYNAMO

Some peculiarities of dynamo theory are very well illustrated by the prototype example of self-exciting dynamo action, viz. the homopolar disc dynamo sketched in figure 7.1. The conducting disc rotates about its axis under the action of an applied torque G. A wire, twisted about the axis in the manner shown, makes sliding contact with the disc at A, and with the axis at B, and carries a current $I(t)$. The magnetic field $\underline{B}$ associated with this current has a flux $\Phi = MI$ across the disc, where M is the mutual inductance between the wire and the rim of the disc. The rotation of the disc in the presence of this flux provides a radial electromotive force $\frac{\Omega}{2\pi}\Phi = \frac{\Omega}{2\pi}MI$ which drives the current I. On this simplistic description, the equation for I is

$$L\frac{dI}{dt} + RI = \frac{M}{2\pi}\Omega I , \qquad (7.1)$$

where R is the total resistance of the circuit and L its self-inductance.

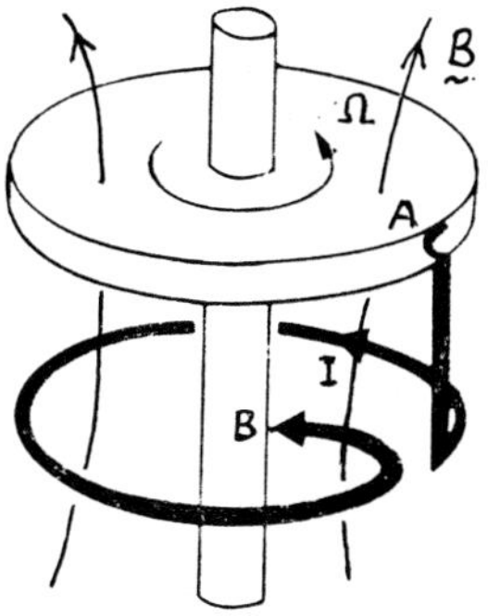

Figure 7.1

Suppose that Ω is maintained constant by suitable adjustment of the driving torque. Then (7.1) has exponential solution $I(t) = I(0)e^{pt}$ where

$$p = L^{-1}\left\{\frac{M}{2\pi}\Omega - R\right\} , \qquad (7.2)$$

and we have exponential growth of $I(t)$ and so of the magnetic field to which it gives rise (i.e. we have dynamo action) provided $M\Omega > 2\pi R$, i.e. provided the disc rotates rapidly enough.

Appealing though this description is in its simplicity, it cannot be correct (although it will be found in many texts and review articles!). For consider the limiting situation of a perfectly conducting disc and wire, in which case $R = 0$. Then, on the one hand, (7.2) gives $p = M\Omega/2\pi L$ so that we still have dynamo action. But on the other hand, the rim of the disc is a closed circuit moving with a perfect conductor, and Alfven's theorem (the most basic theorem in magnetohydrodynamics) tells us that the flux Φ through this circuit must be constant. There is an obvious contradiction. What has gone wrong?

The answer is that we have neglected the currents that flow azimuthally in the disc - i.e. the very currents that are associated with the diffusion of flux across the rim of the disc. These currents become particularly important in the limit $R \to 0$, and they completely invalidate the above description. The paradox can be resolved by supposing that the azimuthal current $J(t)$ is constrained to flow round the rim of the disc (by a suitable distribution of radial insulating strips). Then the fluxes through the I and J circuits are given by

$$\Phi_1 = LI + MJ \tag{7.3}$$

$$\Phi_2 = MI + L'J$$

and the equations governing the current flow are

$$\frac{d\Phi_1}{dt} = \frac{\Omega}{2\pi}\Phi_2 - RI \tag{7.4}$$

$$\frac{d\Phi_2}{dt} = -R'J$$

where R', L' refer to the J-circuit. This system still admits exponential solutions, $(I,J) \propto e^{pt}$, and the criterion for dynamo action is still $M\Omega > 2\pi R$. Now however, $p \to 0$ as $R' \to 0$, and so the description is consistent with Alfven's theorem. Details may be found in Moffatt (1979) where the nonlinear dynamical system (including the equation for $\Omega(t)$ for constant torque G is considered. As shown by Knobloch (1981), a rescaling of the variables for this problem yields the Lorenz system with the now familiar chaotic characteristics. It is noteworthy that this simplest prototype dynamo system already contains the seeds of chaos (provided the formulation is self-consistent).

It is important to note that, while dynamo action requires that the resistance of the circuit R be low, i.e. that the conductivity σ of disc and wire be high, we lose the dynamo if we go to the limit $\sigma \to \infty$, because then the field cannot diffuse into the region in which induction is operative. An efficient dynamo requires a conductivity that is large but not too large.

7.2 THE STRETCH-TWIST-FOLD DYNAMO

The magnetic field $\underline{B}(\underline{x},t)$ evolves in a conducting fluid of diffusivity η moving with velocity $\underline{u}(\underline{x},t)$ according to the induction equation

$$\frac{\partial \underline{B}}{\partial t} = curl\,(\underline{u} \times \underline{B}) + \eta \nabla^2 \underline{B} . \qquad (7.5)$$

In the perfectly conducting limit ($\eta \to 0$), the magnetic lines of force ('$\underline{B}$-lines') are frozen in the fluid, and if the motion is incompressible ($\nabla . \underline{u} = 0$), then stretching of $\underline{B}$-lines implies proportionate intensification. The simplest 'heuristic' dynamo is based on this effect: a magnetic tube of force can be doubled in intensity by the stretch-twist-fold cycle indicated in figure 7.2 (Vainshtein & Zel'dovich 1982).

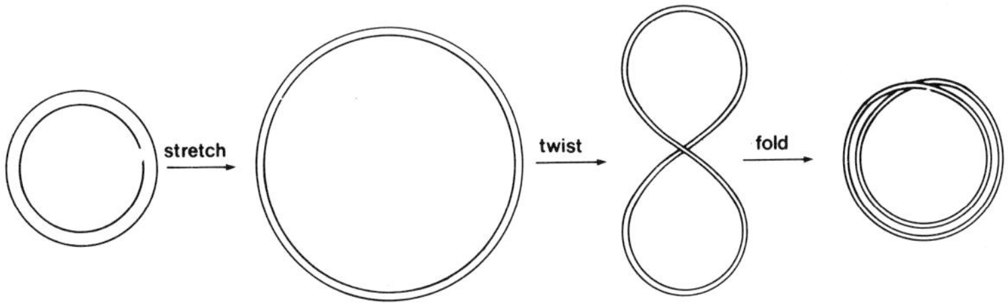

Figure 7.2

Clearly, as recognized by Vainshtein & Zel'dovich, a little diffusion is needed to 'get back to square one', but nevertheless the doubling time for the process does not apparently depend on diffusivity; in this sense the dynamo is a 'fast' dynamo.

Here again, however, there is a danger of over-simplification. When account is taken of the tube structure, and the way that this evolves under repeated application of the cycle of figure 7.2 (see Moffatt & Proctor 1984), a highly complex field structure emerges, and the indications are that the field $\underline{B}(\underline{x},t)$ develops increasingly fine-scale structure as the cycle continues, right down to the diffusive scale $O(\eta^{1/2})$. In the limit $\eta \to 0$, the field becomes non-differentiable everywhere. So here also, although the doubling process of figure 7.2 is non-diffusive in character, the fast dynamo, if it exists, depends in a subtle way on the action of diffusion even in the limit $\eta \to 0$.

7.3 BEHAVIOUR OF THE DIPOLE MOMENT IN A CONFINED SYSTEM

This vital influence of diffusivity in permitting dynamo action is evident also from the classical results of Bondi & Gold (1950) concerning the dipole moment $\underline{\mu}(t)$ associated with electric currents confined to a sphere of radius R of conducting fluid. If $\underline{B}$ is the resulting magnetic field, then two equivalent expressions for $\underline{\mu}(t)$ are

$$\underline{\mu}(t) = \frac{3}{8\pi}\int_{r<R} \underline{B}\, dV = \frac{3}{8\pi}\int_{r=R} (\underline{B}\,.\,\underline{n})\,\underline{x}\, d\underline{S}\,. \tag{7.6}$$

From the second of these expressions, it is easy to obtain an upper bound on $|\underline{\mu}|$, viz

$$|\underline{\mu}| \leqslant \frac{3}{4\pi} R\,\Phi \tag{7.7}$$

where Φ is the total flux of $\underline{B}$ entering the sphere, i.e. the integral of $\underline{B}\,.\,\underline{n}$ over that part of S on which $\underline{B}\,.\,\underline{n} > 0$. If $\eta = 0$, then $\Phi = cst.$ (Alfven's theorem again) and so exponential increase of $\underline{\mu}$ is certainly impossible; no matter what the velocity field $\underline{u}(\underline{x}\,.\,t)$ may be, the inequality (7.7) controls the situation.

Diffusivity however may release this control. Using (7.6), and some elementary manipulation, we have

$$\frac{d\underline{\mu}}{dt} = \frac{3}{8\pi}\int_{r=R} \underline{u}(\underline{n}\,.\,\underline{B})\, dS - \eta\frac{3}{8\pi}\int_{r=R} \underline{n}\times(\nabla\times\underline{B})\, dS\,. \tag{7.8}$$

When $\eta = 0$, the first term redistributes the flux on $r = R$, but respects the inequality (7.7). When $\eta \neq 0$, provided the velocity field is such as to maintain a predominantly positive value of $[-\underline{\mu}\,.\,\underline{n}\times(\nabla\times\underline{B})]$ over the surface $r = R$, diffusion will provide a sustained (and potentially unbounded) increase of $|\underline{\mu}|$. Here therefore the primary mechanism for dynamo action is diffusion, and the growth rate p may be expected to depend on η, with $p \to 0$ as $\eta \to 0$. This is a 'slow' dynamo in the terminology of Vainshtein & Zel'dovich (1982). In fact all known dynamos that have been rigorously established are of the 'slow' variety. Frequently $p = O(\eta^q)$ with $0 < q < 1$, as $\eta \to 0$.

7.4 THE PROS AND CONS OF DYNAMO ACTION

As mentioned in §7.1, dynamo action can occur only if the fluid conductivity is 'sufficiently large', i.e. only if $\eta = (\mu_0\sigma)^{-1}$ is sufficiently small. How small is sufficient? A partial answer is provided by two classical results obtained by manipulation of the equation for magnetic energy associated with electric currents in a sphere $r < R$:

necessary conditions for dynamo action are

$$\eta < e_m R^2 / \pi^2 \qquad \text{(Backus 1958)} \qquad (7.9)$$

$$\eta < U_m R / \pi \qquad \text{(Childress 1969)} \qquad (7.10)$$

where U_m is the maximum value of $|\underline{u}|$ in $r < R$, and e_m is the maximum of the largest principle rate of strain in $r < R$. Frequently $e_m R = O(U_m)$, so that (7.9) and (7.10) are comparable, though not the same. It may happen however that $e_m R \ll U_m$ (if the velocity gradients are everywhere high as in a turbulent flow), and then (7.10) is a much stronger results.

It must be emphasised that (7.9) and (7.10) are necessary for dynamo action, but by no means sufficient. A simple sufficient condition can be formulated only for turbulent flow (see §7.6 below).

The results (7.9) and (7.10), which have been strengthened by Proctor (1977), are the 'pros' of dynamo action. The 'cons' are provided by the various anti-dynamo theorems, mainly variants and generalisations of Cowling's (1934) theorem which states that "steady axisymmetric dynamo action is impossible". A systematic treatment of this class of theorems is provided by the recent work of Hide & Palmer (1982).

7.5 FLUX EXPULSION AND TOPOLOGICAL PUMPING

A further effect which mitigates against efficient dynamo action when η is small is the effect of the expulsion of magnetic flux from any region of closed streamlines. Just as for the homopolar disc dynamo, if magnetic flux cannot penetrate such a region, then any inductive effect in that region will be quite impotent.

Flux expulsion occurs because the velocity field winds up the magnetic field, generally into a tight double spiral, in the region of closed streamlines. Diffusion then acts to eliminate the field from this region. The process is well illustrated by the model problem sketched in figure 7.3: (see Moffatt & Kamkar 1983). Here the initial field $(0, b_0 \cos k_0 x, 0)$ is sheared by the velocity field $\underline{u} = (\alpha y, 0, 0)$. The problem is easily solved in terms of the vector potential $(0, 0, A)$ of $\underline{B}$ which satisfies the convection-diffusion equation

$$\frac{\partial A}{\partial t} + \underline{u} \,.\, \nabla A = \eta \nabla^2 A \,, \tag{7.11}$$

with initial condition

$$A(x,y,0) = -k_o^{-1} b_o \sin k_o x \,. \tag{7.12}$$

The solution here is

$$A(x,y,t) = -k_o^{-1} B_o \ \mathrm{Im} \left[a(t) e^{i\underline{k}(t).\underline{x}} \right] \,, \tag{7.13}$$

where

$$\underline{k}(t) = (k_o \,,\, -\alpha t k_o \,,\, 0) \tag{7.14}$$

and

$$a(t) = \exp\left\{ -\eta k_o^2 (t + \tfrac{1}{3}\alpha^3 t^3) \right\} \,. \tag{7.15}$$

It is the t^3-term in the latter expression which encapsulates the flux-expulsion effect. The time-scale of this field-elimination process is evidently

$$t_{fe} = \alpha^{-1} R_m^{1/3}$$

where $R_m = \alpha / \eta k_o^2$ ($\gg 1$) is the magnetic Reynolds number associated with the shear. This estimate is consistent with that inferred in the pioneering study of Weiss (1966).

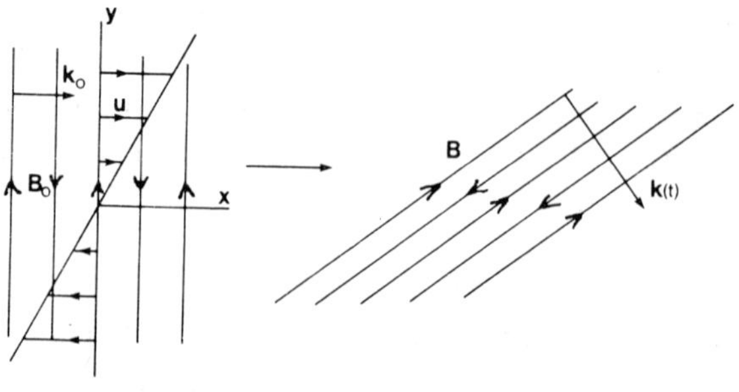

Figure 7.3

If the shear is localised (figure 7.4) then flux expulsion acts only in the region of shear, and reconnection of lines of force is inevitable, as indicated in the figure.

This however is not the whole story. Rhines & Young (1983) have recently studied (7.11) in the context of scalar diffusion, and have observed that a residual field may survive in a region of closed stream-lines over the ordinary diffusive time-scale $t_d = \alpha^{-1} R_m$. It is easy to see how this may occur in the magnetic context considered here. If the $\underline{B}$-lines coincide with the $\underline{u}$-lines in the region of closed $\underline{u}$-lines, then there is no 'winding-up' effect (figure 7.5).

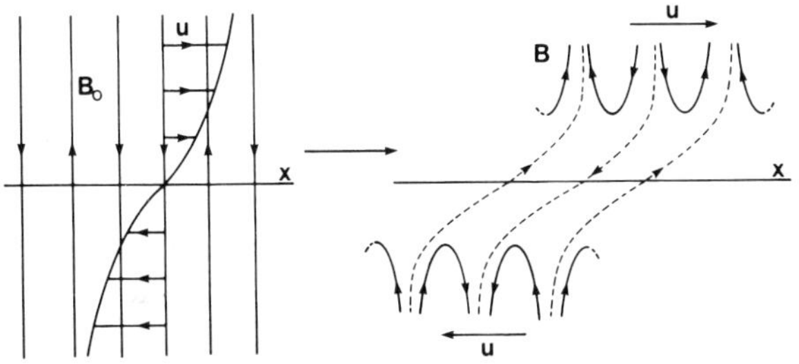

Figure 7.4

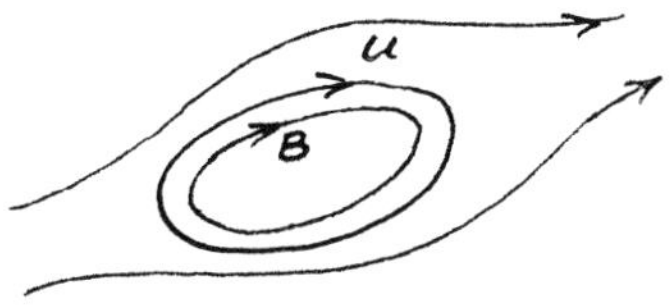

Figure 7.5

A field of this kind will diffuse so that it will not remain exactly aligned with $\underline{u}$; but as shown by Rhines & Young, the strong shearing effect of the $\underline{u}$-field is always such as to maintain a $\underline{B}$-field that is (to leading order) aligned with $\underline{u}$, and this field does indeed survive in the region of closed $\underline{u}$-lines on the long time-scale t_d.

It is an open question whether flux expulsion occurs, or not, in more complex three-dimensional situations. One situation of particular current interest is that in which the $\underline{u}$-lines are ergodic (space-filling) in some region V of R^3. Can a magnetic field survive indefinitely in such a region (when $\eta \neq 0$) or is it expelled by a quasi-two-dimensional mechanism on the $R_m^{1/3}$-timescale? No general answer to this question is as yet known.

An interesting three-dimensional variant of the flux expulsion process is the 'topological pumping' mechanism, identified by Drobyshevski & Yuferev (1974). In the topologically asymmetric motion associated with hexagonal cells in a Benard layer, horizontal $\underline{B}$-lines can be convected downwards, but cannot be convected upwards, since the regions of upward moving fluid are disconnected. There is therefore a net pumping effect downwards which becomes more effective as R_m increases from small values.

Recent computations for larger R_m (100–200) by Galloway and Proctor (1984) and by Arter (1984) have shown that here also the effects are much more subtle than originally realised. Not only is flux apparently pumped downwards, but by some mysterious mechanism that is not altogether clear, reversed flux is generated near the top of the layer. (Could this phenomenon have some bearing on the as yet unexplained appearance of reverse field in the Reversed-Field Pinch (Bodin & Newton 1980) ??).

7.6 MEAN-FIELD ELECTRODYNAMICS

There can be no dispute that the major advances in dynamo theory over the past 20 years have been associated with the development of mean-field electrodynamics, in a turbulent context, whose origins may be traced to the work of Parker (1955), Braginskii (1964) and Steenbeck, Krause & Radler (1966). This theory is fully described by Moffatt (1978) and by Krause & Radler (1980), and it will be sufficient here to discuss certain key points of the theory, and to comment on some weak points which call for further investigation.

The theory is based on a decomposition of the total velocity field $\underline{u}_{tot}$ and total magnetic field $\underline{B}_{tot}$ into mean and fluctuating parts

$$\underline{u}_{tot} = \underline{U} + \underline{u} , \qquad \underline{B}_{tot} = \underline{B} + \underline{b} . \tag{7.16}$$

The mean of the induction equation is then

$$\frac{\partial B}{\partial t} = \nabla \times (\underline{U} \times \underline{B}) + \nabla \times \underline{E} + \eta \nabla^2 \underline{B} , \tag{7.17}$$

where $\underline{E} = \langle \underline{u} \times \underline{b} \rangle$ is the electromotive force associated with the turbulence. Consideration of the equation for the fluctuating field $\underline{b}$ establishes (on quite general grounds) a linear relationship between $\underline{E}$ and $\underline{B}$; and provided there is a scale separation (the scale of the fluctuating fields being small compared with the scale of the mean fields) this linear relationship takes the form

$$E_i = \alpha_{ij} B_j + \beta_{ijk} \frac{\partial B_j}{\partial x_k} + \ldots , \tag{7.18}$$

where α_{ij} and β_{ijk} are pseudo-tensors, determined (in principle) by the statistics of the turbulence, and the parameter η. When the scale of $\underline{B}$ is sufficiently large, the series (7.18) may be expected to converge rapidly; and in practice only the first two terms are retained. It is however quite common in dynamo models to find that the β-term in (7.18) is comparable in importance with the α-term, and one may detect here the seeds of a certain inconsistency: if the first two terms are comparable, then what about the third term, to say nothing of the nth term?

The first problem in mean-field electrodynamics (analogous to transport problems in statistical physics) is then to obtain explicit expressions for α_{ij}, β_{ijk} in terms of η and of statistical properties of $\underline{u}$. The astrophysically interesting situation is that in which $\eta \to 0$ (or, more strictly, in which the turbulent magnetic Reynolds number is large); unfortunately this is the limit in which theoretical analysis is peculiarly difficult! If typical magnitudes of α_{ij}, β_{ijk} are denoted by α and β, and if these are independent of η in the limit $\eta \to 0$, then on dimensional grounds one would expect that

$$\alpha = O(u_0) , \qquad \beta = O(u_0 \ell_0) , \tag{7.19}$$

where $u_0 = \langle \underline{u}^2 \rangle^{1/2}$ and ℓ_0 is a characteristic scale of the turbulence; and indeed the estimates (7.19) are commonly used (with suitable numerical coefficients) in the astrophysical literature. But we have already noted the subtleties of the limit $\eta \to 0$ in the laminar context; and there is no reason to suppose that the behaviour will be any less subtle in the turbulent context. If astrophysical dynamo models have to depend only on the dimensional justification of (7.19), this is a shaky foundation for an enormous superstructure!

There is however some evidence from numerical simulation experiments that (7.19) may, despite the apparent naivety, be essentially correct. Formally exact expressions for α_{ij} and β_{ijk} were obtained by Langrangian analysis by Moffatt (1975) and these were used in a numerical simulation by Kraichnan (1976) who showed that, except possibly in the artificial case of 'frozen' turbulence, α and β do settle down to values of order u_0 and $u_0 \ell_0$ respectively. Current work of Drummond, Duane & Horgan (1984), which incorporates weak diffusion via a Brownian 'jiggle' superposed on the turbulence, finds results so far consistent with Kraichnan's study, and this is at least reassuring. The calculations are however at the limit of available computer power, and one must question whether true asymptotic ($t \to \infty$) conditions are attained in these computations.

The case of isotropic turbulence (statistically invariant under rotations of the frame of reference) deserves particular comment. In this case, α_{ij} and β_{ijk} are isotropic, i.e.

$$\alpha_{ij} = \alpha\,\delta_{ij}\,, \qquad \beta_{ijk} = \beta\,\epsilon_{ijk} \tag{7.20}$$

where, now, α is a pseudo-scalar and β is a scalar. This difference is highly significant: α can be non-zero only in turbulence that 'lacks reflexional symmetry'; β, on the other hand, is generally non-zero, whether the turbulence lacks reflexional symmetry or not.

The simplest measure of the lack of reflexional symmetry in a field of turbulence is the mean helicity

$$H = \langle \underline{u} \, . \, curl\, \underline{u} \rangle \, . \tag{7.21}$$

At <u>low</u> turbulent magnetic Reynolds number, there is a direct relationship between α and H: α is a weighted integral of the spectrum of H (Moffatt 1978, §7.8).

It is known that, when $\underline{U} = 0$ and $\alpha \neq 0$, equation (7.17) admits dynamo solutions provided $|\alpha| R/(\eta+\beta)$ exceeds a critical value dependent only on the shape of the fluid domain, where R is a typical scale of this domain. Hence, a <u>sufficient</u> condition for dynamo action in such a domain is that $|\alpha|$ be non-zero and R be sufficiently large; the former condition is generally satisfied if the turbulence in the domain lacks reflexional symmetry. This is the sufficient condition referred to in §7.4 above.

7.7 SOME PROPERTIES OF THE PSEUDO-TENSORS α_{ij} and β_{ijk}

If the turbulence is not isotropic (and it seldom is!) then there are certain other effects concealed in α_{ij} and β_{ijk} in addition to the simple α-effect and the eddy diffusivity (β-) effect that are present in isotropic conditions. Firstly, α_{ij} need not be symmetric; if we decompose it into symmetric and antisymmetric parts, i.e.

$$\alpha_{ij} = \alpha_{ij}^{(s)} + \epsilon_{ijk}\gamma_k\,, \tag{7.22}$$

then it is evident that $\underline{\gamma}$ is a polar vector which need not vanish in reflexionally symmetric turbulence. The symmetric part $\alpha_{ij}^{(s)}$ does however vanish unless the turbulence lacks reflexional symmetry.

In the 'first-order smoothing approximation' in which terms quadratic in fluctuating quantities are neglected in the fluctuation equation, it turns out that α_{ij} is symmetric, i.e. $\underline{\gamma} = 0$. At the next order, however, 'second-order smoothing', $\underline{\gamma}$ can be expressed as a weighted integral of triple spectra (i.e. Fourier transforms of triple velocity correlations), and is in general non-zero. A more important situation is perhaps that in which the turbulence is inhomogeneous; in this case a contribution to $\underline{\gamma}$ is obtained at the first-order smoothing level, in the direction of decreasing turbulence intensity:

$$\underline{\gamma} = \frac{-k}{\eta} \nabla(\ell_o^2 \langle \underline{u}^2 \rangle) , \tag{7.23}$$

where again ℓ_o is the scale of the turbulence, and k is a dimensionless constant of order unity; the factor η^{-1} is a product of the first-order smoothing approximation. Note that for inhomogeneous turbulence, the vector $\underline{\gamma}$ given by (7.23) will be a function of position, $\underline{\gamma} = \underline{\gamma}(\underline{x})$. When substituted in the mean-field equation, via (7.22) and (7.18), it gives a contribution

$$\frac{\partial \underline{B}}{\partial t} = \nabla \times (\underline{\gamma} \times \underline{B}) + \ldots \tag{7.24}$$

i.e. $\underline{\gamma}$ acts like an effective velocity, transporting the mean field relative to the fluid. It is important however to note that $\underline{\gamma}$ is in general non-solenoidal, i.e. $\nabla . \underline{\gamma} \neq 0$, so that the qualitative effect of $\underline{\gamma}$ is quite different from that of the actual fluid mean velocity $\underline{U}$, which is assumed to satisfy $\nabla . \underline{U} = 0$. In fact, the $\underline{\gamma}$-effect identified here is none other than the flux-expulsion effect (incorporating topological pumping also), reappearing within the mean-field framework.

Turning now to β_{ijk}, a first-order smoothing analysis gives two contributions (Moffatt & Proctor 1982). The first is a weighted integral of the symmetric part of the spectrum tensor of the turbulence, and admits interpretation as an anisotropic eddy diffusivity. The second part is a weighted integral over the helicity spectrum function $H(\underline{k},\omega)$, viz

$$\beta_{ijk}^{(2)} = \iint \left[\frac{\omega}{\omega^2 + \eta^2 k^4}\right] \left[\frac{4\eta^2 k_i k_j k_k}{\omega^2 + \eta^2 k^4} + \frac{k_i \delta_{jk} - k_j \delta_{ki}}{2k^2}\right] H \, d\underline{k} \, d\omega \, . \tag{7.25}$$

This full expression is given here just to indicate the measure of tensorial complexity that arises even at the lowest order of approximation. In the special case of axisymmetric turbulence, it can be shown that the expression (7.25) contains the Radler effect (Radler 1969):

$$\beta^{(2)}_{ijk} \frac{\partial B_j}{\partial x_k} = R(\underline{e} \;\; \underline{J})_i + \ldots . \qquad (7.26)$$

where $\underline{J}$ is the mean current, $\underline{e}$ is a unit vector along the axis of symmetry, and R is the Radler coefficient (a pseudo-scalar). As shown by Moffatt & Proctor (1983), if the turbulence is statistically symmetric about a plane perpendicular to the axis of symmetry, then (at first-order smoothing level), $\alpha_{ij} = 0$ but $R \neq 0$; in this situation the Radler effect may be important for field generation.

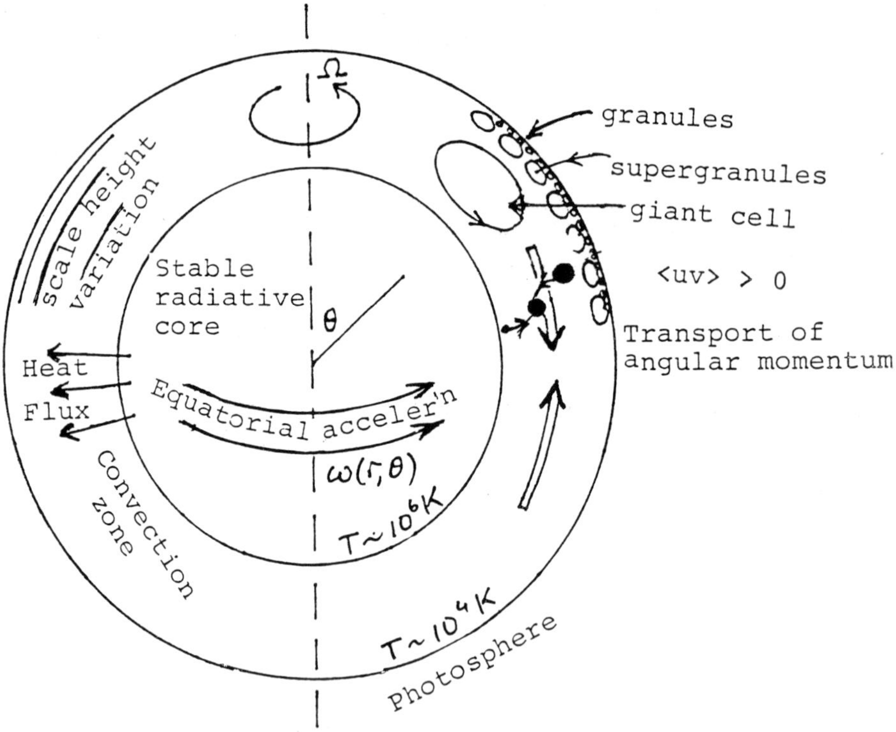

Figure 7.6

7.8 THE SOLAR DYNAMO

Let us now consider some aspects of the solar dynamo problem. The solar scenario for dynamo action is indicated in figure (7.6). The rotation of the Sun has an important double influence on the convective cells in the convection zone: first, Coriolis forces cause a deflection of rising blobs of fluid; this causes the generation of a Reynolds stress distribution, which in turn is believed to be responsible for the differential rotation $\omega(r,\theta)$ of the Sun. Secondly, as blobs rise, they expand and therefore tend to rotate more slowly (conserving their intrinsic angular momentum); this establishes a correlation between vertical velocity and vertical vorticity, i.e. a helicity distribution, which in turn leads to an α-effect. Thus, the two ingredients of an $\alpha\omega$-dynamo, the α-effect and differential rotation, are both a consequence of Coriolis forces; from a dynamical point of view, we are not free to specify $\alpha(r,\theta)$ and $\omega(r,\theta)$ independently - they should both be derived in a self-consistent manner from the governing dynamical equations. This desirable aim has not as yet been attained.

Let us however look at the two processes in a little more detail. The equation of motion, whatever else it may contain, contains a Coriolis force,

$$\frac{\partial u}{\partial t} = -2\underline{\Omega} \times \underline{u} + \ldots \tag{7.27}$$

where, in local Cartesian coordinates (south, east, and vertically up) at colatitude θ,

$$\underline{\Omega} = (-\Omega \sin\theta \, , 0 \, , \Omega\cos\theta) \, . \tag{7.28}$$

With $\underline{u} = (u,v,w)$, and with u and v initially zero, we find an initial tendency (from (7.27))

$$v = -2w \sin\theta \, . \, \Omega t + O(t^3) \, , \tag{7.29}$$

$$u = -2w \cos\theta \sin\theta \, (\Omega t)^2 + O(t^4) \, , \tag{7.30}$$

so that the Reynolds stress is

$$\langle uv\rangle = 4(\Omega t)^3 \langle w^2\rangle \cos\theta \sin^2\theta + O(t^5) \, . \tag{7.31}$$

This suggests that a reasonable approximation in a statistically steady state should be

$$\langle uv\rangle = 4(\Omega t_c)^3 \langle w^2\rangle \cos\theta \sin^2\theta \, , \tag{7.32}$$

where t_c is a coherence time for the rising blobs ($t_c = 3 \times 10^5$ s, $\Omega t_c = 0.2$ for supergranular scales). This generates differential rotation ω whose θ-dependence is given by

$$\nu_T \frac{\partial \omega}{\partial \theta} = \langle uv \rangle \qquad (7.33)$$

where ν_T is an eddy viscosity (80 km^2/s) associated with granular and sub-granular scales. Integrating (7.33) gives

$$\omega(r,\theta) = \frac{4\langle w^2 \rangle}{3\nu_T} (\Omega t_c)^3 (\sin^3\theta - \frac{4}{3\pi}) , \qquad (7.34)$$

where the constant of integration is chosen so that $\langle \omega \rangle = 0$, i.e. $\omega(r,\theta)$ represents the fluctuation about the mean. The expression (7.34) indicates equatorial acceleration, as observed in the Sun, and indeed the difference in ω between equator and poles,

$$\omega(r,\frac{\pi}{2}) - \omega(r,0) = \frac{4}{3\nu_T} \langle \omega^2 \rangle (\Omega t_c)^3 \quad 6.6 \times 10^{-7} s^{-1} , \qquad (7.35)$$

which compares very favourably with the observed value ($7.9 \times 10^{-7} s^{-1}$).

Consider now the mechanism of generation of an α-effect (Steenbeck, Krause & Radler 1966). As a blob rises into a region of decreasing density, the vertical component of $(\underline{\omega} + 2\underline{\Omega})/\rho$ tends to be conserved (where $\underline{\omega}$ is the vorticity). Hence for small t,

$$\omega_3 \approx 2\Omega \cos\theta\, (wt) \frac{d}{dz} (\ln \rho_0(z)) \qquad (7.36)$$

where $\rho_0(z)$ is the basic density stratification, and so the helicity is

$$H = \langle \underline{u} . \underline{\omega} \rangle \approx \langle w\omega_3 \rangle = -(\Omega t_c) \langle w^2 \rangle \cos\theta / H_\rho \qquad (7.37)$$

where H_ρ is the density scale-height. The associated α-effect (on the simplest theory) is

$$\alpha = -\frac{1}{3} H t_c = \frac{1}{3} \Omega t_c^2 \cos\theta \langle w^2 \rangle / H_\rho . \qquad (7.38)$$

Equations (7.34) and (7.38) provide a pair of dynamically consistent expressions for α and ω, which could usefully be employed in numerical investigation of dynamo modes.

7.9 MAGNETIC BUOYANCY AS AN EQUILIBRATION MECHANISM

It is well-known that when R_m is large as in the Sun, that α-effect in conjunction with differential rotation will yield solutions of (7.17) in a spherical geometry having an oscillatory dynamo character, i.e.

$$\underline{B}(\underline{x},t) = \mathrm{Re}\left\{ \hat{\underline{B}}(\underline{x})\, e^{(p_r + ip_i)t} \right\}, \qquad (7.39)$$

where $p_r > 0$, $p_i \neq 0$. The field then grows in intensity from one cycle of its periodic behaviour to the next, and ultimately it must react back upon the dynamical system through some equilibration mechanism. There are three possibilities here: (i) a strong field will tend to suppress the turbulent convection, and thus to decrease the α-effect, an effect studied by Moffatt (1972); (ii) likewise, a strong field will react upon the mean velocity field, and in particular will tend to damp the differential rotation; this mechanism was first studied by Malkus and Proctor (1975), and it has recently been identified by Gilman (1984) in his monumental numerical investigation of the solar dynamo, as a mechanism of crucial importance. The third mechanism, not included in the Gilman model, is probably equally important: this is magnetic buoyancy (Parker 1955). When a strong toroidal magnetic field $\underline{B}_T$ is generated deep in

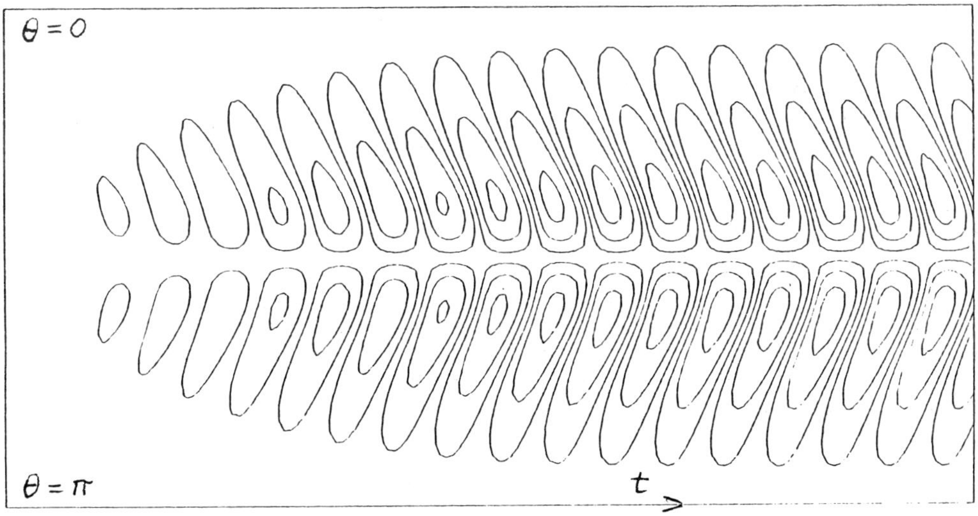

Figure 7.7 (from Nightingale 1985)

the solar convection zone, it is subject to a self-induced instability which causes flux tubes to rise and burst through the photosphere. If downward topological pumping is present, then this magnetic buoyancy instability is what must limit the accumulation of toroidal flux near the bottom of the convection zone. Magnetic buoyancy can be incorporated in an $\alpha\omega$-dynamo via the γ-effect described in §7.7 above, and with $\underline{\gamma}$ a vertical effective velocity proportional to $-\frac{d}{dr}(B_T^2)$ (Nightingale 1985). The boundary condition adopted on the photospheric surface $r = R$ must be such as to allow the toroidal field to escape when it gets there - e.g. a boundary condition of the form

$$B_T + R\frac{\partial B_T}{\partial r} = 0 \qquad \text{on} \qquad r = R \tag{7.40}$$

is one possibility. Figure (7.7) shows contours of $B_T(r,\theta,t)$ at a fixed value of r in the (θ,t) plane (butterfly diagrams), for a particular choice of α, ω and γ. The initial exponential growth is clear, as is the equilibration at constant amplitude induced by the magnetic buoyancy term in the equations. Nightingale's choice of α and ω was based on the previous purely kinematic study of Roberts (1972), and is not dynamically consistent in the sense of §7.8 above - nevertheless it does succeed in establishing that magnetic buoyancy can equilibrate, and it points the way for future studies that should aim in addition at dynamical consistency.

REFERENCES

Arter W.: 1983 Fluid Mech. **132**, 25-48.

Backus, G.E.: 1958 Ann. Phys. **4**, 372-447.

Bodin, H.A.B. and Newton, A.A.: 1980 Nuclear Fusion **20**, 1255.

Bondi, H. and Gold, T.: 1950 Mon. Not. Roy. Astr. Soc. **110**, 607-611.

Braginskii, S.i.: 1964 Sov. Phys. JETP **20**, 726-735.

Childress, S.: 1969 Lectures on Dynamo Theory Inst. Henri Poincare, Paris.

Cowling, T.G.: 1934 Mon. Not. Roy. Astr. Soc. **94**, 39-48.

Drummond, I.T., Duane, S. and Horgan, R.R.: 1984 J. Fluid Mech. **138**, 75-91.

Galloway, D.J. and Proctor, m.R.E.: 1983 Geoph. Astr. Fluid Dyn. **34**, 109-136.

Hide, R. and Palmer, T.N.: 1982 Geoph. Astr. Fluid Dyn. **19**, 301-319.

Knobloch, E.: 1981 Phys. Lett. **82A**, 439-440.

Kraichnan, R.h.: 1976 J. Fluid Mech. **77**, 753–768.

Krause, F. and Radler, K.-H.: 1980 Mean-field magnetohydrodynamics and dynamo theory. Pergamon.

Malkus, W.V.R. and Proctor, M.R.E.: 1975 J. Fluid Mech. **67**, 417–444.

Moffatt, H.K.: 1972 J. Fluid Mech. **53**, 385–399.

Moffatt, H.K.: 1974 J. Fluid Mech. **65**, 1–10.

Moffatt, H.K.: 1978 Magnetic field generation in electrically conducting fluids. Cambridge University Press.

Moffatt, H.K.: 1979 Geophys. Astr. Fluid Dyn. **14**, 147–166.

Moffatt, H.K. and Kamkar, H.: 19183 In Stellar and Planetary Magnetism (ed. A.D. Soward), Gordon & Breach, 91–98.

Moffatt, H.K. and Proctor, M.R.E.: 1983 Geophys. Astr. Fluid Dyn. **21**, 265–283.

Moffatt, H.K. and Proctor, M.R.E.: 1984 J. Fluid Mech. **154**, 493–507.

Nightingale, S.: 1985 Magnetic flux pumping and magnetic buoyancy in mean-field dynamos Ph.D. Thesis, Cambridge University, in preparation.

Parker, E.N.: 1955 Astrophys. J. **122**, 293–314.

Rhines, P.B. and Youngs, w.R.: 1983 J. Fluid Mech. **133**, 133–145.

Roberts, P.H.: 1972 Phil. Trans. Roy. Soc. **A 272**, 663–698.

Steenbeck, M., Krause, F. and Radler, K.-H.: 1966 Z. Naturforsch. **21a**, 1285–1296.

Vainshtein, S. and Zel'dovich, Ya.B.: 1978 Sov. Phys. Usp. **15**, 159–172.

Weiss, N.O.: 1966 Proc. Roy. Soc. **A293**, 310–328.

CHAPTER 8

SOLAR WIND AND THE EARTH'S BOW SHOCK

Steven J. Schwartz
Theoretical Astronomy Unit
School of Mathematical Sciences
Queen Mary College
Mile End Road
London E1 4NS
United Kingdom

In this chapter I shall review the basic phenomena associated with the solar wind from both the macroscopic (e.g. fluid) point of view and also from the microscopic (e.g. particle) perspective. The solar wind provides a fascinating laboratory for studying a wide range of plasma physical processes, and has been extensively observed by *in situ* satellite measurements for some 25 years. In addition to its intimate connection to the sun and to its interaction with the planets, the solar wind provides a unique opportunity to study and attempt to understand many important topics common to other astrophysical plasmas, e.g. particle acceleration and propagation, stellar winds, turbulence, the role of magnetic fields and shocks. The third section of the chapter, in fact, is devoted to the Earth's bow shock, which has provided us in recent years with the kind of detail necessary to understand and model the way in which a *collisionless* plasma supports a macroscopic shock, and its attendant zoo of particle phenomena and turbulence.

It is not possible in the space provided here to do justice to any of these subjects. More details can be found in many excellent review articles and books. In the case of the solar wind these include Hundhausen (1972), Kennel et al. (eds.) (1979), and the proceedings of recent solar wind conferences (Rosenbauer, 1981; Neugebauer 1983). Also of interest for the stellar connection is Bonnet & Dupree (1981). The bow shock material has had less time to be digested. Nonetheless, special issues of the *Journal of Geophysical Research* (June 1981; 1985, in press) provide useful reference points, the latter including many excellent invited lectures from a recent conference.

8.1 THE SOLAR WIND AS A FLUID

The solar wind is basically an extension and expansion of the solar atmosphere into interplanetary space. As such, it maps in a sometimes complicated way the spatial features on the solar surface (loops, coronal holes, etc.) to the Earth's orbit and beyond, while temporal features (flares and transients) give rise to variations and shocks which also reach the far corners of the heliosphere. In brief, the solar wind is a super-sonic (and super-Alfvénic) expansion of the solar corona. As we shall see below, the solar wind flows essentially because the corona is so hot that the local instellar medium could not contain the higher pressure associated with a static atmosphere (although this argument, originally put forward by Parker (1958), was the subject of some historical debate). The basic process, then, is the conversion of the corona's thermal energy into the kinetic energy of more-or-less radial bulk flow of the solar wind. In the first sub-section I shall go on to demonstrate this in a more mathematical way. The interest in this subject comes firstly from its impact on solar and stellar physics through the obvious loss of mass and angular momentum. It also has a profound effect on planetary magnetospheres and planetary formation. Finally, the solar wind is a good medium for studying a variety of fundamental plasma physical topics, including the nature of collisionless plasmas, a nearly homogeneous plasma permeated by a magnetic field, particle beams and turbulence, and shocks.

8.1.1 Fluid Models of the Solar Wind

The simplest model of the solar wind treats it as a single fluid. Although we know that the solar wind is actually made up of several particle species (protons, electrons, alpha particles and others) not in thermal equilibrium, the collective interaction amongst these and the fields makes this approach more fruitful than a back of the envelope demonstration of its invalidity (see §8.2.1 below) might suggest. Since the one-fluid equations basically express the conservation of mass, momentum and energy, it is perhaps not so surprising that the overall existence of the solar wind is accessible via this approach. We shall assume for the moment that the flow is steady, purely radial and spherically symmetric with no magnetic field. Then the conservation of mass and momentum are expressed by

$$\frac{d}{dr}(\rho u r^2) = 0 \tag{8.1}$$

and

$$u\frac{du}{dr} = -\frac{1}{\rho}\frac{dp}{dr} - \frac{GM_\odot}{r^2} \tag{8.2}$$

where ρ is the mass density, u the radial velocity and p the total pressure of the flow and r is the radial distance from the sun's centre. For simplicity, I shall assume that the energy equation can be represented by a polytropic equation of state, namely

$$\frac{d}{dr}(p\rho^{-\gamma}) = 0 \tag{8.3}$$

rather than some more complicated energy equation whose transport coefficients are poorly known at best. These equations possess two straightforward integrals, i.e. the mass flux, ρur^2=constant, and an energy integral,

$$\frac{1}{2}u^2 + \frac{\gamma}{\gamma-1}\frac{P}{\rho} - \frac{GM_\odot}{r} = E = \text{constant} \tag{8.4}$$

which is just Bernoulli's theorem for this problem. The quantity E in (8.4) is the total energy flux (which is also constant) divided by the total mass flux, i.e. the energy per unit mass carried in the wind. It is clear from (8.4) that E must be positive if there is any flow as $r \to \infty$, and that this flow can accelerate as the enthalpy decreases. Using these results, we can rearrange the momentum equation (8.2) into one which involves only the Mach number, $M = \sqrt{(\rho u^2/\gamma p)}$, which is the ratio of the flow velocity to the local speed of sound, c_s (1.14). After some tedious algebra, this produces

$$\frac{M^2-1}{M^2}\frac{dM^2}{dr} = \frac{2}{r}\left[(\gamma-1)M^2+2\right]\left[1-\frac{\gamma+1}{4(\gamma-1)}\frac{(GM_\odot/r)}{E+(GM_\odot/r)}\right] \tag{8.5}$$

The first term in brackets on the right hand side of (8.5) is always positive (for $\gamma>1$) while it is easy to show by examining the large and small r limits that the second term passes through zero for some r provided only that $\gamma<5/3$. This value of r, known as the critical radius r_c, is easily found to be

$$r_c = \frac{3}{4}\frac{(\frac{5}{3}-\gamma)}{\gamma-1}\frac{GM_\odot}{E} \tag{8.6}$$

At $r=r_c$, then, the left hand side of (8.5) must also go through zero. This occurs if M^2 passes through a local maximum or minimum at this radius, or if $M^2=1$ there, with dM^2/dr non-zero. This latter case corresponds to a smooth sub- to super-sonic transition or vice versa. The full set of solution topologies can now be inferred, as sketched in figure 8.1. For completeness I have also included the double-valued solutions which pass through $M^2=1$ with infinite slope at some $r \neq r_c$.

The actual solution relevant to the solar wind problem is determined by the boundary conditions (1) $M^2<1$ at $r=R_\odot$ as observed and (2) the pressure must tend to zero as $r\to\infty$ in order to match the interstellar medium. Only the heavy curve shown in figure 8.1, which passes smoothly from sub-sonic to super-sonic at the critical radius and remains super-sonic thereafter, satisfies both these conditions. In reality, the outer boundary is probably marked by a shock transition back to the sub-sonic regime as the interstellar medium is encountered at distances believed to be around 100 astronomical units. Note that this transition must be in the form of a shock, as the only continuous possibility occurs at $r=r_c$. Typically, the critical radius is located in the vicinity of 5 $R_\odot$ for solar parameters.

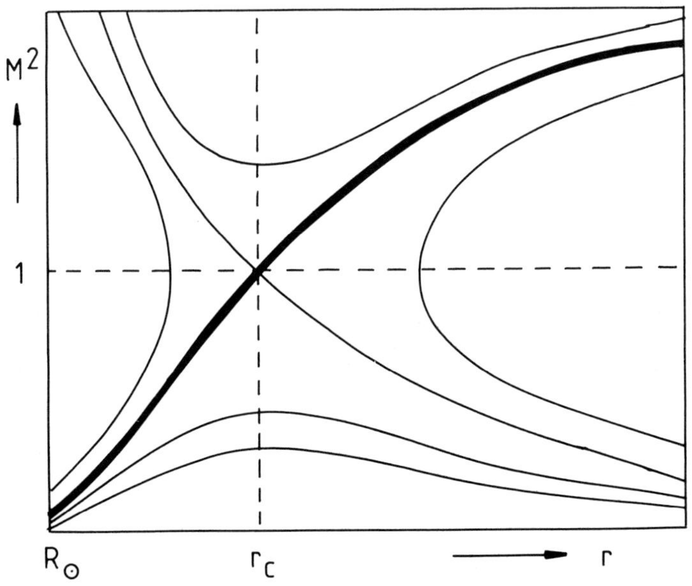

Figure 8.1 Solution topologies for the solar wind problem. Only the heavy solution, which passes through a sub-sonic to super-sonic transition at the critical radius, r_c, satisfies the necessary boundary conditions.

8.1.2 Solar Wind Magnetic Fields

The simple fluid model of the solar wind described in the previous section adequately confirms the necessity of a supersonic wind. From the plasma physics aspects, however, there are numerous refinements to make. Let's begin here by investigating the fate of solar magnetic fields in such a wind. Due to the fact that the solar surface is an excellent electrical conductor, field lines are firmly anchored there. In order to remove the induced electric field arising from the time variation of $\underset{\sim}{B}$ at $r=R_\odot$ resulting from the solar rotation, I shall work in a frame co-rotating with the sun, at an angular frequency Ω.

The frozen flux condition requires that the field threading a fluid parcel convects with that parcel. Thus, if the field and the flow are parallel at the solar surface ($\underset{\sim}{E}=0$ there), as the parcel moves it carries that field with it, so that the field and the flow are parallel to one another (see figure 8.2). Since, in the co-rotating frame the fluid velocity is given by

$$\underset{\sim}{V} = u\,\hat{\underset{\sim}{r}} - \Omega r\,\hat{\underset{\sim}{\Phi}} \tag{8.7}$$

the field components are therefore in the ratio

$$\frac{B_\Phi}{B_r} = -\frac{\Omega r}{u} \tag{8.8}$$

while $B_r \propto r^2$ to preserve $\nabla \cdot B=0$. For u=constant, which is not too bad an approximation over much of the heliosphere, (8.8) corresponds to an Archimedean spiral pattern, as sketched in figure 8.2. Note that the purely radial flow assumption only applies if the field is passive and plays negligible dynamical role. Mathematically, this holds provided the magnetic pressure, $B^2/2\mu_0$, is much less than the dynamical pressure, $\rho u^2/2$, so that the flow is unaffected by the field. Re-arranging gives the condition

$$u > v_A \tag{8.9}$$

where v_A is the Alfvén speed (1.9). Equality in (8.9) can be used to define an Alfvénic radius, r_A, which is typically 50 $R_\odot$, or 1/4 of an astronomical unit, inside of which the field is strong enough to affect the flow and forces near co-rotation with the sun, altering (8.7), while beyond this distance, the field wraps up in the spiral fashion described by (8.8). At 1 astronomical unit, the two field components are comparable in magnitude. In any event, the frozen flux condition always implies, in the steady state, that the field lines trace out the paths of the fluid parcels or, equivalently, the field lines connect solar wind parcels which originated from the same location on the solar surface.

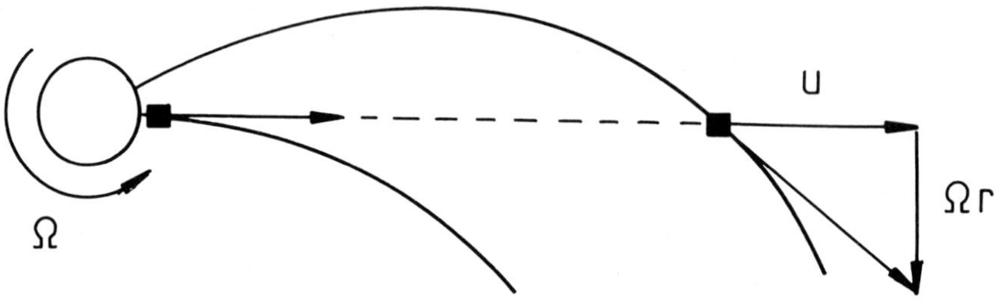

Figure 8.2 The spiral field configuration caused by frozen field lines. Shown is the position of a fluid element as it leaves the sun and at a later time, during which the sun has rotated. The field line is frozen at the sun. In a co-rotating frame, this curve thus has the velocity as its tangent.

8.1.3 Mass and Angular Momentum Loss

Since the sun is the best observed of any star, it is interesting to ask what impact the solar wind has as far as solar, and stellar, evolution are concerned. The most obvious question is whether the mass lost due to the wind is significant or not. Using typical measured values (see Table 8.1) the sun is apparently losing mass at a rate of 3×10^{-14} $M_\odot$ per year, implying a timescale for significant mass loss of 3×10^{13} years, or some 6,000 times longer than the estimated 5×10^9 year age of the sun. Clearly, the sun needn't worry about the mass it's losing. Of course, it is possible that the sun went through an earlier period during which the wind was much stronger/denser and could have been an important factor in its evolutionary track. Many stars appear to show very high mass loss (see, e.g. Hartmann, 1981).

Equally interesting is the angular momentum carried away by the wind, which could play a major role in the spindown of, e.g. stars which formed by the collapse of a rotating cloud. Here the presence of a magnetic field plays a crucial role. As noted above, the atmosphere and wind co-rotate as a rigid body out to the Alfvénic radius, r_A. Thus the angular momentum carried away by a particle of mass m in the wind is $\Omega r_A{}^2 m$ rather than $\Omega R_\odot{}^2 m$. The angular momentum timescale is therefore shorter than the mass loss timescale by a factor $(R_\odot/r_A)^2 \approx 1/2500$, making the angular momentum timescale comparable to the solar lifetime.

More rigorous calculations include the angular momentum carried away by the fields as well. In principle, the azimuthal velocity of the solar wind can be measured and the angular momentum loss (and hence an estimate of r_A) determined experimentally. In practice, however, this component is so small compared to the super-sonic radial flow that an enormous amount of care and effort is required to obtain meaningful results. This has only recently been done with any confidence (Pizzo et al., 1983; Marsch & Richter, 1984a).

8.1.4 Refinements of Fluid Models

There have been a multitude of attempts at improving our description of the solar wind by increasing the level of sophistication of the fluid model under consideration. Almost invariably this has involved an often difficult numerical solution to the governing equations. Some of the "extra" physics used includes:

(i) Additional forces, such as wave or turbulent pressure, viscosity, rotational forces, radiation pressure, etc. Some of these, particularly the acceleration by hydromagnetic waves, have been used in an attempt to model the "high speed" solar wind (see Table 8.1) which simpler theories just could not explain. Of course, if you add enough free parameters, it is usually possible to find a solution which matches at least some of the observations!

(ii) More realistic energy equations. This is the weakest aspect of the solution shown in section 8.1.1, and also the hardest to get right. Additions include heat conduction (although the conductivity in a collisionless plasma such as the solar wind is not well understood), energy addition due to wave dissipation and/or viscous heating, and others. Conductivity, like viscosity, makes the equations of higher order and much more unstable from the numerical point of view.

(iii) Two-fluid equations, one for electrons and one for protons, each with their own temperatures. This is much more realistic since the electrons and protons are collisionally decoupled throughout most of the interplanetary medium.

(iv) Anisotropic pressure tensors. In the near absence of collisions, the particle distributions do not remain isotropic Maxwellians. Within a fluid description, this is accommodated by allowing the temperature of the species to be different along and across the magnetic field (see §8.2.2 below).

(v) Non-spherical geometry, modeling the interaction of the field and the flow in inhomogeneous regions, such as coronal hole boundaries.

(vi) Numerous other refinements.

Basically, the simple models, such as the one displayed above, do OK to within factors of 2-4. Some of the above refinements help improve the agreement between model and observation, while many puzzles remain. Some of these puzzles are undoubtedly related to our lack of understanding of the microphysics and how, if possible, to build it into the fluid picture. For example, we shall see in the next few sections that we really don't know what determines the level of heat which is conducted by the solar wind plasma. There does not seem to be any simple relationship between the fluid parameters (temperature, pressure, velocity, etc.) and the heat flux. Certainly, the classical conduction law based on frequent particle collisions doesn't hold. Since the solar wind is in essence a conversion of thermal energy to bulk flow, these factors affecting the thermal properties are crucial in any attempt to obtain more detailed agreement than the simple models do with spacecraft observations.

Table 8.1 Some properties of low and high speed solar wind at 1AU

	Low Speed	High Speed
Typical velocity (km/s)	300	700
Density (part./cm^{-3})	10	4
Temperatures	$T_p < T_e \approx 10^5$K	$T_p > T_e \approx 10^5$K
Proton Distributions	$\approx$ Isotropic	$T_{p\perp} > T_{p\parallel}$ plus beam
Collisions per scalelength	a couple	none
Solar magnetic field	closed loops, sectors	open, coronal holes
Qualitative character	variable, clumpy	smooth, structureless

To make matters worse, the solar wind is not the nice, spherically symmetric, steady state flow envisaged in earlier sections. What is now termed "slow" solar wind was originally thought to be the quiescent state. Now we know that the "high speed" solar wind actually is much smoother and more homogeneous, because it comes to us from magnetically open regions known as coronal holes. The low speed wind probably originates from smaller, at least partially closed regions which result in more variability and structure in this wind. Table 8.1 provides a brief contrast of these two solar wind states, although even within them there is considerable variation. For a more statistical account of solar wind parameters see Feldman et al. (1977) and Schwenn et al. (1983)

8.2 THE SOLAR WIND AS A PLASMA

8.2.1 Why a Plasma Description is Needed

The fluid approximation for any gas requires frequent particle collisions to ensure that the gas or plasma is in equilibrium and the particle distributions nearly Maxwellian. Thus the crucial parameter is the ratio of the collisional mean free path, λ_{mfp}, to a typical dimension over which the macroscopic parameters vary (retaining our interest in steady state, lengths are more appropriate to compare than times). The usual variable of interest is the temperature, so the non-dimensional parameter used is

$$|B_T| = \lambda_{mfp}{}^e \, |\underset{\sim}{\nabla} \ln T_e| \tag{8.10}$$

The electrons are used for this purpose since they usually dominate the transport processes. For $|B_T| \ll 1$, the plasma is collision-dominated and the transport processes (heat conduction, etc.) are calculated by considering the small deviations of the distribution function from Maxwellian (e.g. Spitzer, 1956). This yields the standard result that the heat flux, $\underset{\sim}{q}_e$ is proportional to ∇T_e with the proportionality constant, the conductivity, taking its "classical" value.

If $|B_T| \gtrsim 1$, the distributions show strong deviations from Maxwellian and the classical conduction law is inappropriate. In the solar wind, $|B_T|$ increases as $r^{3/7}$ from small values in the corona to ≈ 1 at 1AU. Indeed, the electron distributions show strong deviations from a simple Maxwellian, with a central core component, and a much hotter, less dense halo which is drifting in such a way as to carry most of the actual heat flux (Feldman et al., 1975). It has often been postulated that these non-equilibrium distributions are unstable to a variety of microinstabilities (Forslund, 1970), and that the microturbulence so generated regulates the heat flux (see Schwartz, 1980 for a review of the vast literature on this subject). However, the observational evidence does not seem to indicate any strong correlation with the expected parameters. Recent interesting attempts (Scudder and Olbert, 1979) to study the effects of the rare Coulomb collisions suffered by an electron during its travels to the outer heliosphere and back (the electron thermal speed is larger than the solar wind velocity, so that the solar wind is not super-"sonic" as far as electrons are concerned) show some promise, although not all the details agree with the data (Feldman et al., 1982).

Table 8.2 provides a brief summary of the various plasma parameters in the solar wind. Again, Feldman et al. (1977) provides a more complete, statistical survey. One of the interesting features to emerge from Table 8.2 is the fact that although the solar wind is a good plasma ($N_D \gg 1$), the Debye length is larger than typical spacecraft dimensions.

This means that a space probe effectively sits in a vacuum and can collect particles one-by-one as they enter, in marked contrast to the usual laboratory situation in which the probe often violently disturbs the plasma and forces the use of more indirect diagnostic tools. Solar wind (and magnetospheric) plasmas provide a unique opportunity to measure and study the detailed particle populations of a plasma. I have already discussed briefly the nature of solar wind electrons. Let's turn now to the ion distributions.

Table 8.2 Average Solar Wind Plasma Parameters at 1AU ($=215R_{\odot}=1.5\times10^{11}$m)

Parameter	Value	Parameter	Value
Macroscopic			
density(n)	5cm^{-3}	magnetic field ($\|B\|$)	6γ
velocity(v_{SW})	300-800km/s		$=6\times10^{-5}$G
temperature(T)	10^5K		$=6\times10^{-9}$Tesla
Velocities (km/s)			
$v_{thermal}$(proton)	30	$v_{thermal}$(electron)	1000
$v_{Alfvén}$	50		
Frequencies (rad/s)			
ion plasma(ω_{pi})	3000	electron plasma(ω_{pe})	10^5
ion cyclotron(Ω_i)	0.5	Ω_e	1000
ion collision($2\pi\nu_{cp}$)	$2\pi\times3\times10^{-7}$		
Timescales (s)			
ion gyroperiod	12	electron gyroperiod	6×10^{-3}
collision time	3×10^6	expansion (r/v_{SW})	3×10^5
Lengthscales			
Debye length (λ_D)	10m	mean free path(λ_{mfp})	0.6AU
Ion Larmor (r_{Li})	60km		$=9\times10^{10}$m
Electron Larmor	1km		
Non-dimensional parameters			
Debye number($N_D=n\lambda_d^3$)	5×10^9	ion plasma beta(β_i)	0.5-1

8.2.2 Solar Wind Protons

The slow, dense solar wind is observed to be nearly isotropic and Maxwellian. In fact, Marsch and Goldstein (1983) have shown that when a careful treatment of Coulomb collisions is made, most of the protons in this type of solar wind have suffered one or two collisions between the sun and us. This then accounts for their near-Maxwellian character.

The high speed solar wind, however, is less dense and practically free from collisions. Collisionless plasma theory predicts that the proton distribution should conserve two adiabatic invariants,

$$c_{\perp} = T_{\perp p} / B_0 = \text{constant} \tag{8.11}$$

which results from magnetic moment conservation, and a longitudinal invariant

$$c_{\parallel} = T_{\parallel p} B_0^2 / n^2 = \text{constant} \tag{8.12}$$

where subscripts $\parallel$ and $\perp$ refer to directions parallel and perpendicular to the background magnetic field $\tilde{B}_0$ respectively. Combining (8.11) and (8.12) predicts that the ratio of parallel to perpendicular temperatures should increase with radial distance according to

$$\frac{T_{\parallel p}}{T_{\perp p}} \propto \frac{n^2}{B_0^3} \propto r^2 \tag{8.13}$$

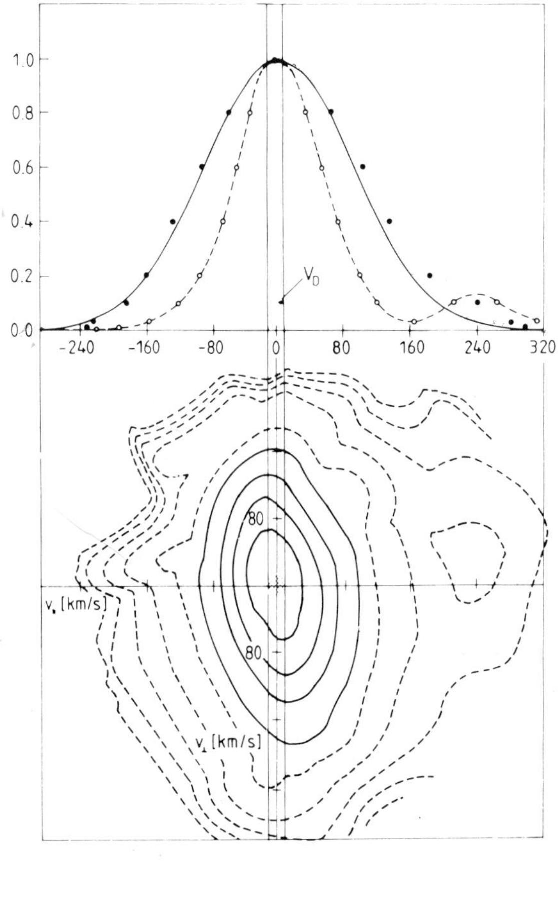

Figure 8.3 Typical high speed solar wind proton distribution. The top panel shows cuts along (dashed) and across (solid) the magnetic field, while the bottom panel shows the contours of the distribution. The vertical lines and stippled region indicate the extent of velocity-space which is expected to be collision dominated (and hence isotropic). Note the strong $T_{\perp p} > T_{\parallel p}$ anisotropy of the main protons, and the beam component moving at a higher velocity (from Marsch & Goldstein, 1983)

Although this strong r dependence is based on an assumed purely radial field, which must decrease $\propto r^{-2}$, even taking into account the spiral field of figure 8.2 still leads us to the conclusion that the distribution at 1AU ought to be considerably hotter (i.e. have a broader full width at half maximum) along the field than across it. A typical observed distribution is shown in figure 8.3. The distribution clearly is anisotropic, but in the opposite sense to that predicted! In addition, there is evidence for a separate beam component of protons travelling faster than the peak of the main protons. This beam is accompanied by the solar wind alpha particles (see the next section).

Various attempts have been made to explain these features of the ion population in terms of a cylcotron interaction between the protons and the ever-present spectrum of magnetic fluctuations, predominantly Alfvénic (i.e. torsional and incompressive) in character (Schwartz et al., 1981; Marsch et al., 1982b; Isenberg & Hollweg, 1982). The debate continues over the detailed mechanism(s) responsible for the nature of the proton distributions in the high speed solar wind, and has spurred a number of observational (e.g. Schwartz & Marsch, 1983) and theoretical (Marsch & Richter, 1984a) efforts.

8.2.3 Minor Ions in the Solar Wind

The solar wind is predominantly made up of protons (and accompanying electrons). However, other heavier ions are also present, of which by far the most significant are alpha particles. Being some 4-5% of the total number density, they contribute 16-20% of the total mass in the wind. Two interesting features emerge. Firstly, the alpha particles have a higher temperature, by a factor ~ 4, than the protons. Similarly, heavier ions show temperatures in rough proportion to their masses, i.e. the thermal *velocities* of all solar wind ions are nearly the same. Secondly, the bulk speed of the alpha particles is typically higher than the protons, by an amount roughly equal to, and certainly scaling with, the local Alfvén speed. These features are shown in the solar wind data presented in figure 8.4. It is not yet understood why these heavier particles, which should have more difficulty in leaving the corona, are accelerated more than the protons. Possibly this is related to their higher thermal speed, but the causal relationship one way or the other is not established. The apparent proportionality to the Alfvén speed suggests that the flow, and possibly the heating, is in some way related to a wave process (Isenberg & Hollweg, 1982; Isenberg, 1984; Marsch et al. 1982b), but the ability of theoretical ideas to fit the observations is not yet convincing.

8.2.4 Waves in the Solar Wind

A plasma is able to support a wide range of wave modes, from low-frequency MHD waves to high-frequency plasma oscillations. A sketch of the dispersion curves for waves propagating in a warm plasma at an oblique angle to the magnetic field is shown in figure 8.5. Many other modes are also possible, and those shown can be considerably modified, under different plasma conditions and special propagation

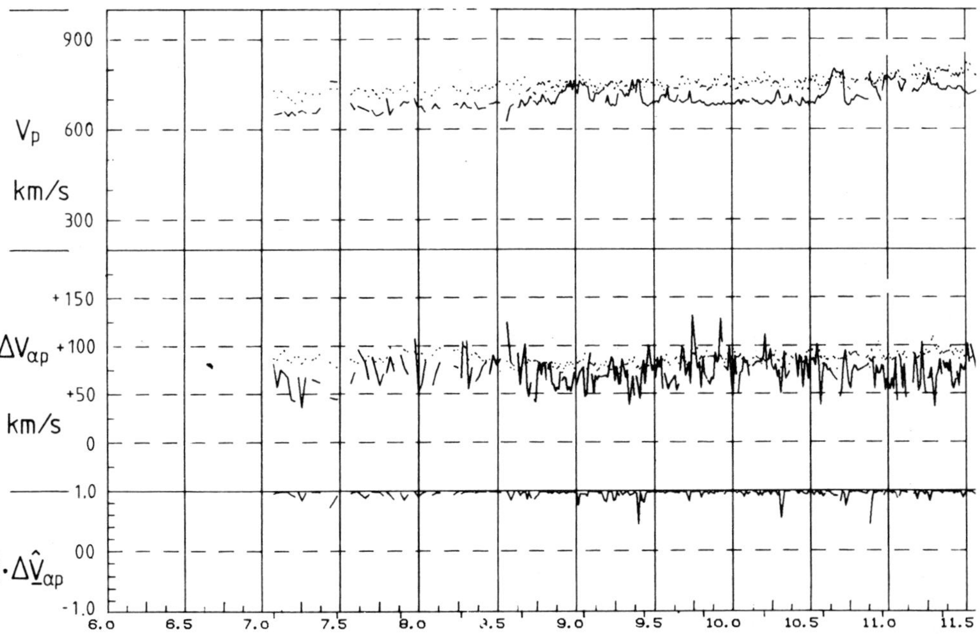

Figure 8.4 Some solar wind data from the Helios satellite. Top: proton bulk speed (dots = α particle speed) Middle: Differential speed between alphas and protons (dots = v_A) Bottom: cosine of angle between the magnetic field and the differential velocity showing the field-aligned nature of the relative motion. Notice the clear tendency of the alpha particles to flow faster than the protons by an amount which closely follows the local Alfvén speed (from Marsch et al., 1982a)

directions. The presence of other species can also influence the wave characteristics.

Most of the wave modes shown in figure 8.5, and a whole host of others, have been observed (or claimed to be so) in the solar wind. One of the difficulties in such observations is the large Doppler-shift due to the super-sonic solar wind flow. This often masks the underlying rest-frame wave frequency and complicates the identification of particular wave modes. Another difficulty arises from the single point observations of a spacecraft which thus mixes temporal and spatial variations. In many cases, though, the plasma parameters, wave polarization, etc. are sufficient to allow a unique determination of the wave mode.

The higher frequency modes are often cited for their importance in wave-particle interactions and scattering. They are more "bursty" in nature, and are often associated with features in the particle distributions. The lower frequency, MHD, modes can be observed in both the field data and the particle parameters. The classic example is shown in figure 8.6, which compares the components of the magnetic field with the corresponding fluctuations in the solar wind flow velocity. Notice the excellent correlation between the field and velocity, and the relative absence of variation in the total field strength and number

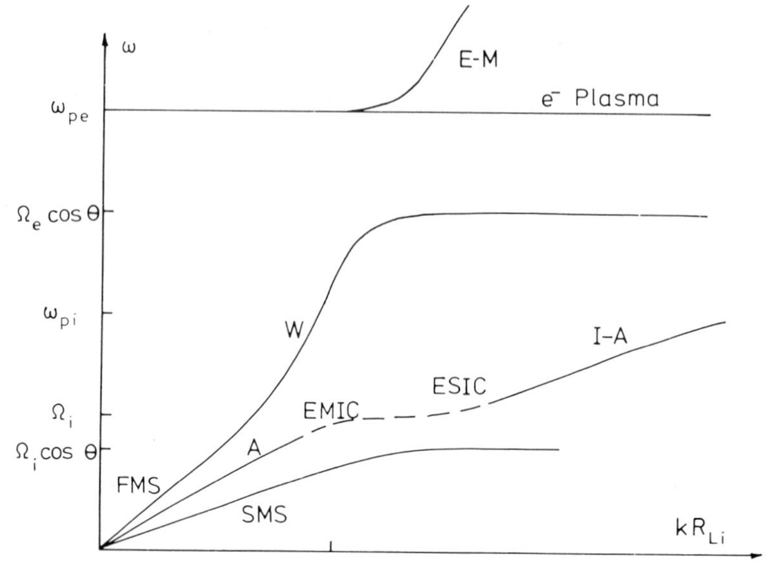

Figure 8.5 Dispersion curves (angular frequency ω vs. wavenumber k) for obliquely propagating waves in a warm plasma. (W=whistler, I–A=ion acoustic, ESIC=electrostatic ion cyclotron, EMIC=electromagnetic ion cyclotron, FMS=fast magnetosonic, A=Alfvén, SMS=slow magnetosonic) (from Schwartz, 1980)

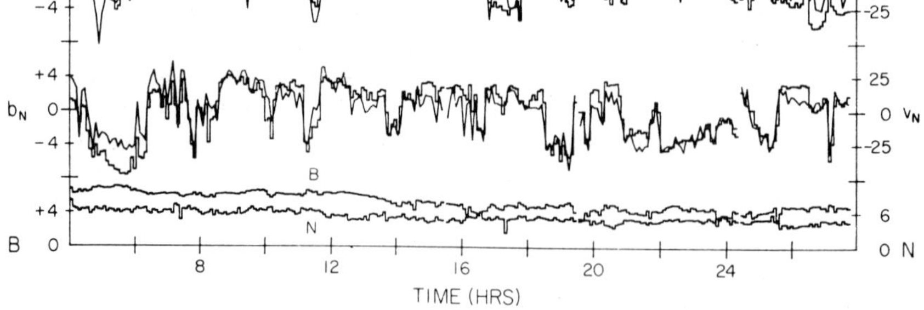

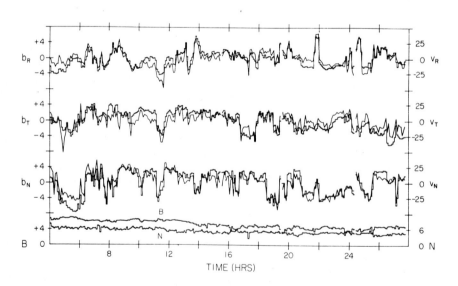

Figure 8.6 Solar wind Alfvén waves. Plotted are the three components of the magnetic field and flow velocity variations, along with the field strength and plasma number density (from Belcher & Davis, 1971).

density (bottom two curves). The vector proportionality between $\underset{\sim}{v}$ and $\underset{\sim}{B}$ is precisely that expected for Alfvén, or shear MHD, waves. These waves are, in fact, an exact non-linear solution of the ideal MHD equations (e.g. Hollweg, 1974). This is somewhat comforting, since an inspection of figure 8.6 reveals that these waves have amplitudes $|\delta \underset{\sim}{B}|$ which are comparable to the background field.

Notice, however, that these waves are not the simple sinusoids used in textbooks. Moreover, apart from the Alfvén wave case, most theoretical treatments of plasma waves, either based on MHD equations or kinetic theory, have resorted to a linear analysis, perturbing about a homogeneous background state. However, the role of plasma waves in modifying the particle distributions and/or bulk parameters, as discussed, for example, in connection with the electron heat flux, requires a non-linear description. Some theoretical progress has been made along these lines, and computer simulations have added insight into their validity, but our basic understanding of nonlinear plasma physics remains in its infancy. Nonetheless, the kind of details observable in the solar wind have made it one of the major testing grounds of modern ideas (e.g. Mattheaus & Goldstein, 1982; Ovenden et al., 1983). For a more thorough discussion of solar wind plasma wave observations and related theory, see, for example, Barnes (1979), Hollweg (1978), Schwartz (1980), Gurnett (1981), and references therein.

8.3 THE EARTH'S BOW SHOCK

8.3.1 Why a Shock is Needed

We now turn to address what happens when the super-sonic solar wind meets an object, such as the Earth's magnetosphere, in its path. In considering the general problem of flow past an obstacle, three possibilities exist. If the flow is *sub*-sonic, information about the object can be transmitted via sound waves or pressure waves against the flow and reach the regions upstream of the object. The flow then responds to this information and is deflected around the obstacle in a smooth, laminar fashion.

If the flow is *super*-sonic, on the other hand, such signals get swept downstream faster than they can propagate, so that they cannot "inform" the upstream flow about the presence of the obstacle. In some cases, the flow simply impinges on the object and is absorbed by it. Alternatively, a shock is launched which then stands in the upstream flow. The shock effects a super- to sub-sonic transition. This sub-sonic flow then is capable of being deflected around the obstacle. I shall proceed in this third and final part of the chapter to discuss the many details about the Earth's bow shock which have come to light in recent years through a combination of dedicated spacecraft studies (notably the ISEE series of two satellites spaced close enough together (100-200 km) to separate the spatial and temporal variations) and associated theoretical work. Simultaneous development of computer simulations has also been a prime input into this field (Leroy et al., 1982; Quest et al., 1983). Those unfamiliar with the gross structure of the outer magnetosphere may find reference to the summary figure 8.15 helpful. I begin here with a purely local description of an idealized planar shock.

8.3.2 General Shock Considerations

Although shocks are a nearly discontinuous jump in the bulk parameters of the flow and vary considerably in nature depending on a variety of factors (magnetic field strength and direction, plasma beta, Mach number, etc.) all shocks have four basic properties in common: They must conserve mass, momentum, and energy. These conservation laws fix the shock jump conditions, known as the Rankine-Hugoniot relations, which specify the values of downstream parameters, given the upstream ones, independent of the structure within the shock itself (Tidman & Krall, 1971). And shocks must be dissipative, that is, entropy must increase as the flow traverses the shock. This last property prohibits, for example, a sub-sonic flow from spontaneously undergoing a shock transition and becoming super-sonic.

The major problem in shock physics is in understanding and modelling this dissipation. In an ordinary hydrodynamic shock, the dissipation usually takes the form of viscosity, and the characteristic thickness is then related to the viscous, or collisional, scalelength. Clearly, the discussion in §8.2.1 and table 8.2 imply that the Earth's bow shock cannot take the form of such a *collisional* shock, since it would then occupy a large fraction of the region between us and the sun. To contemplate an idealized discontinuity whose thickness spans a wide variation in solar wind parameters is somewhat absurd.

The inclusion of a magnetic field changes the picture slightly. In fact, the relevant compressional signals in the solar wind are fast magnetosonic waves rather than ordinary sound waves. These waves, in their small amplitude limit, propagate at a speed v_{fms}, where

$$v_{fms}^2 = \frac{1}{2}(c_s^2+v_A^2)\left[1+\sqrt{1-\frac{4c_s^2v_A^2\cos^2\theta}{(c_s^2+v_A^2)^2}}\right] \qquad (8.14)$$

and the relevant Mach number is therefore the magnetosonic Mach number, $M_{ms} = v_{SW}/v_{fms}$. Often, however, the Alfvén Mach number, $M_A = v_{SW}/v_A$ is used instead, since it does not depend on the angle of propagation, and is usually close to M_{ms} in value.

Such a fast shock involves the compression of both the plasma and the magnetic field, and thus has associated with it the current layer needed to effect the jump in the field across the shock. For low Mach number fast shocks ($M_A \lesssim 3$), electron resistivity is capable of providing the dissipation required by the shock. Of course, such resistivity must be "anomalous", i.e. the scattering must be caused by microturbulence, or we would have the same problem of shock thickness discussed above. There is indeed ample evidence for the existence of microinstabilities and turbulence within the Earth's bow shock (for a review, see, e.g., Wu et al., 1984; Winske, 1985). However, there is a limit, regardless of the value of the resistivity, to the amount of this dissipation, which essentially stems from the fact that the total current is fixed by the shock jump conditions.

Above this critical Mach number (which depends on plasma parameters and the orientation of the magnetic field) the ions must participate in the dissipation process. Since the shock is considerably thinner than a collisional mean free path, this collisionless process can severely distort the ion distributions. A by-product of such non-equilibrium configurations is the possible collisionless interaction and acceleration of at least some of the incident ions. Indeed, although shocks have often been postulated as likely cosmic particle accelerators (Axford, 1981), the bow shock has provided the first real evidence that this is the case. Eventually, thermalization occurs in the downstream region and the shock is seen, on a larger scale, to have provided the necessary deceleration of the super-Alfvénic solar wind, converting the incident kinetic energy flux associated with the bulk flow into predominantly thermal energy in the downstream region.

In the absence of particle collisions to provide a gradual deceleration, the individual solar wind ions must be decelerated by a combination of electric and magnetic fields accompanying the shock.

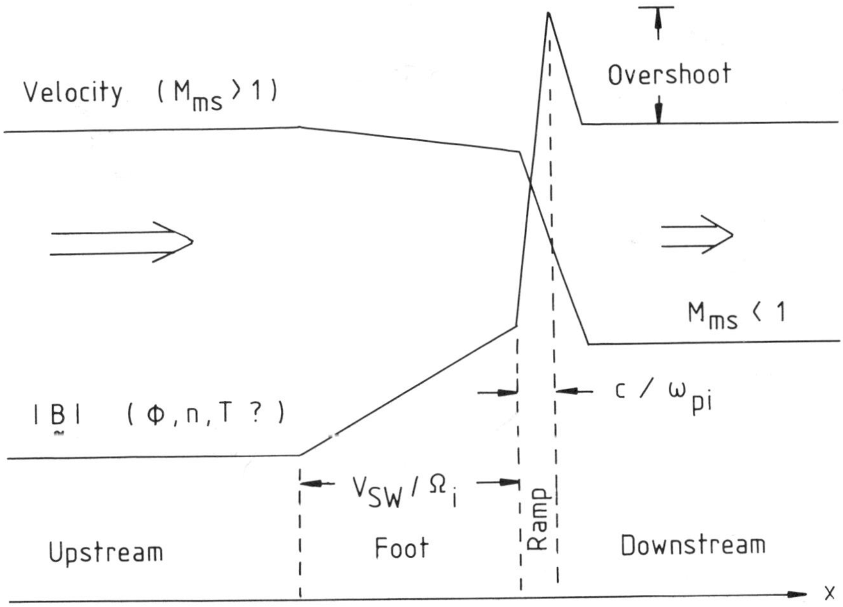

Figure 8.7 General shock structure and nomenclature of a fast, super-critical, collisionless shock such as the Earth's bow shock. The shock, as sketched here, is assumed to be planar. The variation of the plasma parameters, which have not all been observed in this kind of resolution, and fields is shown for illustrative purposes, and includes an overshoot. Real shocks show considerably more structure than this figure suggests.

The slower moving parts of the distribution may even suffer some reflection at the shock itself, and subsequently gyrate into the upstream region before passing downstream. Thus, there are at least two length scales relevant to a fast, super-critical collisionless shock: the Larmor radius, $r_{Li}=v_{SW}/\Omega_i$, associated with these reflected particles and scaling with the solar wind speed (not the thermal speed), and the ion inertial length, c/ω_{pi}, which corresponds to the likely thickness of the current layer. Typical parameter values yield lengths of 1000 km and 100 km respectively for these two scales. The general shock structure is sketched in figure 8.7.

8.3.3 Macroscopic Fields at Collisionless Shocks

In this section I shall display the properties of the macroscopic electromagnetic fields and potentials at collisionless shocks, and discuss how these behave under transformation between various frames in which the shock is at rest. These are important for the particle considerations of the following sections. CGS units are used throughout this and subsequent sections.

The notation used here is relatively simple. Subscripts "u" and "d" denote quantities measured in the upstream and downstream regions respectively. The shock is in the y-z plane, and the upstream-pointing normal is $\hat{n} = -\hat{x}$. The magnetic field lies in the x-z (coplanarity) plane upstream and downstream, and, as we shall see, may have a $\hat{y}$ component through the shock. Although there are an infinite number of frames in which the shock is at rest, I shall concentrate on the two most popular: the de Hoffman-Teller frame, denoted "HT" (de Hoffman and Teller, 1950; Schwartz et al., 1983), in which the bulk flow velocity is aligned with the magnetic field in the upstream and downstream region, and the normal incidence frame, denoted "N", in which the upstream flow is anti-aligned with the upstream-pointing shock normal. The upstream and downstream velocity-space configurations are shown in figure 8.8 for the fast shock case, including the transformation velocity $\underset{\sim}{V}_{HT}$ between these two frames.

The general de Hoffman-Teller transformation is given by equation 4 of Schwartz et al. (1983). This transformation removes the motional $-\underset{\sim}{V}\times\underset{\sim}{B}/c$ electric field, leaving only that due to charge-separation, which has only an $\hat{x}$ component under the planar symmetry assumed here. The HT frame moves with a speed $\underset{\sim}{V}_{HT}$ with respect to the frame in question. In the case of the normal incidence frame, reference to figure 8.8 shows that

$$\underset{\sim}{V}_{HT} = - V_{xu} \tan \theta_{Bnu} \hat{z}$$

$$= - V_{HT} \hat{z} \qquad (8.15)$$

where V_{xu} is the upstream normal velocity, which is the same in all frames, and θ_{Bnu} is the acute angle between the upstream magnetic field and the shock normal. Typical solar wind velocities reveal that this transformation is non-relativistic unless θ_{Bnu} is very close (within ~ 1°)

to 90°. The non-relativistic transformation laws $\underset{\sim}{E}' = \underset{\sim}{E} + \underset{\sim}{V}\times\underset{\sim}{B}/c$, $\underset{\sim}{B}' = \underset{\sim}{B}$ (assuming $E = O(VB/c)$), where $\underset{\sim}{V}$ is the velocity of the primed frame with respect to the unprimed, thus give

$$E_x^N = E_x^{HT} - \frac{V_{HT}}{c} B_y \tag{8.16}$$

$$E_y^N = + \frac{V_{HT}}{c} B_x \tag{8.17}$$

$$E_z^N = E_z^{HT} = 0 \tag{8.18}$$

and

$$\underset{\sim}{B}^N = \underset{\sim}{B}^{HT} = \underset{\sim}{B} \tag{8.19}$$

where I have used the fact that $\underset{\sim}{E}^{HT}$ has only an $\hat{\underset{\sim}{x}}$ component.

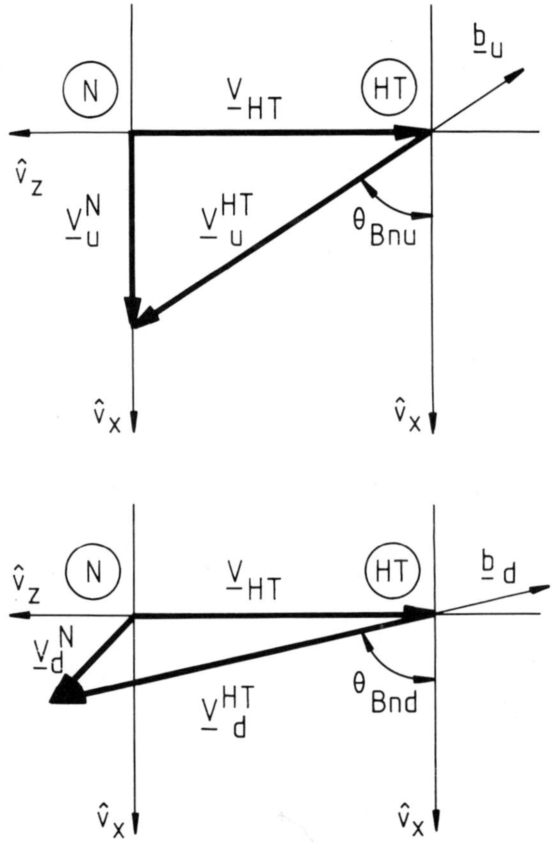

Figure 8.8 Transformation between the normal incidence frame ("N") and the de Hoffman-Teller frame ("HT") for a fast magnetosonic shock. Top: Upstream, Bottom: Downstream. In both frames the shock is at rest, with upstream-pointing normal $\hat{\underset{\sim}{n}} = -\hat{\underset{\sim}{x}}$. From an arbitrary shock rest frame, transformation to the HT frame also requires a $\hat{y}$-component to $\underset{\sim}{V}_{HT}$ (see Paschmann et al., 1980).

We now need a set of potentials (A, Φ), which form a four-vector, and give $B = \nabla \times A$, $E = -\nabla\Phi - c^{-1}\partial A/\partial t$. It is convenient to choose the electric field to be purely electrostatic, i.e. $E = -\nabla\Phi$, in one frame. We shall see that the transformation does not preserve this property. I shall avoid the common practice of defining a "potential" solely by the integral of E_x^N while leaving E_y^N as the "motional electric field" (e.g., Tidman and Krall, 1971), although this may be convenient for some purposes.

A suitable set of potentials in the HT frame is

$$\Phi^{HT}(\underset{\sim}{x},t) = -\int_{-\infty}^{x} E_x^{HT} dx \tag{8.20}$$

$$A_x^{HT}(\underset{\sim}{x},t) = 0 \tag{8.21}$$

$$A_y^{HT}(\underset{\sim}{x},t) = -z^{HT} B_x + \int_{-\infty}^{x} B_z \, dx \tag{8.22}$$

and

$$A_z^{HT}(\underset{\sim}{x},t) = -\int_{-\infty}^{x} B_y \, dx \tag{8.23}$$

Using the transformation properties of the four-vector (A,Φ) and noting that the coordinates obey $z^{HT} = z^N + V_{HT} t$ gives

$$\Phi^N = \Phi^{HT} + \frac{V_{HT}}{c} \int_{-\infty}^{x} B_y \, dx \tag{8.24}$$

$$A_x^N = A_x^{HT} = 0 \tag{8.25}$$

$$A_y^N = -(z^N + V_{HT} t) B_x + \int_{-\infty}^{x} B_z \, dx \tag{8.26}$$

$$A_z^N = -\int_{-\infty}^{x} B_y \, dx + O(V_{HT}\Phi/c) \tag{8.27}$$

where the second term in (8.27) is small in the non-relativistic limit, given our earlier assumptions about the relative orders of magnitude of E and B. It is straightforward to show that the potentials (8.20-8.27) and the fields (8.16-8.19) are consistent. Note, though, that the vector potential (8.26) is necessarily time-dependent in the normal incidence frame, even for the stationary shock problem. It is also possible to define the electric field in the normal incidence frame to be purely potential. Then the vector potential in the HT frame would be time-dependent. Moreover, the electric potential in the normal incidence frame would also be a function of the y coordinate since E_y^N is non-zero.

Finally, we can relate the electric fields seen in the various frames to the charge densities in order to illuminate the origins and role of the charge separation at the shock. Taking the divergence of $\tilde{E}^N$ in (8.16–8.18), using Gauss' Law for each frame, and relating B_y to the current density j_z via Ampere's Law yields

$$\rho^N = \rho^{HT} - \frac{V_{HT}}{c^2} j_z \tag{8.28}$$

which is simply the transformation of the four-vector $(\tilde{j},\rho c)$. Note that if there is a significant B_y, and hence j_z, the charge separation as seen in the two frames is different. In all cases, it is related only to E_x via $dE_x{}^{N,HT}/dx=4\pi\rho^{N,HT}$.

The downstream electron heating can be used to determine the size of the potential in the HT frame (Axford, 1981). Goodrich & Scudder (1984) show this by combining the relativistic invariance of $\tilde{E}\cdot\tilde{B}$ with the generalized Ohm's law. An equivalent approach is to consider the electron fluid momentum equation:

$$m_e V_x \frac{d\tilde{V}}{dx} = -\frac{1}{n_e}\tilde{\nabla}\cdot\tilde{\tilde{P}}_e - e\left[\tilde{E} + \frac{\tilde{V}\times\tilde{B}}{c}\right] + \frac{1}{n_e}\tilde{F}_{fric} \tag{8.29}$$

where $\tilde{F}_{fric}$ is any frictional force (resistivity, etc.) acting on the electrons. Working in the HT frame, where $\tilde{E} = E_x{}^{HT}\hat{x}$, and neglecting both any pressure anisotropy and $\tilde{F}_{fric}$, dotting (8.29) with $\tilde{V}$ leads to

$$-\frac{d\Phi^{HT}}{dx} = E_x^{HT} = -\frac{1}{e}\left[\frac{d}{dx}\left(\frac{1}{2}m_e V^2\right) + \frac{1}{n_e}\frac{dp_e}{dx}\right] \tag{8.30}$$

in which the last term dominates the right hand side by virtue of the small m_e.

Observationally, (8.30) leads to a potential which is considerably smaller than the incoming ion bulk kinetic energy, $m_i V_{xu}{}^2/2e$, although computer simulations suggest that the integral of $E_x{}^N$ is comparable to this upstream ion energy. Simulations also show a substantial B_y component (Leroy et al., 1982; Forslund et al., 1984; Leroy and Winske, 1983; Quest et al., 1983) which reconciles the two electric fields via (8.16). We shall see below how these fields affect the particle dynamics through the shock transition.

8.3.4 Particle Dynamics at Collisionless Shocks - Electrons

The sign of the potential jump in the HT frame, as given by (8.30), is such that the ions are decelerated. Thus the upstream electrons are *accelerated* across the shock, resulting in a shift in the peak of the electron distribution with respect to the centre of mass frame. This process is sketched in figure 8.9. Since there can be no net current through (i.e. normal to) the shock, the trailing part of the distribution must be filled in by hot downstream electrons, yielding the dashed portion of the sketch. Although Liouville's Theorem would imply that the height of the peak is the same upstream and downstream, some scattering occurs and lowers it as the electrons become thermalised. In fact, a typical set of distributions through the shock is shown in figure 8.10, and shows that the peak is completely dissipated, and the downstream distribution is very flat-topped. It is not yet clear whether this shape results from global considerations of the electrons behind a curved bow shock or, more probably, reflects the relaxation of the unstable distribution containing the shifted peak due to scattering by excited microturbulence. There is no shortage of candidate unstable wave modes (Winske, 1985; Wu et al., 1984). In this section I shall provide a few simple theoretical predictions of these transmitted electrons.

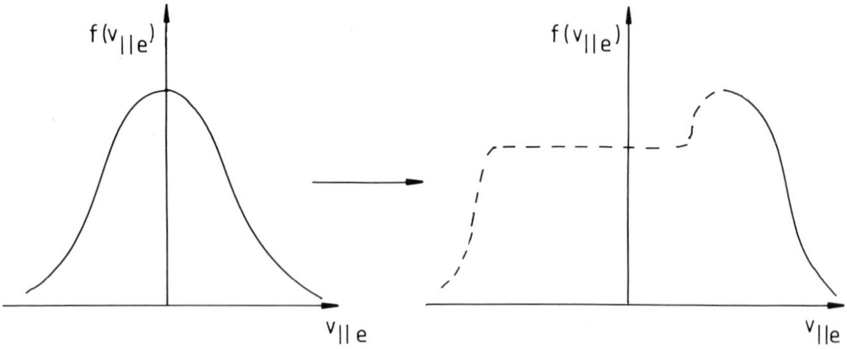

Figure 8.9 Sketch showing the formation of an offset peak in the electron distribution related to the acceleration by the shock potential drop in the HT frame. The dashed portions of the distribution must be filled in by scattered and/or heated downstream electrons in order to preserve the zero current condition in the downstream region (i.e. $j_x=0$ everywhere).

One of the easiest ways of calculating the energetics of particles which traverse a shock is to restrict the calculation to the HT frame. Since E_y is zero here, away from the shock a particle's *kinetic* energy is constant. Moreover, this energy is independent of the details of the shock traversal and is simply

$$|\underset{\sim}{v}_d|^2 = |\underset{\sim}{v}_u|^2 - \frac{2q}{m}\,\Delta\Phi^{HT} \tag{8.31}$$

where $\underset{\sim}{v}_{u,d}$ is the particle's velocity as measured in the HT frame in the upstream and downstream regions respectively, and $\Delta\Phi^{HT} = \Phi_d^{HT} - \Phi_u^{HT} = \Phi^{HT}(x{=}+\infty)$ from (8.20) is the potential difference across the shock in the HT frame. It is often convenient to decompose the particle velocity into its guiding centre motion, $v_{\parallel}\hat{\underset{\sim}{b}}$, and its gyromotion, $\underset{\sim}{v}_{\perp}$. The motion in the N frame is found by adding the de Hoffman-Teller velocity, $\underset{\sim}{V}_{HT}$, to the guiding centre motion (see, e.g., Schwartz et al., 1983,

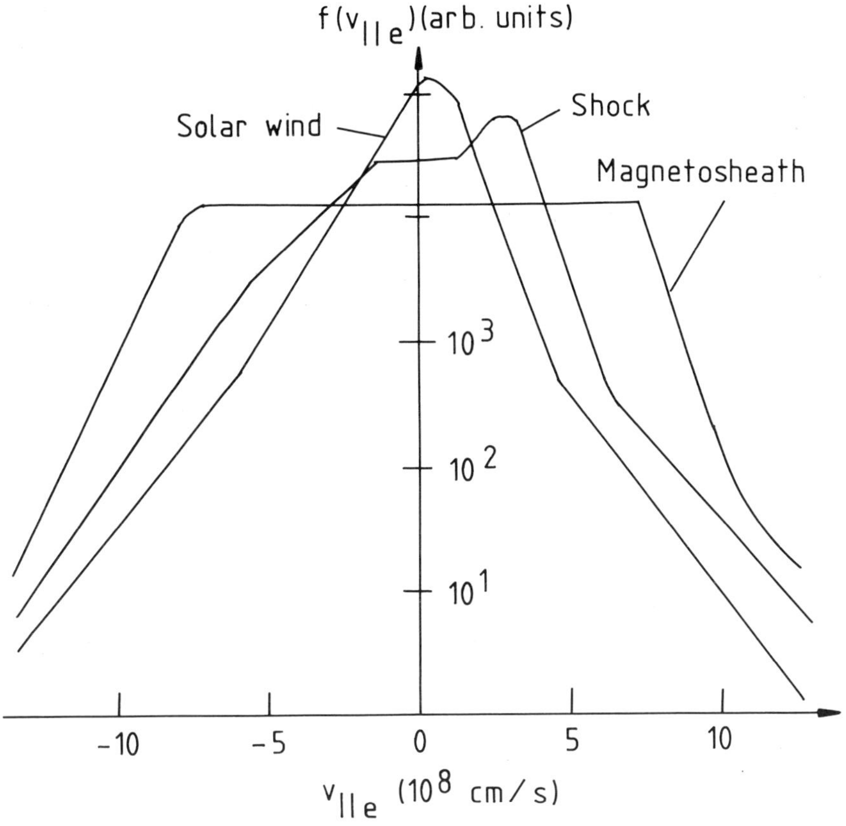

Figure 8.10. Cuts through an observed electron velocity distribution parallel to the magnetic field direction in the upstream region, within the shock, and in the downstream region during a typical bow shock crossing (after Feldman et al., 1983). Note the presence of an offset peak, as sketched in figure (8.9), the scattering implied by its decrease with increasing penetration through the shock, and the flat-topped nature of the downstream distribution.

equations (9)-(10)).

In order to perform this calculation directly in the N frame, the detailed particle motion is required in order to take account of the energy gained by displacements along E_y^N. Goodrich & Scudder (1984) have shown that, when these are properly included, the results of the separate calculations in the two frames are indeed related by the Galilean kinematics assumed here. We note here that (8.31), and the transformation to the N frame, hold even if $B_y = 0$. It is not possible to deduce, as previous authors have claimed, that $B_y \neq 0$ on the basis of an analytical discussion of particle dynamics. Rather, the fact that B_y is non-zero within the shock transition can be shown to follow from, e.g. the linearized resistive shock evolution equations (Tidman and Krall, 1971). It is possible to deduce this fact *observationally* by noting that the electron and ion energetics in the N frame are not the same due to the $\hat{y}$-drifts. Using the observed electron heating to estimate $\Delta\Phi^{HT}$, Goodrich & Scudder point out essentially that $E_x^{HT} \neq E_x^N$, from which (8.16) immediately reveals that B_y does not vanish everywhere.

For electrons, the energy gain in traversing the shock results in a velocity which is typically larger than the fluid flow. Thus it is convenient to decompose their velocities as

$$\underset{\sim}{v} = (V_x \sec \theta_{Bn} + v_{\parallel p})\, \hat{\underset{\sim}{b}} + \underset{\sim}{v}_{\perp} \tag{8.32}$$

where $v_{\parallel p}$ is the particle's peculiar velocity along the field relative to the (field-aligned) bulk flow (as determined, say, by the Rankine-Hugoniot relations), and $\underset{\sim}{v}_{\perp}$ is its gyromotion. Given the parameters of the particle upstream, and assuming the electrons conserve their magnetic moments in traversing the shock, (8.31) can be rearranged to yield

$$v_{\parallel pd} = \left[(v_{\parallel pu} + V_{xu} \sec \theta_{Bnu})^2 + v_{\perp u}^2 \left[1 - \frac{B_d}{B_u} \right] + \frac{2e\Delta\Phi}{m_e} \right]^{1/2} - V_{xd} \sec \theta_{Bnd} \tag{8.33}$$

Some electrons will mirror at the shock and return to the upstream region, while those for which $v_{\parallel pu} + V_{xu} \sec\theta_{Bnu}$ is negative are travelling away from the shock to begin with. A full description of this process requires knowledge of the profiles of B and Φ through the shock, since the acceleration by the shock can reduce the particle's pitch angle if it occurs before the rise in B is reached, making mirroring less likely. Any particle for which the expression inside the square root in (8.34) is negative, however, must mirror and cannot enter the downstream region. This mirroring is probably responsible for electron beams seen in the earth's foreshock (Anderson et al., 1979; Feldman et al., 1983; Wu, 1984). We can use (8.34) to follow, say, the peak of the upstream electron distribution ($v_{\parallel pu} = 0 = v_{\perp u}$). These electrons also form the peak of the downstream distribution, in the absence of scattering, via Liouville's theorem. If the HT potential is a fraction f_Φ of the upstream ion kinetic energy, $m_i V_{xu}^2/2$, then the potential term inside the square root bracket dominates provided

$$\cos^2\theta_{Bnu} > \frac{m_e}{f_\Phi m_i} \tag{8.34}$$

For $f_\Phi \approx 1/10$ (Goodrich & Scudder, 1984) the potential term dominates for all geometries $\theta_{Bnu} \lesssim 86^\circ$. Thus in all but very nearly perpendicular shocks, we expect the peak of the distribution function downstream to be offset from the bulk flow by an amount

$$v_{\parallel pd} \approx \left[\frac{2e\Delta\Phi}{m_e} \right]^{1/2} - V_{xd} \sec \theta_{Bnd} \qquad \theta_{Bnu} \lesssim 86^{\circ} \tag{8.35}$$

where the second term is a small, but in the fast shock case non-negligible, correction which can only be found by solving the Rankine-Hugoniot relations for V_{xd} and θ_{Bnd}. Equations (8.33-8.35) provide a relatively straightforward method for interpreting and quantitatively assessing the offset electron "beams" shown in figures 8.9 and 8.10.

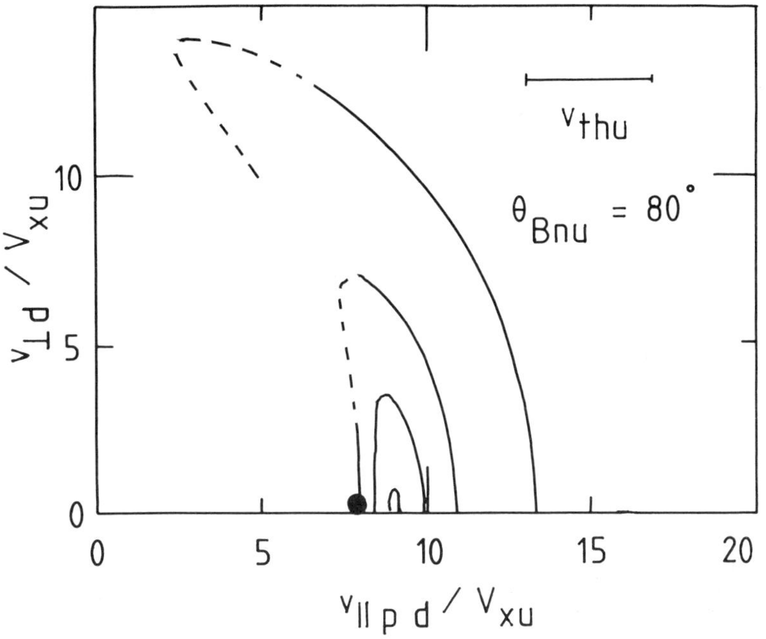

Figure 8.11 Contours of the distribution function of transmitted electrons at a fast shock using (8.33), the Rankine-Hugoniot relations and Liouville's theorem. The shock parameters used are: Alfvén Mach number, M_A = 8, electron and ion plasma beta both = 1, cross-shock de Hoffman-Teller potential = 1/10 the incident ion energy (i.e., f_Φ = 1/10) and θ_{Bnu} = 80°. The upstream electrons were assumed to be isotropic. Contour levels correspond to 0.1, 0.5, 1 and 2 upstream thermal speeds, v_{thu}. The dot on the $v_{\parallel pd}$ axis gives the approximation (8.35). The dashed portions correspond to particles whose transmission depends on the relative positions of the rise in B and Φ.

Figure 8.11 shows an application of these results for parameters typical of the bow shock. The downstream distribution function corresponding to an upstream isotropic Maxwellian is shown, calculated using the general result (8.33) and Liouville's Theorem, along with the approximate expression (8.35). Notice that the effect of the potential is to reduce the parallel temperature of the beam. This is always the case for an accelerated distribution, and can be easily demonstrated by setting $v_{\perp u}=0$ in (8.33) and differentiating with respect to $v_{\parallel pu}$. For example, compare the distance from the peak to the third contour in figure 8.11 with v_{thu}. Additionally, the adiabatic behaviour of the electrons results in an increased perpendicular spread. The resulting $T_{\perp}>T_{\parallel}$ anisotropy of the beam may be important in the growth rates, etc., of whistler or other microinstabilities which are responsible for the dissipation, relaxation and thermalization of the beam energy (Tokar et al., 1984). Note the incomplete contour arizing from upstream electrons which are either travelling away from the shock from the outset or mirror at the shock and return to the upstream region. The dashed portion corresponds to electrons which *might* mirror if the rise in B occurs before the rise in potential. Such contours, and indeed a large fraction of the velocity space extending from the beam peak to negative parallel peculiar velocities, must be filled in by scattered or trapped electrons from the downstream region.

Ignoring the difficulties associated with dissipative processes, which probably act to reduce the overall beam speed, the electron beams seen in the earth's bow shock provide a direct measure of the shock potential in the HT frame. A complementary approach using the electron fluid momentum equation (8.30) gives similar values, probably because the electron heating, as characterized by the ledge energy of the flat-topped magnetosheath, is mainly due to the relaxation of the free energy associated with the beam (Winske, 1985, and references therein).

8.3.5 Particle Dynamics at Collisionless Shocks - Ions

Turning now to ions, let us begin by looking at some observations. Figure 8.12 shows a series of 2-D ion distributions at high time resolution during a transition from the solar wind through the shock and into the downstream, magnetosheath region. Although it is rare to find all the features shown in this figure during a single pass through the shock, it does illustrate many features which are seen. Figure 8.13 is a summary of theoretical ideas using the N and HT frame diagram of figure 8.7 (rotated through 90° to facilitate comparison with figure 8.12). I shall begin with a discussion of backstreaming ion beams seen well ahead of the shock, returning later to the more complicated ion signatures seen within the shock transition.

Upstream of the shock (figure 8.12A) is a field-aligned beam of particles. These ions are able to travel against the E×B drift and fill a region known as the ion foreshock (see figure 8.15 below). There are two main hypotheses for the source of these particles, which have energies (in the spacecraft frame) of ~4 - 25 times the solar wind energy. Sonnerup (1969) noted that these particles could be explained by taking some of the solar wind particles and reflecting them adiabatically (i.e. conserving their magnetic moments). In the HT frame, where there is no electric field apart from the electrostatic shock potential, such particles would simply reverse their velocity and remain field-aligned, as indicated in the diagram of figure 8.13. This hypothesis agrees well with both the field-aligned nature and the energy of these beams (Paschmann et al., 1980; Thomsen et al., 1983a; Schwartz & Burgess, 1984; figure 8.14 below), but does not provide any argument as to why particles should behave adiabatically at a shock whose thickness is less than a gyroradius. Perhaps the answer lies in the complicated multiple shock traversals solar wind ions may suffer (Burgess & Schwartz, 1984). In any event, if too much of their energy is in gyromotion, particles will not be able to escape from the shock, so that well upstream of the shock one could only expect to find particles with small pitch angles.

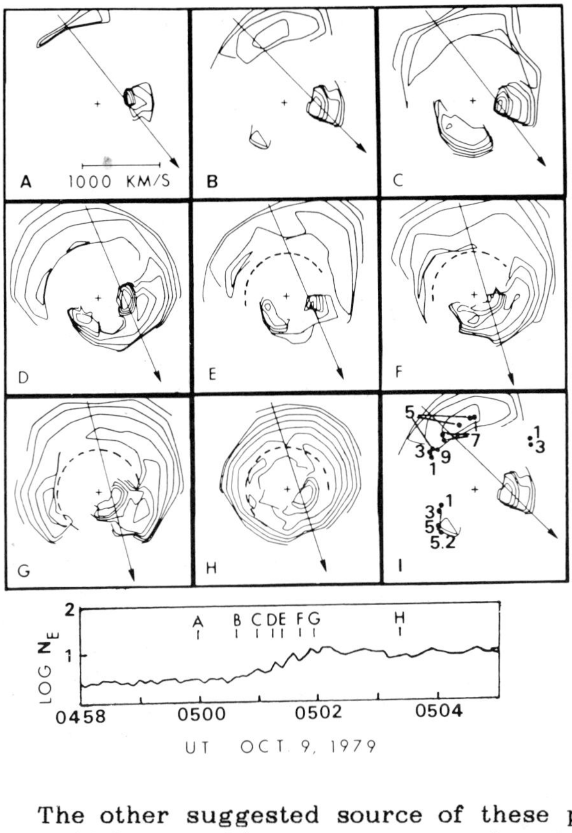

Figure 8.12 High time resolution ion velocity space contours during a bow shock crossing. The plus indicates zero velocity in the spacecraft frame, which is almost the same as the normal incidence frame. The shock normal points approxmiately to the left. The small set of contours immediately to its right in, e.g., panel A is the solar wind, which is not fully resolved by this instrument. The arrow which passes through this component is aligned with the instantaneous direction of the magnetic field. The lower panel shows the electron density variation (which mimics the magnetic field) and indicates the positions of the various panels (from Thomsen et al., 1983b).

The other suggested source of these particles is hot, magnetosheath ions which are able to return to the shock and leak across into the upstream region (Edmiston et al., 1982; Tanaka et al., 1983). In order to reach the shock, ions must have a velocity greater than zero in the HT frame, i.e. a speed > V_{HT} in the N frame. This can be thought of as a combination of the E×B drift and a field-aligned guiding centre motion which is sufficient to overcome the downstream drift component in the N frame. In leaking across the shock, these particles gain an energy $e\Delta\Phi^{HT}$ in the HT frame, which is ~ 1/10 the solar wind energy based on only the normal component of solar wind velocity. If this acceleration is unmagnetized, as seems likely, such particles pick up some gyromotion

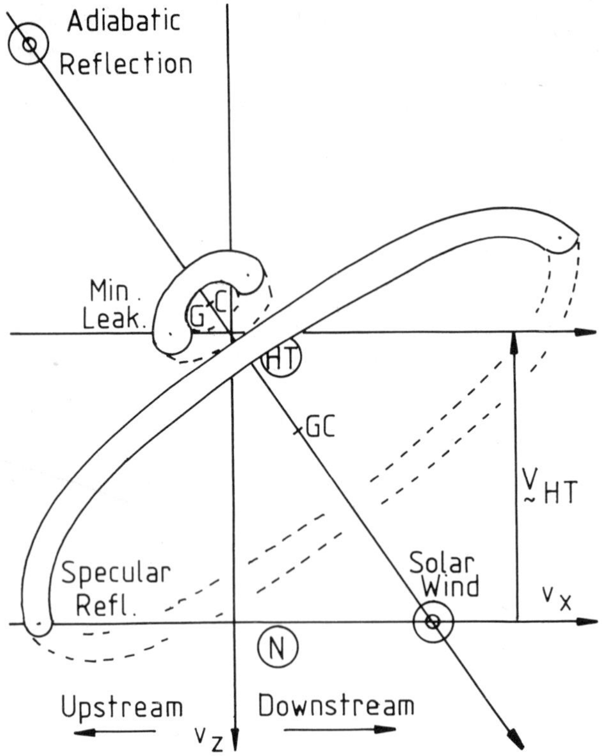

Figure 8.13 Theoretical sketch of upstream ions at the bow shock. This velocity space diagram uses the N and HT frame diagram of figure 8.8 (rotated through 90° for comparison with the data shown in figure 8.12). Shown are the positions of the incident solar wind, specularly reflected solar wind ions (whose subsequent motion in velocity-space follows the torus centred about the guiding centre, GC, as indicated), adiabatically reflected solar wind ions, and minimally escaping leaked magnetosheath ions which are accelerated by the shock potential and also gyrate about their guiding centres. Only a segment of the toruses would be observed at any one spatial location due to gyrophase bunching.

in addition to an upstream guiding centre motion, as shown in figure 8.13 (Schwartz et al., 1983). A slightly higher velocity in the HT frame is obtained if they preserve their magnetic moments and emerge field-aligned (Edmiston et al., 1982; Tanaka et al., 1983). In any case, these minimally escaped ions are less energetic than the adiabatically reflected ones. Note, though, that the difference is less in the N frame due to the addition of the large frame transformation, V_{HT}. As the geometry becomes more nearly perpendicular, V_{HT} increases and the energies in the N frame become dominated by V_{HT}, independent of the production mechanism. This problem led Schwartz & Burgess (1984) to transform the observed velocities into the HT frame before making the comparison with theoretical predictions. The results of this comparison are shown in figure 8.14. Although some theoretical predictions include particles which have more than just the minimal speed V_{HT} in the downstream region due to energization at other portions of the curved bow shock (Tanaka et al., 1983), and the observations shown in figure 8.12H confirm the presence of such ions (the dotted circle has a radius = V_{HT} in that diagram), many of the observed beams seem to fit the adiabatic reflection prediction quite well and cannot be explained by currently available leakage predictions. Of course, there are inherent errors in the data, in particular in the angles which are a crucial part of the analysis. And the more complicated theories, e.g. that of Tanaka et al., have only been undertaken for a limited range of parameters. Moreover, figure 8.12I shows the predictions of minimally escaping leaked ions as the set of dots clustered near the field-aligned direction, after folding in the instrument response, superimposed on the data from 8.12B. The numbers are distances upstream of the shock in units of 100km. This strongly suggests that at least some of the beams seem to be coming directly from the downstream distribution.

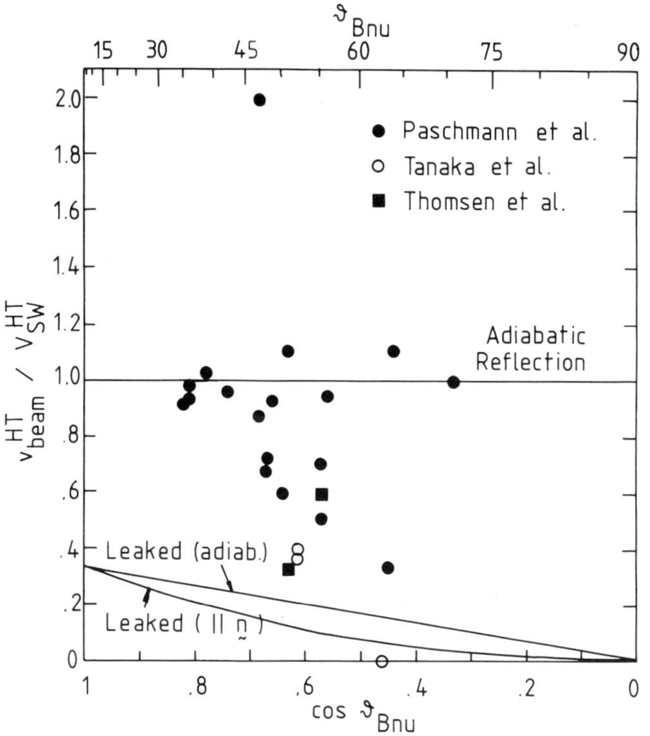

Figure 8.14 Theoretical predictions vs. observations of upstream ion beams. The solid circles are data from Paschmann et al. (1980) while the squares are the two events shown in Thomsen et al. (1983) (one of which is figure 8.12 above). The lines show the adiabatic reflection and the leakage predictions, while the open circles are the beams resulting from the computations of Tanaka et al. (1983).

Before leaving this topic of backstreaming ion beams, I wish to discuss the energetics as seen in various frames and the resulting implications for the determination of the shock potentials $\phi^{N,HT}$. It has been argued (Goodrich & Scudder, 1984) that the important magnetic forces associated with B_y within the shock in the HT frame (required to make up the difference between the prodominantly electrostatic force in the N frame and the lower HT potential) necessitate the use of shock models with finite thickness in order to accurately compute ion dynamics at the shock. Of course, the energetics are particularly simple in the HT frame, and these must come out right, but the velocity space details may alter. An equally important problem arises in the N frame. Here, the energetics depend in part on any displacements along the shock front parallel to the motional $-\tilde{v}\times B$ electric field, which is in the $\hat{\tilde{y}}$ direction. The problem is exactly analogous to the electron case discussed above. Without detailed knowledge of the trajectory, it is impossible to compute the resulting particle energy in this frame. This is true even if assumptions about the particle motion, e.g. magnetic moment conservation, are made. Knowledge of the normal electric field in this frame, E_x^N, is not sufficient.

As an example, let's look at the simple case of minimally escaping leaked ions, namely those which have zero speed in the HT frame before falling down the potential hill. In the normal incidence frame, these particles are just able to return to the shock against the convective electric field drift. For the sake of brevity, I treat here two extreme possibilities: (1) that ions leaked into the upstream region conserve their magnetic moments (Edmiston et al., 1982; Tanaka et al., 1983) and (2) that leaked ions behave unmagnetized and suffer an acceleration along the shock normal in the HT frame (Schwartz et al., 1983).

If the ion's magnetic moment is conserved, then $v_{\perp u}^{HT} = 0$, where $v_{\perp,\parallel u}^{HT}$ are the components of an ion's velocity perpendicular and parallel to the magnetic field in the HT frame. Transforming to the normal incidence frame by adding $\tilde{V}_{HT}$ to the parallel beam speed $v_{\parallel u}^{HT} = \sqrt{(2e\Delta\Phi^{HT}/m)}$ gives the energy gain as seen in the normal incidence frame as

$$|\tilde{v}_u^N|^2 - |\tilde{v}_d^N|^2 = \frac{2e\Delta\Phi^{HT}}{m_i} + 2\,V_{xu}\left[\frac{2e\Delta\Phi^{HT}}{m_i}\right]^{1/2}\frac{\sin^2\theta_{Bnu}}{\cos\theta_{Bnu}} \qquad (8.36)$$

where $\tilde{v}_d^N = \tilde{V}_{HT}$ and V_{xu} is the normal component of the solar wind velocity. The second term in (8.36) is usually larger than the first if the HT potential is a small fraction of the upstream bulk kinetic energy.

On the other hand, for particles leaked and accelerated along the shock normal, a similar analysis leads to an energy gain given by

$$|\tilde{v}_u^N|^2 - |\tilde{v}_d^N|^2 = \frac{2e\Delta\Phi^{HT}}{m_i} + 2V_{xu}\left[\frac{2e\Delta\Phi^{HT}}{m_i}\right]^{1/2}\sin^2\theta_{Bnu}(1-\cos\Phi_g) \qquad (8.37)$$

where $\Phi_g(t)$ is the gyrophase of the ensuing motion. The energy gain (8.37) is in general not equal to (8.36). Indeed, for $\theta_{Bnu} > 60^\circ$ it is always smaller, regardless of gyrophase. A similar analysis show that the guiding centre (i.e. beam) energy corresponding to (8.37) is always smaller than (8.36).

Thus different assumptions about the particles' behaviour (magnetized conserving their magnetic moments, unmagnetized acceleration along the shock normal, etc.) lead to different energy gains in the N frame. These differences are due to the unspecified $\hat{\tilde{y}}$ drifts along E_y^N. Indeed, if $\Delta\Phi^{HT}$ were known, (8.36–8.37) could be used to calculate the extent of these drifts. Knowledge of the N-frame potential, Φ^N, is not sufficient to determine the final energetics, even if magnetic moment conservation, etc. is assumed, since it does not determine $\Delta\Phi^{HT}$ without knowledge of the shock structure via (8.24). Previous leakage calculations in the N frame have apparently neglected this point (e.g. Edmiston et al., 1982; Tanaka et al., 1983).

In the de Hoffman-Teller frame, all the energy gains are the same, since $E_y^{HT} = 0$; only the pitch angle predictions are different. Thus, while the detailed field profiles are required in the de Hoffman-Teller frame in order to study the velocity space features of variously produced backstreaming ions, similarly detailed particle trajectory studies are required in the normal incidence frame in order to get the energetics right. In both cases the field profiles and overall field magnitudes need to be specified. Similar considerations arise in the interaction of upstream ions with the shock.

Returning now to figure 8.12, panels B–E show the clear presence of a third ion component below and to the left of the centre cross. These are particles within the shock transition which have been reflected by the shock. Figure 8.13 shows the predicted positions of such particles on the assumption that the reflection is specular, i.e. mirror-like. The normal component of velocity is reversed while the others remain unchanged. The resulting bunch of ions subsequently gyrate about their guiding centre, analogous to the accelerated leaked ions discussed above, as indicated by the large torus in figure 8.13. All such ions leave the shock with the same gyrophase, and subsequently gyrate coherently as they travel. Thus, only sections of this torus are observable at any one location, a phenomenon known as *gyrophase bunching*. Figure 8.12I also shows the excellent agreement when these predictions are superimposed on the data.

Simple geometry using figure 8.13 reveals that the guiding centre motion for specularly reflected ions is directed downstream for $\theta_{Bnu} > 45^\circ$, as is the case in the diagram. These particles return to the shock and are thus seen only within one Larmor radius upstream (Paschmann et al., 1982) (i.e. in the "foot" region of figure 8.7) and downstream as distinct bunches which eventually contribute to the ion thermalization (Sckopke et al., 1983). This reflection is the first step in the way a thin, collisionless shock "heats" the ion distribution, i.e. spreads it in velocity-space.

For $\theta_{Bnu} < 45^\circ$, the guiding centre motion of specularly reflected ions is directed *upstream*. Such backstreaming gyrating ions have also been observed (Gosling et al., 1982).

Finally, other types of shock associated ion distributions are also seen upstream of the earth's bow shock. The diffuse ions (Gosling et al., 1978) are an energetic, nearly isotropic shell, with energies typically higher than the field-aligned beams. Intermediate distributions are thought to provide the connection between these two types, as they appear to be a smeared out beam. A popular scenario involves the formation of intermediate distributions via a beam-driven instability due to the original beam (Lee, 1982; Lee & Skadron, 1985). Further pitch-angle scattering eventually returns these ions to the shock by filling in an entire shell. The higher energies found in the diffuse component implies, in this scenario, that some ions have returned to the shock more than once, acquiring more energy each time by what is termed first-order Fermi acceleration (i.e. bouncing between two converging mirrors - the scattering centres in the solar wind and the slower ones at or behind the shock). This picture provides the first direct evidence of a cosmic particle accelerator.

8.3.6 The Global Structure of the Earth's Bow Shock and Foreshock

The previous sections dealt primarily with idealized, planar shocks. While much of the detailed shock structure and dynamics occur over sufficiently small scales that the curvature can be neglected, this is not entirely the case. Moreover, the global aspects are important in understanding the solar-terrestrial interaction. A summary sketch is shown in figure 8.15. The shape of the bow shock, based on numerous crossings, is adequately described by a hyperboloid whose axis is aberrated from the sun-Earth line by the ratio of the Earth's orbital motion to the solar wind speed, and whose focus is located some 3 Earth radii in front of the Earth along this axis (Slavin & Holzer, 1981). In standard form,

$$\frac{L}{r} = 1 + \epsilon \cos \theta \tag{8.38}$$

where $L=23.3\ R_E$ and $\epsilon=1.16$. The distances scale as $(P_{SW}/2.1\times10^{-8})^{-1/6}$, where $P_{SW}=1.16 n_p m_p V_{SW}^2$ is the solar wind dynamic pressure, assuming a 4% number density of alpha particles. As discussed in Chapter 9, this scaling results from the compressibility of the Earth's dipolar magnetic field.

As is evident from figure 8.15, the bow shock provides the entire range of shock geometries, from quasi-perpendicular on the dusk flanks to quasi-parallel on the dawn side. Of course, the interplanetary field does not always assume the nominal spiral angle shown here, and fluctuations in the solar wind pressure cause the shock to move. Further complications arise from the high level of turbulence found at and upstream of quasi-parallel shocks. It is not yet clear whether this turbulence is a general feature of quasi-parallel shocks, which cannot rely on efficient field compression in the same way that perpendicular ones do, or is due to the debris of ion beams and associated turbulence which originate at the quasi-perpendicular shock and are convected with the solar wind to arrive at the quasi-parallel side. It is generally thought that both these factors contribute to the relatively poorly understood mess depicted in the figure.

Directly behind the bow shock, the hot magnetosheath plasma, which is just shocked and deflected solar wind plasma, flows around the magnetosphere, separated from it by the magnetopause. Further downstream of the Earth's location, the hyperboloid shape (8.38) is replaced by a nearly cylindrical tail, held together by the transverse solar wind pressure, which attempts to fill in the void left behind the Earth.

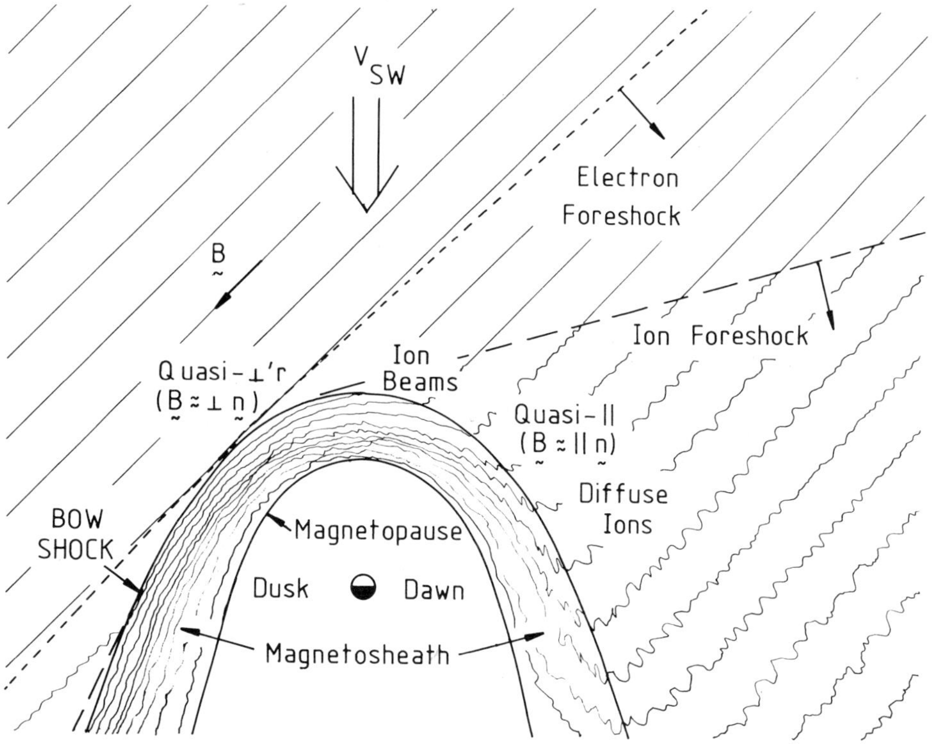

Figure 8.15 The gross structure of the Earths' bow shock and associated foreshock. The foreshock boundaries are found by following a field-aligned particle as it travels away from the shock and is convected perpendicular to B by the ExB solar wind drift. Since the electron velocities are much greater than the solar wind, their foreshock is bounded approximately by the tangential field line. The approximate locations of the various ion distributions are shown. Notice how the shock is nearly perpendicular to the magnetic field on the dusk side, and becomes more nearly parallel toward dawn.

8.4 CONCLUSION

In this chapter, I have reviewed several aspects of the solar wind plasma. It seems clear that we have learned a great deal about plasmas in general by taking advantage of the detailed particle and field measurements which are feasible in space plasmas. Equally clear are several key processes, e.g. electron thermal conduction and minor ion acceleration, which as yet have no solid theoretical framework.

Similar comments apply to the Earth's bow shock. It is fair to say that the bow shock studies over the past several years have resulted in an exciting explosion of data and understanding of collisionless shocks. As is inevitable, however, the questions raised per question answered has yet to reach even the steady state value of unity.

Acknowledgements Parts of §8.3.2 and §8.3.3 are the result of collaboration with Bill Feldman, with partial support from the Royal Society. Much of the material in this chapter has been clarified for me, or taught to me, by Bill Feldman, Peter Gary, Jack Gosling, Michelle Thomsen and Eckart Marsch. I am also indebted to my colleagues at QMC, notably Ian Roxburgh, David Burgess, David Rowse and Cliff Singer.

REFERENCES

Anderson K A, R P Lin, F Martel, C S Lin, G K Parks and H Reme (1979) *Geophys. Res. Lett.* 6, 401.

Axford W I (1981) *Proc. International School and Workshop on Plasma Astrophysics*, Varenna, Italy, 27 Aug.–7 Sept. 1981, ESA SP-161, p425.

Barnes A (1979) in Kennel et al. (1979) p249.

Belcher J W and L Davis (1971) *J. Geophys. Res.* 76, 3534.

Bonnet R M and A K Dupree (eds) (1981) *Solar Phenomena in Stars and Stellar Systems*, NATO ASI, D. Reidel.

Burgess D and S J Schwartz (1984) *J. Geophys. Res.* 89, 7407.

de Hoffman F and E Teller (1950) *Phys. Rev.* 80, 692.

Edmiston J P, C F Kennel and D Eichler (1982) *Geophys. Res. Lett.* 9,531.

Feldman W C, J R Asbridge, S J Bame, M D Montgomery and S P Gary (1975) *J. Geophys. Res.* 80, 4181.

Feldman W C, J R Asbridge, S J Bame and J T Gosling (1977) in *The Solar Output and its Variations*, O R White (ed), Colorado Univ. Press, p351.

Feldman W C, J R Asbridge, S J Bame and J T Gosling (1982) *J. Geophys. Res.* 87, 7355.

Feldman W C, R C Anderson, S J Bame, S P Gary, J T Gosling, D J McComas and M F Thomsen (1983) *J. Geophys. Res.* 88, 96.

Forslund D W (1970) *J. Geophys. Res.* 75, 17.

Forslund D W, K B Quest, J U Brackbill and K Lee (1984) *J. Geophys. Res.* 89, 2142.

Gosling J T, J R Asbridge, S J Bame, G Paschmann and N Sckopke (1978) *Geophys. Res. Lett.* 5, 957.

Gosling J T, M F Thomsen, S J Bame, W C Feldman, G Paschmann and N Sckopke (1982) *Geophys. Res. Lett.* 9, 1333.

Goodrich C C, and J D Scudder (1984) *J. Geophys. Res.* 89, 6654.

Gurnett, D A (1981) in Rosenbauer (1981) p286.

Hartmann L (1981) in Bonnet & Dupree (1981) p331.

Hollweg J V (1974) *J. Geophys. Res.* 79, 1539.

Hollweg J V (1978) *Rev. Geophys. Sp. Phys.* 16, 689.

Hundhausen A J (1972) *Coronal Expansion and SolarWind*,Springer-Verlag

Isenberg, P A (1984) *J. Geophys. Res.* 89, 6613.

Isenberg P A and J V Hollweg (1982) *J. Geophys. Res.* 87, 5023.

Isenberg P A and J V Hollweg (1983) *J. Geophys. Res.* 88, 3923.

Kennel C F, L J Lanzerotti and E N Parker (eds) (1979) *Solar System Plasma Physics*, North-Holland.

Lee M A (1982) *J. Geophys. Res.* 87, 5063.

Lee M A and G Skadron (1985) preprint, University of New Hampshire.

Leroy M M, and D Winske (1983) *Annal. Geophys.* 1, 527.

Leroy M M, D Winske, C C Goodrich, C S Wu and K Papadopoulos (1982) *J. Geophys. Res.* 87, 5081.

Marsch E and H Goldstein (1983) *J. Geophys. Res.* 88, 9933.

Marsch E and A K Richter (1984a) *J. Geophys. Res.* 89, 5386.

Marsch E and A K Richter (1984b) *J. Geophys. Res.* 89, 6599.

Marsch E, K-H Mühlhäuser, H Rosenbauer, R Schwenn and F M Neubauer (1982a) *J. Geophys. Res.* 87, 35.

Marsch E, C K Goertz and K Richter (1982b) *J. Geophys. Res.* 87, 5030.

Mattheaus W H and M L Goldstein (1982) *J. Geophys. Res.* 87, 6011.

Neugebauer M (ed) (1983) *Solar Wind 5*, NASA CP-2280.

Ovenden C R, H Shah and S J Schwartz (1983) *J. Geophys. Res.* 88, 6095.

Parker E N (1958) *Astrophys. J.* 128, 664.

Paschmann G, N Sckopke, J R Asbridge, S J Bame and J T Gosling (1980) *J. Geophys. Res.* 85, 4689.

Paschmann G, N Sckopke, S J Bame and J T Gosling (1982) *Geophys. Res. Lett.* 9, 881.

Pizzo V, R Schwenn, E Marsch, H Rosenbauer, K-H Mühlhäuser and F M Neubauer (1983) *Astrophys. J.* 271, 335.

Quest K B, D W Forslund, J U Brackbill and K Lee (1983) *Geophys. Res. Lett.* 10, 471.

Rosenbauer H (ed) (1981) *Proc. Solar Wind 4*, Max-Planck-Institut Report MPAE-W-100-81-31.

Schwartz S J (1980) *Rev. Geophys. Sp. Phys.* 18, 313.

Schwartz S J and E Marsch (1983) *J. Geophys. Res.* 88, 9919.

Schwartz S J, W C Feldman and S P Gary (1981) *J. Geophys. Res.* 86,541.

Schwartz S J, M F Thomsen and J T Gosling (1983) *J. Geophys. Res.* 88, 2039.

Schwartz S J and D Burgess (1984) *J. Geophys. Res.* 89, 2381.

Schwenn R (1983) in Neugebauer (1983) p489.

Sckopke N, G Paschmann, S J Bame, J T Gosling and C T Russell (1983) *J. Geophys. Res.* 88, 6121.

Scudder J and S Olbert (1979) *J. Geophys. Res.* 84, 2755.

Slavin J A and R E Holzer (1981) *J. Geophys. Res.* 86, 11401.

Sonnerup B U Ö (1969) *J. Geophys. Res.* 74, 1301.

Spitzer L (1956) *Physics of Fully Ionized Gases*, Interscience.

Tanaka M, C C Goodrich, D Winske and K Papadopoulos (1983) *J. Geophys. Res.* 88, 3046.

Thomsen M F, S J Schwartz and J T Gosling (1983a) *J. Geophys. Res.* 88, 7843.

Thomsen M F, J T Gosling, S J Bame, W C Feldman, G Paschmann and N Sckopke (1983b) *Geophys. Res. Lett.* 10, 1207.

Tidman D A and N A Krall (1971) *Shock Waves in Collisionless Plasmas*, J. Wiley, New York.

Tokar R L, D A Gurnett and W C Feldman (1984) *J. Geophys. Res.* 89, 105.

Winske D (1985) *J. Geophys. Res.*, in press.

Wu C S (1984) *J. Geophys. Res.* 89, 8857.

Wu C S, D Winske, Y M Zhou, S T Tsai, P Rodriguez, M Tanaka, K Papadopoulos, K Akimoto, C S Lin, M M Leroy and C C Goodrich (1984) *Sp. Sci. Rev.* 37, 63.

CHAPTER 9

PLANETARY MAGNETOSPHERES

Frances Bagenal
Space Physics Group
The Blackett Laboratory
Imperial College of Science and Technology
London SW7 2BZ

Of the six planets visited by spacecraft, four (Mercury, Earth, Jupiter and Saturn) are known to have internally-generated magnetic fields. By the end of the decade we shall know if Uranus and Neptune are similarly endowed. The regions of space influenced by these planetary magnetic fields range in size from just a layer around the planet in the case of Mercury to the giant Jovian magnetosphere which occupies a volume at least 300 times the volume of the sun. Exploration of the plasma environments near these magnetised planets and studies of their interaction with the solar wind has led to the development of a comparative theory of magnetospheres. Although the word magnetosphere strictly implies that the central body has an intrinsic magnetic field, there are several objects in the solar system which are not magnetised but nevertheless strongly interact with the solar wind. I shall therefore stretch the definition of magnetosphere to include them. I shall also consider some of the satellites that are embedded in a planetary magnetosphere and their interaction with the surrounding plasma flow.

The first section of this chapter is an outline of a comparative theory of magnetospheres. The second section is a survey of the magnetospheres of the solar system starting with the innermost planet Mercury and working radially outwards to Pluto, considering how each case fits into the generalised theory. Comparative studies of magnetospheres have only recently been developed. Since the ground work was laid down by Siscoe(1979) there have been considerable changes in our understanding of the magnetospheres of Venus, Jupiter and Saturn. Recent reviews, taking rather different approaches to the subject, are given by Stern and Ness(1982), Vasyliunas(1983a) and McNutt(1984).

9.1 COMPARATIVE THEORY OF MAGNETOSPHERES.

I shall begin with a simple description of the different ways in which objects interact with a flowing plasma. The nature of the interaction depends on the characteristics of the central body (e.g. its electrical conductivity, the presence of an atmosphere or whether the object has an internally-generated magnetic field) and the properties of the external plasma flow (e.g. the sonic and Alfvenic Mach numbers of the flow). Then I consider the possible sources of plasma and the dominant plasma motions in the case where the central object has a magnetic field of sufficient strength to form a magnetosphere.

9.1.1. Obstacles in a flowing plasma

By the time plasma from the sun reaches the planets its kinetic energy is largely bulk motion (i.e. the flow is supersonic) and, as the interplanetary magnetic field (IMF) is weak, the solar wind is super-Alfvenic (see Chapter 8). The fact that the solar wind is supersonic* means that generally there is a bow shock upstream of the obstacle. The plasma is slowed down and heated as it passes through the shock and hence the flow around the obstacle is then sub- or trans-sonic. In contrast to the solar wind, plasma flows in planetary magnetospheres span wide ranges of sonic and Alfvenic Mach numbers, a consideration that must be kept in mind when comparing the plasma interactions of different planetary satellites.

With regard to the characteristics of the obstacle, it is convenient to consider first the two extreme situations where the object is taken to either a perfect insulator or a perfect conductor. It is then necessary to consider the effects of finite conductivity since planetary bodies probably have effective conductivities in a range between these extremes.

(i) Non-conducting object

A magnetic field diffuses through an object with a timescale $\tau_d \sim \mu_0 \sigma L^2$, where L is the size and σ the electrical conductivity of the body. If this diffusive timescale is much less than the timescale for changes in the ambient magnetic field then the field passes through the body largely unperturbed. For a magnetic field 'frozen' into a plasma flowing at characteristic speed V the object sees the field change over the convective timescale $\tau_c \sim L/V$. Hence the magnetic interaction is weak for a non-conducting body with low magnetic Reynolds number ($R_m = \tau_d/\tau_c = \mu_0 \sigma VL \ll 1$).

In the case of a supersonic flow, although the magnetic field readily diffuses through the non-conducting body the plasma particles obviously cannot penetrate the body and are therefore absorbed. Because the flowing plasma is absorbed on the upstream surface there is a cavity behind the object and a wake is formed downstream as the plasma expands into the low pressure region (Figure 1a).

For subsonic flow, the interaction depends on whether the plasma is dominated by the particles' kinetic pressure or the magnetic pressure. Thus for a plasma where the ratio $\beta=(nkT)/(B^2/2\mu_0)$ is small the magnetic field controls the flow and the plasma is absorbed by the non-conductor. For high-β plasmas the subsonic flow is deflected around the object in the smooth, laminar flow of hydrodynamics, taking the weak magnetic field with it.

* with respect to the fast magnetoacoustic speed.

(ii) Perfectly-conducting object

When there is a relative motion $\underline{V}'$ between a magnetised plasma and a conducting body the Lorentz electric field $\underline{E} = - \underline{V}' \times \underline{B}$ drives a current $\underline{J}=\sigma\underline{E}$ in the object (Figure 1b). The current in the body in turn produces a perturbation in the background magnetic field. Since the magnetised plasma is highly anisotropic (with $\sigma_{\parallel} \gg \sigma_{\perp}$) the current is carried away from the flanks of the object along the magnetic field.

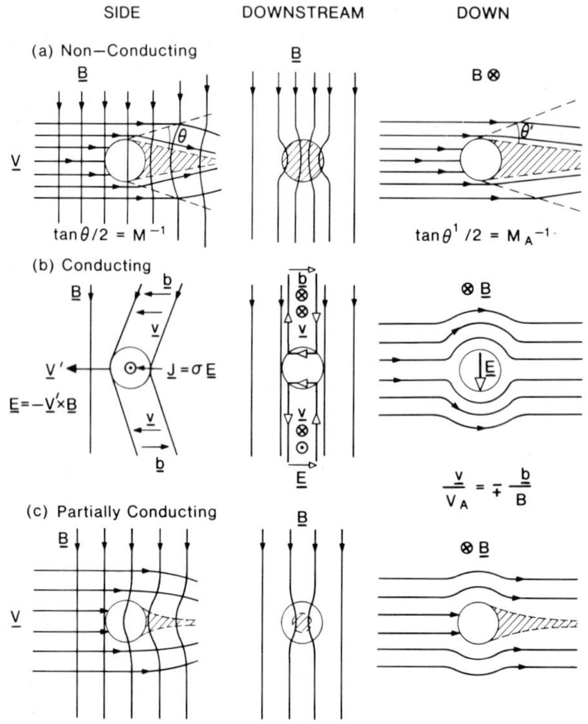

Figure 9.1 The interaction of a magnetised plasma with a (a)non-conducting; (b) conducting; (c) partially-conducting object. The hatching indicates the region of low density in the object's wake.

In the magnetohydrodynamic (MHD) regime the plasma flow is coupled to the magnetic field and hence the plasma flow is also perturbed by the conducting body. In the case of a perfect conductor (i.e. $R_m \gg 1$) the resulting motion of the plasma in the tube of magneti c flux that intersects the body exactly matches the motion of the conductor. The surrounding plasma is then deflected around the body in a manner similar to incompressible hydrodynamic flow around a cylinder with essentially no wake downstream (Drell et al.,1965). The perturbations in the magnetic field ($\underline{b}$) and plasma flow ($\underline{v}$) are an Alfven wave which propagates along the ambient magnetic field with a characteristic speed $V_A=B/\sqrt{(\mu_o \rho)}$ and satisfies the Alfven relation $\underline{v}/V_A=+\underline{b}/B$ (where the sign corresponds to propagation parallel or anti-parallel to the ambient magnetic field). One can consider these Alfven waves to be carrying the field-aligned currents.

(iii) Partially- conducting object

When the object has a finite conductivity the flow perturbation is insufficient to allow all of the surrounding plasma to flow around the body and hence some is absorbed (Figure 1c). At the same time a finite amount of the kinetic energy of the plasma flow is dissipated in the body as Joule heating. Thus finite resistivity leads to deceleration of the plasma flow near the object and the magnetic field being "hung-up" in the stagnant flow. Since the magnetic field far from the object continues to be convected in the unperturbed flow the magnetic field lines become bent or "draped" over the object to form a magnetic tail or wake downstream.

(iv) Object with an atmosphere

On the dayside of all objects having an atmosphere the sun's ultraviolet emissions ionise some of the neutral atoms. For objects that are embedded in a dense plasma, ionisation by particle impact may also be significant. In any case, the degree of ionisation determines the conductivity of the upper atmosphere (ionosphere) and thus affects the nature of the interaction of the object with the plasma in which it is immersed. In the case of a dense ionosphere ($\sigma \rightarrow \infty$) the solar wind magnetic field is excluded and the flow is diverted around the flanks of the body. The boundary between the ionosphere and the surrounding

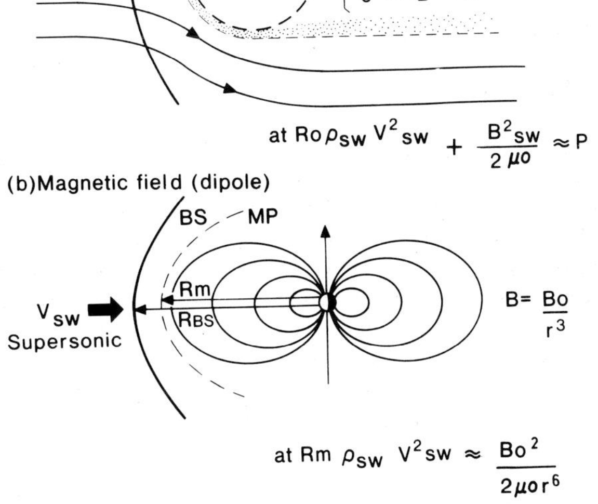

Figure 9.2 The interaction of a magnetised plasma with an object posessing (a) an atmosphere of surface pressure P_o and a scale height H; (b) a magnetic field. The bowshock(BS) and magnetopause(MP) are shown.

(a) Atmosphere

$P \sim P_o \exp \frac{-r}{H}$
$\sigma = 0 \quad B \rightarrow B_{sw}$
$\sigma \rightarrow \infty \quad B \rightarrow 0$

at $R_o \quad \rho_{sw} V^2_{sw} + \frac{B^2_{sw}}{2\mu_0} \approx P$

(b) Magnetic field (dipole)

$B = \frac{B_o}{r^3}$

at $R_m \quad \rho_{sw} V^2_{sw} \approx \frac{B_o^2}{2\mu_0 r^6}$

plasma, the ionopause, is located (at say R_o) where the combined magnetic and ram pressure of the external plasma is balanced by the particle pressure (P) in the ionosphere (Figure 2a). When the ionisation is weak ($\sigma \to 0$) the magnetic field and plasma flow are dragged through the resistive ionosphere causing a substantial downstream wake. We must further consider the consequences of the ionisation of any neutral material extending out into the streaming plasma. On ionisation the particle 'sees' the Lorentz electric field due to its motion relative to the plasma and is accelerated up to the ambient flow. The momentum gained by the newly-created ions comes from the surrounding plasma which correspondingly loses momentum. This effect is called 'mass loading' and contributes to the draping of field lines over an object with a substantial atmosphere (e.g. comets, Venus and Titan).

(v) Magnetised object

Well before Biermann(1957) provided cometary evidence of a persistent solar wind, Chapman and Ferraro (1931) considered how a strongly magnetised body (the Earth) would deflect a flow of particles from the Sun. They proposed that a dipolar magnetic field (of strength B_0 at the planet's equatorial radius R_p) would stand off the flow to a distance R_{cf} where the ram pressure of the flow balances the Maxwell stress of the magnetic field

$$\rho_{sw} V_{sw}^2 = \frac{1}{2} \frac{B_0^2}{\mu_0} \left(\frac{R_p}{R_{cf}} \right)^6 . \quad (9.1)$$

Hence

$$\frac{R_{cf}}{R_p} = \left(\frac{\frac{1}{2} B_0^2}{\mu_o \rho_{sw} V_{sw}^2} \right)^{1/6} . \quad (9.2)$$

This not only assumes that $B_{sw} \lll B_0 (R_p/R_{cf})^3$ (which is generally reasonable but also that the particle pressure inside the magnetosphere is negligible. In reality, the observed stand-off distances, R_m are between 1 and 2 times R_{cf}.

Thus, to first approximation, the magnetic field of the object deflects the plasma flow around it, carving out a cavity in the solar wind. The boundary between the magnetosphere and the solar wind is called the magnetopause (MP) and the layer of deflected solar wind flow behind the bowshock(BS) is called the magnetosheath. The solar wind generally pulls out part of the planetary magnetic field into a long cylindrical magnetotail, extending far downstream behind the planet.

9.1.2. Plasma Sources

It is rather misleading to describe the region dominated by a planetary magnetic field as an empty cavity from which the solar wind is excluded. Magnetospheres contain considerable amounts of plasma which have "leaked in" from various sources (Figure 3). Firstly, the magnetopause is not entirely "plasma-tight". Whenever the IMF has a component anti-parallel to the planetary field, magnetic reconnection is likely to occur and solar wind plasma will leak into the magnetosphere across the magnetopause (see Chapter 5). Secondly, although ionospheric plasma is generally cold and gravitationally bound to the planet, a small fraction has sufficient energy to escape up magnetic field lines and into the magnetosphere. Thirdly, the interaction of magnetospheric plasma with any natural satellites that are embedded in the magnetosphere can generate significant quantities of plasma. Possible mechanisms for these satellite sources are (a)ionisation of the outermost layers of any satellite atmosphere; (b) energetic particle sputtering of the satellite surface producing less energetic ions directly or an extensive cloud of neutral atoms which are eventually ionised far from the satellite.

Figure 9.3 Sources of magnetospheric plasma.

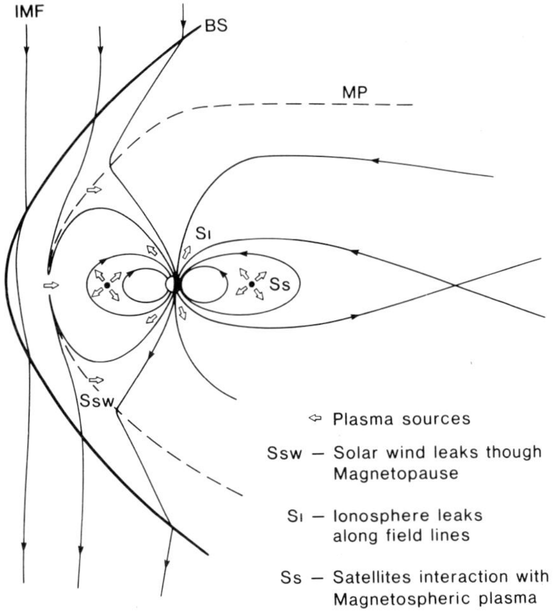

9.1.3 Magnetospheric Flows

The two largest sources of momentum in planetary magnetospheres are the planet's rotation and the solar wind. The nature of any large scale plasma circulation (often called convection) in the magnetosphere depends on which momentum source is tapped. For magnetospheric plasma to rotate with the central body there are two basic requirements: (i) the planet's neutral atmosphere must corotate with the planet: (ii) the neutral atmosphere must be closely coupled to the ionosphere by collisions (i.e. the electrical conductivity of the ionosphere perpendicular to the field, $\sigma_{\perp}$ is large). In a corotating ionosphere (with velocity $\underline{V}^i$) any perpendicular (horizontal) currents are given by Ohm's law

$$\underline{J}_{\perp}^{i} = \sigma_{\perp}^{i}(\underline{E}^{i} + \underline{V}^{i} \times \underline{B}) \,. \tag{9.3}$$

Just above the ionosphere the perpendicular conductivity in the (collision-free) magnetosphere $\sigma_{\perp}^{m}$ is essentially zero. Thus

$$\underline{J}_{\perp}^{m} = \sigma_{\perp}^{m}(\underline{E}^{m} + \underline{V}^{m} \times \underline{B}) = 0 \tag{9.4}$$

and

$$\underline{E}^{m} = - \underline{V}^{m} \times \underline{B} \,. \tag{9.5}$$

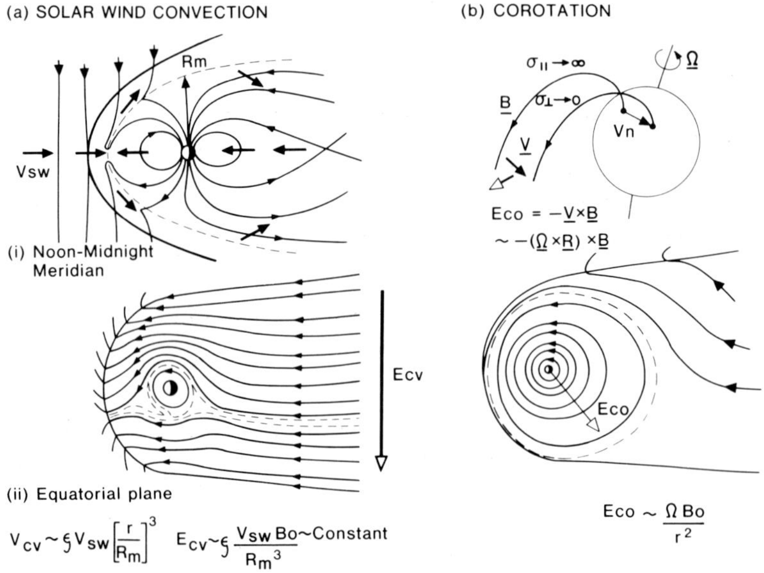

Figure 9.4 Large scale magnetospheric circulation driven by (a) the solar wind; (b) planetary rotation. The magnetic field directions are given corresponding to (a) Earth and (b) Jupiter.

Because the plasma particles are far more mobile in the direction of the local magnetic field the parallel conductivity, $\sigma_{\parallel}{}^{m}$ is large and the field lines can be considered to be equipotentials. Thus the electric field in the magnetosphere can be mapped into the ionosphere. Since the ionosphere is relatively thin the electric field E^{m} obtained by evaluating (9.5) just above the ionosphere is the same as E^{i}. Substituting for $\underline{E}$ in (9.3) gives

$$\underline{J}_{\perp}^{i} = \sigma_{\perp}^{i}(\underline{V}^{i} - \underline{V}^{m}) \times \underline{B} \,. \tag{9.6}$$

The condition for corotation of the magnetospheric plasma is that J^{i}/σ^{i} is sufficiently small that

$$\underline{V}^m = \underline{V}^i = \underline{\Omega} \times \underline{r} \ . \qquad (9.7)$$

For a dipolar magnetic field the corotational electric field is therefore radial with magnitude $E_{co} \sim \Omega B_0/r^2$ (Figure 4b). It is clear that large magnetospheric conductivities facilitate corotation. The large σ^m also means that any currents in the magnetosphere (J^m) that result from mechanical stresses on the plasma (due to the solar wind interaction, for example) are directly coupled by field-aligned (Birkeland) currents to the ionosphere (discussed in Chapter 2). The ionospheric current density $\underline{J}^i$ is therefore proportional to $\underline{J}^m$ which is governed by the momentum equation in the magnetosphere

$$\rho \frac{d\underline{V}}{dt} + \nabla P = \underline{J}^m \times \underline{B} \ . \qquad (9.8)$$

Thus corotation breaks down when mechanical stresses on the magnetospheric plasma drive ionospheric currents which are sufficiently large that J^i/σ^i becomes significant.

Where the magnetospheric plasma may be coupled to the planet's rotation via the ionosphere, the momentum of the solar wind is harnessed by processes occurring near the magnetopause. Although theories of a viscous drag between the solar wind and magnetospheric plasma rivalled theories of magnetic reconnection for many years, there is now evidence that the merging of the IMF with the planetary field on the dayside magnetopause is the dominant coupling process. The resulting convection pattern is caused by the reconnected magnetospheric flux tubes being pulled by the solar wind over the poles and back into an extended magnetotail (Figure 4a). The plasma then drifts towards the equatorial plane and eventually returns in a sunward flow to the dayside magnetopause. Cowley has described in Chapter 5 how the reconnection rate can be considered to be equivalent to an electric field, E_{cv} which is assumed to be roughly constant with time

$$\underline{E}_{cv} = -\underline{V}_{cv} \times \underline{B} \sim \text{constant} \ , \qquad (9.9)$$

where V_{cv} is the corresponding convection speed. V_{cv} can then be estimated from

$$V_{cv} \sim \xi \, V_{sw}(r/R_M)^3 \ , \qquad (9.10)$$

where R_M is the magnetopause distance and ξ is the efficiency of the reconnection process in harnessing the solar wind momentum ($\sim$ 0.1 for the Earth).

An indication of whether magnetospheric circulation is driven by the solar wind or the planet's rotation can be found by comparing the corresponding electric fields. Since $E_{co} \propto r^{-2}$ and $E_{cv} \propto r^3$ it seems reasonable to expect that corotation would dominate close to the planet while solar wind convection would dominate outside a critical distance R_c. Thus the fraction of the magnetosphere that corotates is

$$\frac{R_c}{R_M} = \left(\frac{R_M \, \Omega}{\xi \, V_{sw}} \right)^{\frac{1}{2}} . \qquad (9.11)$$

This simply says that magnetospheres of rapidly rotating planets with strong magnetic fields will be dominated by rotation while the solar wind will control the plasma flows in smaller magnetospheres of slowly rotating planets.

Now that we have covered the basic ideas underlying generalised magnetospheric theories let us consider the various planets in the solar system and see how well these simple ideas match observations.

9.2 PLANETARY MAGNETOSPHERES

In the preceding discussion magnetospheres were characterised using a few key properties of the central object and the plasma flow in which it is embedded. Figure 5 shows values of the corresponding parameters for the nine planets in the solar system. In the top panel there is a logarithmic plot of the size of the planetary system against distance from the sun. The regular spacing illustrates Bode's empirical law.

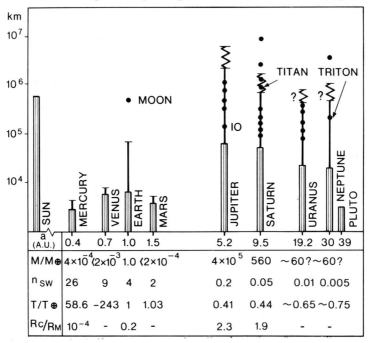

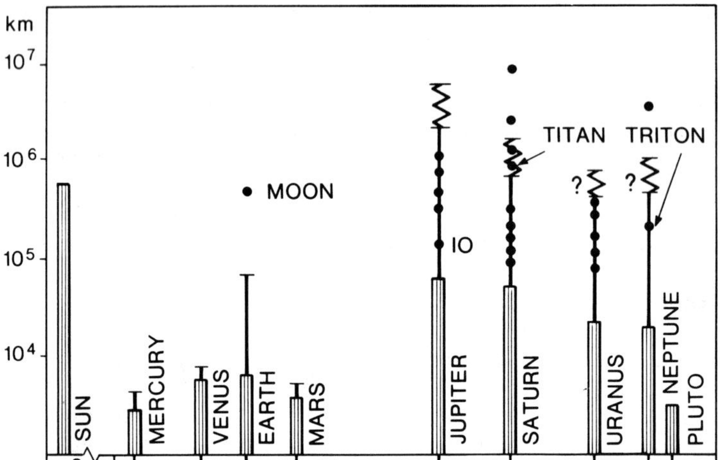

	Mercury	Venus	Earth	Mars	Jupiter	Saturn	Uranus	Neptune	Pluto
a (A.U.)	0.4	0.7	1.0	1.5	5.2	9.5	19.2	30	39
$M/M_\oplus$	4×10^{-4}	$<2\times10^{-3}$	1.0	$<2\times10^{-4}$	4×10^{5}	560	~60?	~60?	
n_{sw}	26	9	4	2	0.2	0.05	0.01	0.005	
$T/T_\oplus$	58.6	-243	1	1.03	0.41	0.44	~0.65	~0.75	
R_c/R_M	10^{-4}	-	0.2	-	2.3	1.9	-	-	

Figure 9.5 Planetary magnetospheres (after Siscoe, 1979). Each hatched bar corresponds to the planet's radius. The barred line shows the extent (variable for the outer planets) of the planet's magnetosphere. Note the scales are logarithmic. The locations of the major satellites are indicated with dots. An astronomical unit (A.U.) is 1.5×10^8 km. The magnetic moment of the Earth, $M_\oplus$ is 8×10^{25} Gauss cm^3 (8×10^{15} Tesla m^3). The Earth's rotation period, $T_\oplus$ is 24 hours.

It is traditional to regard the Asteroid belt as dividing the smaller inner planets from the outer giants (Pluto is the awkward exception). This crude division also holds more or less for the size of the planets' magnetospheres. Of the inner planets Earth has by far the strongest magnetic field (moment $M_{\oplus} = 8\times10^{15}$Tesla m^3) with the only other positive measurement being that of Mercury's much weaker field. Upper limits of $2\times10^{-3}M_{\oplus}$ and $2\times10^{-4}M_{\oplus}$ have been placed on the intrinsic magnetic fields of Venus and Mars respectively. Thus the plasma environments around these inner planets are strongly influenced by the solar wind and only the Earth has an appreciable magnetosphere.

In the outer solar system a combination of strong planetary magnetic fields and a more tenuous solar wind produces very large magnetospheres. In fact the magnetosphere of Jupiter is the largest 'object' in the solar system, occupying over 300 times the volume of the sun. At present it is not known if Uranus, Neptune or Pluto have intrinsic magnetic fields and their inclusion in the group of giant magnetospheres is speculative (and probably wrong in the case of Pluto). The magnetospheres of Jupiter and Saturn are sufficiently large to envelop most of their natural satellites. These satellites in turn provide most of the magnetospheric plasma.

With the exceptions of Mercury and Venus the rotation periods of the planets do not differ greatly from an Earth day. Nevertheless, when one considers the effect of planetary rotation on magnetospheric circulation relative to the influence of the solar wind, it is clear from the values of R_c/R_M in Figure 5 that rotation dominates throughout the giant magneospheres, for only $\sim$ 20% of the Earth's and to a negligible extent in the rest.

With these generalizations in mind we shall now embark on a tour of the solar system and consider each planetary magnetosphere in turn.

9.2.1 Mercury

Mariner 10 was the only spacecraft to visit Mercury, making three flybys between March 1974 and March 1975. Two of these were close to the nightside of the planet where the magnetospheric signatures of a bowshock, magnetopause and magnetotail were detected by the Mariner 10 magnetometer and particle detectors (reviewed by Ness,1979). Nevertheless the magnetic field was found to be weak ($M_{Me} \sim 4\times10^{-4}M_{\oplus}$) and the magnetosphere small with a stand-off (noon magnetopause) distance $R_M \sim 1.3 \pm 0.2\ R_{Me}$, (Figure 6). Siscoe and Christopher (1975) calculated the variation in the stand-off distance with expected changes in the solar wind ram pressure and predicted that the solar wind would directly impinge on the surface of Mercury a little under 1% of the time. With a negligible atmosphere and no satellites the solar wind is the only source of plasma for the Mercury's magnetosphere. However there are no stably-trapped particles; scaling from the Earth's magnetosphere (discussed in section 9.2.4) would put any plasmasphere well inside the body of the planet. Scaling from the Earth also predictsthat the region on Mercury's dayside where solar wind is funnelled into the magnetosphere (the 'polar cusp') to extend down to 50-60° latitude (compared with $\sim 77°$ for the Earth's cusp). Similarly, on the nightside, Mercury is expected to have a larger auroral region (extending to 25-35° compared with 70°for the Earth) where material from the magnetotail precipitates on to the surface of the planet. Mercury's weaker magnetic field also means the planet is not shielded from energetic solar particles or cosmic rays.

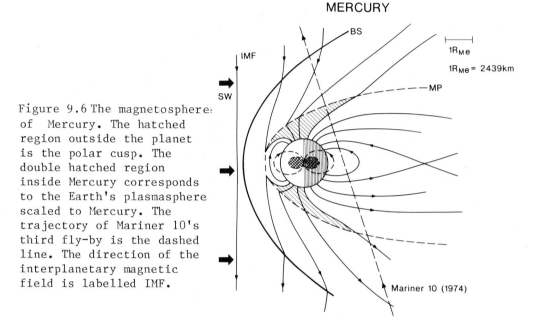

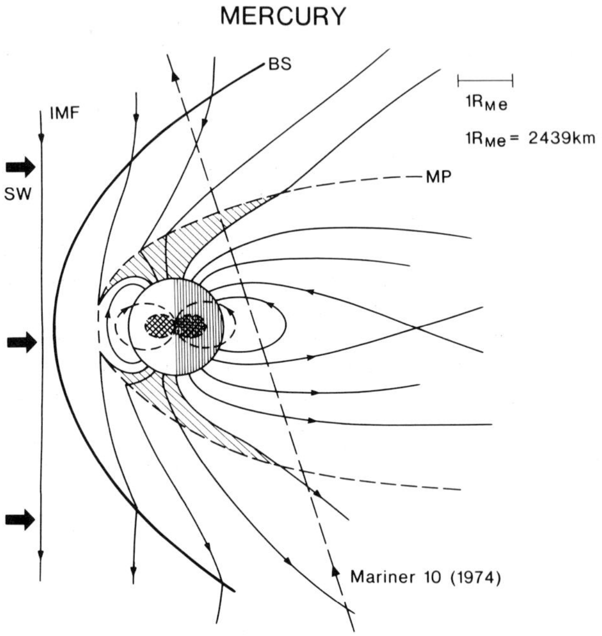

Figure 9.6 The magnetosphere of Mercury. The hatched region outside the planet is the polar cusp. The double hatched region inside Mercury corresponds to the Earth's plasmasphere scaled to Mercury. The trajectory of Mariner 10's third fly-by is the dashed line. The direction of the interplanetary magnetic field is labelled IMF.

Moreover, in Mercury's magnetotail the particle instruments on Mariner 10 detected 4 bursts of energetic particles which lasted for about 1 minute, separated by about 6 minutes. Siscoe et al.(1975) proposed that these bursts are evidence of nightside reconnection events (substorms), the timescale being about 1/20 that at the Earth because of the small spatial scale of Mercury's magnetosphere and the lack of an ionosphere to dampen the motion of magnetic fluxtubes. Thus, although Mercury's magnetosphere is very small and contains no trapped plasma populations its very simplicity could well make it more suitable for investigating nightside magnetic reconnction than the Earth's magnetosphere.

9.2.2 Venus

Although early missions to Venus (Venera 4 and 6, Mariner 5 and 10) detected and later extensively mapped (Venera 9 and 10) an upstream bowshock, the Pioneer Venus Orbiter (PVO) was the first spacecraft to explore deep into the Venusian magnetosphere with a full complement of particle and field instrumentation. From the random orientation of the magnetic field measured by the PVO magnetometer behind the planet and the lack of correlation between magnetic signatures and surface features, Russell et al.(1980) concluded that any dipole moment of Venus is less than 5 x 10^{-5} that of the Earth ($M_V = 0 \pm 5.5 \times 10^{11}$ Tesla m^3). Thus the 'magnetosphere' of Venus is the result of the interaction of the solar wind with the planet's atmosphere rather than with an intrinsic magnetic field (reviewed and compared with cometary interactions by Russell et al.,1982).

The fact that the bowshock upstream of Venus is relatively weaker than the terrestrial bowshock suggests that some ($\leqslant$1%) of the solar wind flow is absorbed rather than deflected by the planet. Behind the bowshock the subsonic magnetosheath flow is compressed in front of the planet (Figure 7). This 'squeezing' of the tubes of interplanetary magnetic flux causes the constituent plasma to evacuate along the field and the magnetic pressure to increase. The degree of plasma depletion depends on the relative timescales for plasma evacuation compared with convection of the flux tube. The flow comes close to stagnation ($\leqslant$ 20 km s^{-1}) at the sub-solar point which should be compared with a thermal speed of about 100 km s^{-1} for shocked solar wind protons. Thus the dynamic pressure of the solar wind is essentially replaced by the Maxwell stress in the compressed flux tubes. With the upper atmosphere of Venus being strongly ionised by solar ultraviolet radiation a boundary (the ionopause) is formed between the two plasma populations at the altitude where the thermal pressure of the ionosphere balances the magnetic pressure of the compressed solar wind (i.e. at R_o $(nkT)_{ionoshpere} \simeq \frac{1}{2}B^2/\mu_o \simeq (\rho V^2)_{solar\ wind}$). The peak ionospheric

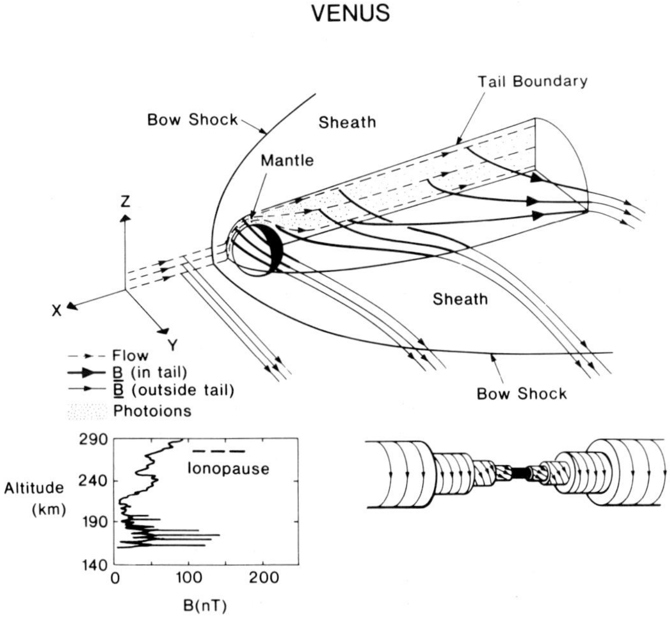

Figure 9.7 (Top) The interaction of the solar wind with Venus (after Saunders and Russell,1985). (Bottom left) An example of the magnetic field measured by PVO inside Venus' ionosphere (Phillips et al.,1984). The small regions of strong magnetic field at low altitudes are thought to correspond to magnetic flux ropes, illustrated bottom right (from Russell and Elphic, 1979).

pressure is large enough to stand off the average solar wind to an ionopause height of 250-300 km. Therefore, with exception of rare occasions when the solar wind dynamic pressure exceeds the peak ionospheric pressure ($\leqslant$ 15% of the time), one would expect the IMF to be excluded from Venus by its conducting ionosphere and the solar wind to be deflected around the planet.

Although PVO indeed measured average magnetic fields close to zero in the ionosphere, the spacecraft frequently encountered small regions of high magnetic fields even under quiet solar wind conditions. The magnetic field in these regions tended to be twisted up into 'fluxropes' as illustrated in Figure 7 (Russell and Elphic, 1979). The origins of these flux ropes are unknown though it seems plausible that small bundles of magnetic flux are either "sheared-away" (by the Kelvin-Helmholtz instability) at the ionopause or just dragged through the ionosphere by the magnetic tension along the field lines exerted ultimately by the solar wind.

The neutral atmosphere of Venus extends beyond the ionopause into the magnetosheath. Photoionisation (and/or charge exchange with the solar wind protons in the sheath) produces H^+, H_2^+, H_e^+ and O^+ ions which are 'picked-up' in the sheath flow and carried downstream past the planet. Moreover, the PVO plasma data showed evidence of blobs of ionospheric plasma being 'scavenged' in some manner and carried away by the sheath flow. The momentum gained by these new or scavenged ions must come from the background plasma flow which is correspondingly decelerated. Thus the interplanetary field lines are 'caught up' and draped over the planet forming a wake or tail downstream. Similar to the Earth's magnetotail, the smaller magnetotail of Venus has two lobes separated by a current sheet. The orientation of the Venus current sheet, however, is determined solely by the direction of the IMF rather than a planetary field (Saunders and Russell, 1985).

Thus the Venus magnetosphere is confined to a relatively small ($\leqslant$ 0.3 R_V) boundary layer where the solar wind interacts with the Venusian ionosphere. In the process, plasma is added to the deflected solar wind flow and the resulting mass-loaded flux tubes are then draped over the planet to form a thin tail ($\sim 4\ R_V$ across) downstream.

9.2.3 Earth

The Earth's magnetosphere has been explored by many spacecraft over the past two decades. Although the general morphology is well-understood, the present task is to investigate details of the physical processes occurring in different regions (see Chapter 2). In the hierarchy of planetary magnetospheres the Earth comes between the small, solar-wind dominated magnetospheres of the inner planets and the large, rotation-dominated magnetospheres of the giant planets.

Balancing the Earth's dipolar magnetic field pressure with the ram pressure of the solar wind puts the Chapman-Ferraro standoff distance R_{cf} at 8.8 to 10.4 R_E. The fact that the average noon position of the Earth's magnetopause is 10.8 R_E suggests that a finite pressure of magnetospheric particles should also be included in the pressure balance. These particles have two sources: plasma escaping from the

Earth's ionosphere populates the inner region of the magnetosphere (plasmasphere) while the solar wind leaks in across the magnetopause on reconnected field lines (see Chapter 5) and populates the outer region of the dayside magnetosphere and the magnetotail(Figure 8). The energetic particle population in the Van Allen radiation belts consists of protons and helium ions (presumably from the solar wind) mixed with oxygen ions from the ionosphere. The relative proportion of radiation belt particles with solar wind or ionospheric origins is not known and probably varies with solar wind conditions.

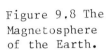

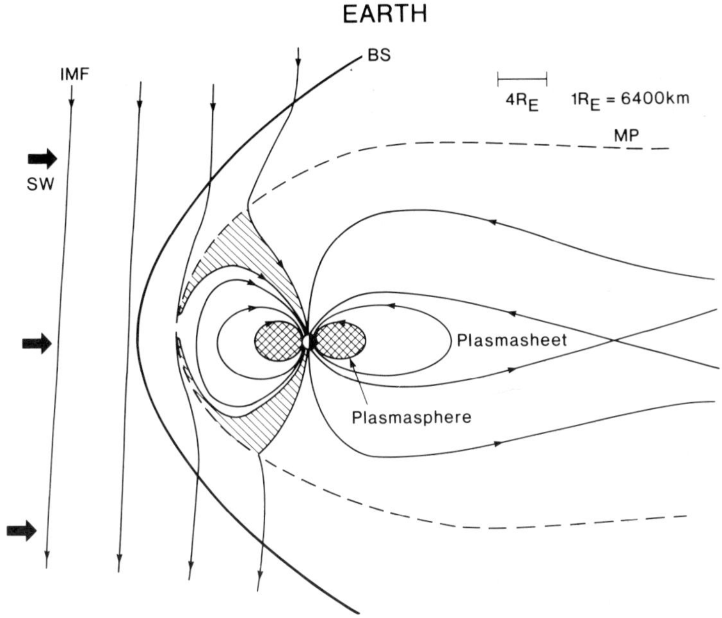

Figure 9.8 The Magnetosphere of the Earth.

The low energy plasma inside the plasmasphere largely corotates with the Earth. The plasmapause is the boundary (at $\sim 4R_E$) where the magnitude of electric field associated with reconnection exceeds the corotational electric field. Thus outside the plasmasphere the plasma circulation is driven by the solar wind (Chapter 5).

The Moon, $66R_E$ from the Earth, spends less than 10% of its orbit inside the Earth's magnetosphere. The early spacecraft Explorers 33 and 35 carried magnetometers and plasma detectors to investigate the interaction of the solar wind with the Moon. First they discovered there was no bow shock upstream of the Moon. When the spacecraft later passed behind the Moon, downstream of the object, the perturbations in magnetic field (Figure 9), the plasma density rarefaction and deflection of the solar wind flow,were found to be consistent with supersonic flow around a non-conducting body (Taylor et al.,1968). Models of the solar wind interaction with a non-conducting body (Figure 1a) such as the Moon have been developed by Spreiter et al.(1970).

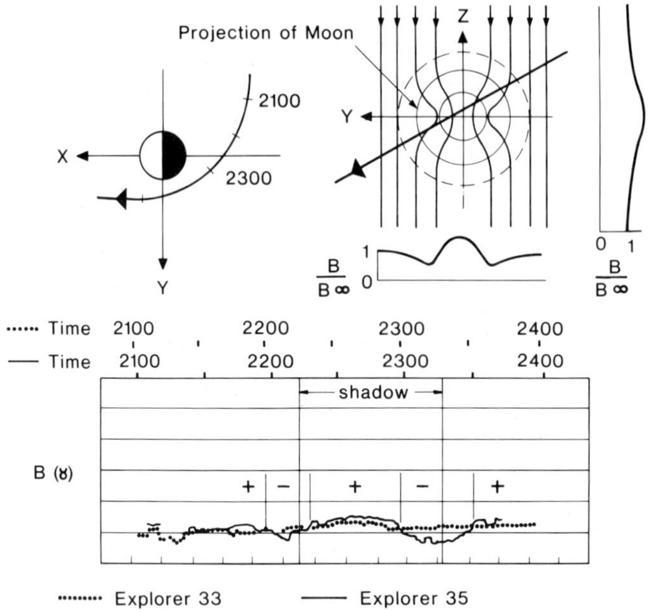

Figure 9.9 The interaction of the solar wind with the Moon. Top right shows the magnetic field perturbations that would be expected if the Moon is a non-conductor. The arrowed line is the spacecraft trajectory. (Below) The magnetic field signatures observed by Explorers 33 and 35 in 1967 (from Spreiter et al.,1970).

9.2.4 Mars

Although Mariner 4 indicated the presence of a bow shock as early as 1965, there has been very limited exploration of the solar wind interaction with Mars. Soviet spacecraft Mars 2,3 and 5 carried magnetometers and confirmed the presence of a bow shock (at 1.5±0.15 R_{Ma}). Unfortunately, none of these spacecraft reached altitudes less than 1000km and the interpretation of the scant data sets has become a controversial subject. While some put forward evidence of a small intrinsic magnetic field (Dolginov,1978; Slavin and Holzer,1982) others claim there is no need to invoke entry into a Martian magnetosphere to explain the data (Russell,1978; Russell et al.,1984). The most optimistic upper limit on the planetary field (2×10^{-3} $M_\oplus$) was given by Dolvinov (1978) by assigning a wake boundary as a magnetopause. On the other hand Russell(1978) suggested that the wake boundary is due purely to draping of the IMF and brought the limit down to ($2.5 \times 10^{-5} M_\oplus$). More recently, Slavin and Holzer (1982) have suggested an intermediate value of $1.8 \pm 0.8 \times 10^{-4}$.

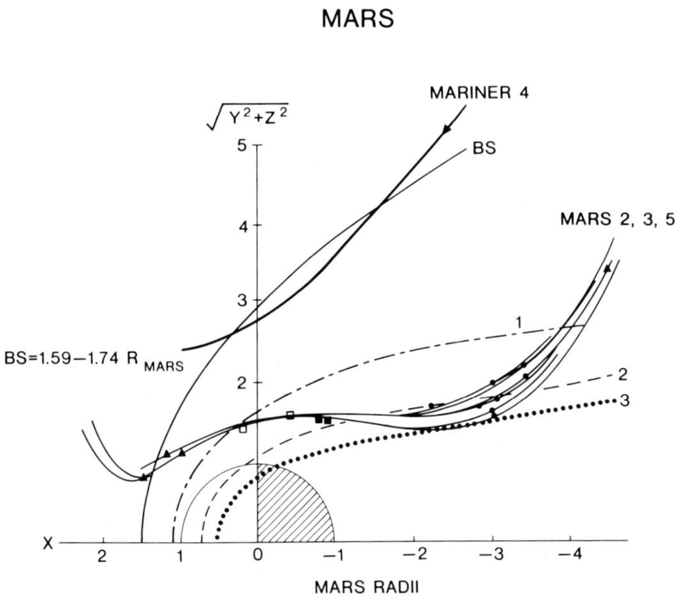

Figure 9.10 The interaction of the solar wind with Mars. The locations of Mars 5 'magnetospause' crossings, as identified from magnetometer data, are shown by squares and dots. The triangles are bowshock crossings. Line 1 is the maganetopause location scaled from the Earth to be consistent with the observed bowshock location. Line 2 is the best fit to the magnetopause crossings using a scaled terrestial magnetopause. Line 3 is the largest scaled terrestial maganetopause consistent with the observations under the assumption that none of the magnetopause encounters are in fact encounters with the planetary magnetic field. (After Russell, 1978).

In any case, the interaction with the solar wind will be confined to a small region very close to planet. The atmosphere of Mars is much less dense than that of Venus and therefore it is unlikely that the pressure of ionospheric material could stand off the solar wind. Thus it is expected that the majority of the solar wind slips around a conducting outer layer with limited entrainment of ionospheric plasma and the IMF is weakly draped over the planet in a small downstream wake.

9.2.5 Jupiter

When Burke and Franklin (1955) discovered, somewhat by accident, that Jupiter is a source of radio emission, it was soon realised that this radio emission must come from energetic charged particles in a strong magnetic field. This remarkable discovery came before Van Allen's detection of the Earth's radiation belts and the in situ verification of the solar wind (Neugerbauer and Snyder, 1962). A more puzzling discovery came a few years later when Bigg (1964) revealed that the low frequency (decametric) component of the Jovian radio emission was influenced by Io, the innermost of the four large Galilean satellites.

The radio emission from Jupiter provided information about the planet's magnetic field; the high frequency cutoff indicated the great strength of the Jovian magnetic field ($M_J \simeq 4 \times 10^5 M_\oplus$). Moreover, from the periodic variations in the high frequency (decimentric) component of the radio emission it was possible to determine Jupiter's rotation period accurately and the 9.6° tilt of the magnetic dipole with respect to the planet's rotation axis. The low frequency emission was assumed to be synchrotron radiation from electrons with energies of ∿ 10 MeV that gyrate around dipolar magnetic field lines at a radial distance of a few times the radius of Jupiter ($1\ R_J = 71400$ km). This basic picture of a strong magnetic field trapping a large, energetic particle population was confirmed by the Pioneer 10 and 11 spacecraft which reached Jupiter in December 1973 and 1974 respectively. The Pioneers also revealed that farther from the planet the magnetic field is considerably stretched out so that the Jovian magnetosphere is shaped more like a disc than a sphere (Figure 11). Although the large size ($R_M = 50{-}100\ R_J$) and radial distension implied the presence of a substantial amount of plasma at lower energies, the Pioneer plasma detector provided little information of the thermal population. Nevertheless, the theorists had already come out strongly in favour of a magnetosphere dominated by the planet's rotation (Gledhill,1967; Melrose,1967; Brice and Ioannidis,1970).

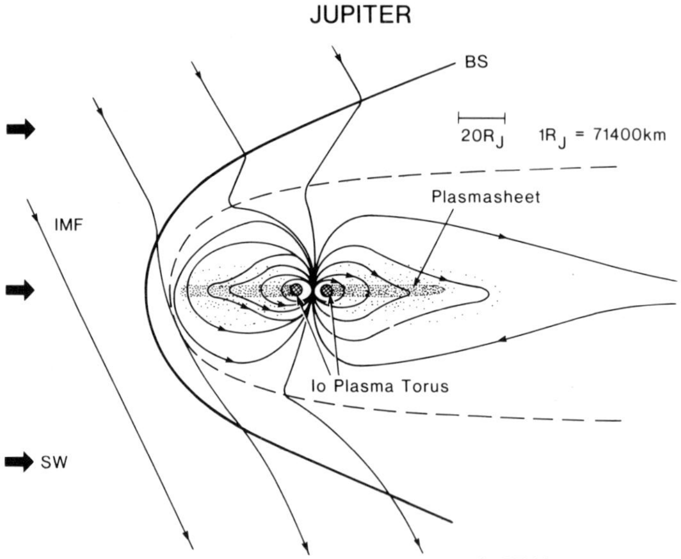

Figure 9,11 The magnetosphere of Jupiter.

A few months before the Pioneer 10 encounter, Brown (1974) detected optical (d-line) emission from a cloud of neutral sodium atoms in the vicinity of Io using a ground-based telescope. The first direct evidence of the presence of ionised material at low energies near Jupiter came with the discovery by Kupo et al. (1976) of optical emission S^+ ions. Brown (1976), borrowing techniques from studies of more remote astronomical gaseous nebulae, concluded that the S^+ emission came from a dense ring of cold plasma inside the orbit of Io. The Voyager 1 and 2 spacecraft confirmed that Io is the major source of plasma in the Jovian magnetosphere, a fact that seemed less surprising when the Voyager cameras revealed the satellite's active volcanos.

(i) The Io-plasma interaction

Bigg's observation that Io modulates the intensity of the Jovian decametric radio emission initiated many early models of the satellite's interaction with the magnetospheric plasma (Marshall and Libby,1967; Piddington and Drake,1968; Goldreich and Lynden-Bell,1969; Gurnett,1972; Geortz and Deift,1973). These early studies assumed Io was a perfect conductor and the ambient plasma density very low. They examined how Io's motion in the planetary magnetic field might cause the satellite to act as a unipolar generator and investigated the possibility that large field-aligned currents might directly connect the satellite to the planet. Following Drell's description of a large conducting body generating Alfven waves as it moves through a magnetic field (Drell et al.,1965), Marshall and Libby (1967) were the first to propose that Io might generate large amplitude Alfven waves that propagate along the magnetic field to the ionosphere ofJupiter where the radio bursts are triggered. However, in applying the theory to Io, the early theorists were hampered by the fact that very little was known about the properties of Io and the surrounding plasma.

The large perturbations of the magnetic field that were measured in the vicinity of Io when Voyager 1 passed beneath the satellite confirmed the theoretical expectations of a strong interaction between Io and the magnetospheric plasma. Indeed, further analysis of the Voyager 1 observations indicated that an Alfvenic disturbance was radiated by Io, carrying a $\sim 10^6$ amp field-aligned current towards the ionosphere of Jupiter (Belcher et al.,1981). Moreover, the observed high plasma densities implied that the propagation speed of Alfven waves is small in the torus. This means that by the time an Alfven wave has travelled from Io to the ionosphere (where it is reflected) and back, Io has moved along its orbit so that the field-aligned currents do not form a closed loop as first suggested by Goldreich and Lynden-Bell(1969) but rather form open-ended Alfven wings similar to Drell's model. Neubauer(1980), Geortz(1980) and Southwood et al.(1980) have developed theoretical models of Io's interaction with the magnetospheric plasma which incorporate some of the basic Voyager results and involve the generation of large amplitude Alfven waves.

Although it seems that to first approximation Io is a good conductor, in detail the Io-plasma interaction is complicated; for example, one should consider the presence of a neutral atmosphere, local ionisation of neutral material by the corotating plasma and the sputtering of energetic charged-particles on the satellite's surface which in turn generates more plasma.

Concerning the Io-modulation of the decametric radio emission, there have been attempts to relate features of the emission to models of the Io-generated Alfven waves (Gurnett and Goertz,1981; Bagenal,1983) but it has not yet been proved that there is a causal relationship between the two phenomena.

(ii) The Io plasma torus

As the Voyager spacecraft approached Jupiter the ultraviolet spectrometer detected powerful ($2\text{-}3\times10^{13}$ watts) emission from a

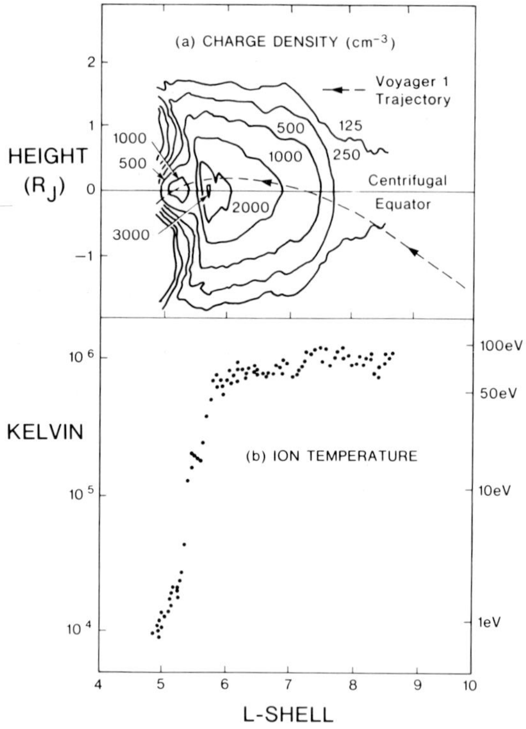

Figure 9.12 Voyager 1 measurements of positive ions in the Io plasma torus. (a) Contours of positive charge density (equal to the electron density). (b) Ion temperature.

toroidal region encompassing the orbit of Io. The observed spectrum indicated the emission was coming from ions of sulphur (S^+, S^{2+}, S^{3+}) and oxygen (O^+, O^{2+}). When the Voyager spacecraft later flew through the plasma torus, the Plasma Science instrument made in situ measurements of both the electrons and the various positive ionic species. From the Voyager plama measurements (Figure 12) it is clear that the Io plasma torus is divided into two distinct regions with a sharp boundary at 5.7 R_J, inside Io's orbit (5.9R_J)(Bagenal and Sullivan,1981;but see Bagenal et al.,1985). It is the large outer region of warmer (∿80eV) plasma that produces the UV emission while the optical emissions come mainly from the much colder (∿1eV) plasma inside 5.7 R_J.(The spectrophotometry of the Io plasma torus is reviewed by Brown et al.,1983). It is estimated that at least 10^{28} ions must be produced by Io per second to maintain the plasma torus but the actual source mechanism is not known. Two possibilities are that (i) plasma is produced directly in the interaction between the satellite and the magnetospheric plasma; or (ii) neutral material is sputtered off the satellite's surface and escapes to form a large neutral cloud that is later ionised. The tenuous atmosphere of Io and the lack of enhancd UV emission near Io limit the source strength from the first mechanism and recent detection of emission from extended neutral clouds of oxygen (Brown, 1981) and sulphur (Durrance et al., 1983) favour the second mechanism.

In either case, when the neutrals are ionised they experience a Lorentz force due to their motion relative to the local magnetic field; this force causes the ions to gyrate about the magnetic field at a speed equal to the magnitude of the neutral's initial velocity relative to the surrounding plasma (Figure 13). The ion is accelerated until its guiding centre motion matches the plasma rest frame, corotating with Jupiter. Because a particle's gyro-radius (R_g) is mass-dependent, the new ion and its electron are separated after ionisation. Hence there is a radial current due to the ions being 'picked-up' by the magnetic field. This radial current across the torus is linked by field-aligned (Birkeland) currents to the ionosphere of Jupiter where the $\underline{J}\times\underline{B}$ force is in the opposite direction to the planet's rotation. Thus the planet's essentially-limitless source of angular momentum is tapped electrodynamically by the newly ionised plasma.

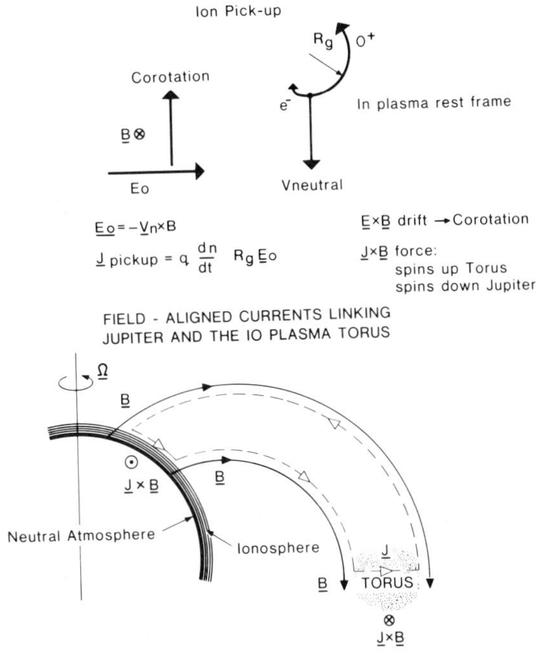

Figure 9.13 Ion pick-up ion the Io plasma torus.

Oxygen and sulphur ions 'picked up' by the magnetic field gain gyro-energies 170 and 540 eV respectively. The initial velocity distributions are highly anisotropic and the different ionic species and electrons are not in thermodynamic equilibrium. Coulomb collisions between ions or with electrons will change the distribution to a more stable one (Maxwellian). The effects of collisions over increasing timescales are: firstly to make the velocity distribution of each ionic species isotropic; secondly, to produce equipartition of energy for each species; and thirdly, to produce equipartition of energy between ions of different mass (including eventually the electrons). However, except in the cold inner torus, complete thermal equilibrium is unlikely because the timescales for collisions are of the same order as the timescale for transport (i.e. days).

The distribution of plasma along magnetic field lines is limited by the strong centrifugal forces which tend to confine the plasma to the region of the field line farthest from the rotation axis, the centrifugal equator. To first approximation the plasma density, n, decreases exponentially with distance from the centrifugal equator

$$n(z) = n(0)\ \exp-(z/H)^2, \tag{9.12}$$

where the scale height H is given by $H=(2kT_i/3m_i\Omega^2)^{\frac{1}{2}}$ for ions of mass m_i and temperature T_i. Thus the warmer ions in the outer region of the torus have a larger scale height and are more spread out along the field than the cold ions inside Io's orbit (see Figure 12).

(iii) Radial transport

In the region dominated by rotation in the Earth's magnetosphere plasma is re-distributed by episodic escape of material through the dayside magnetopause followed by thermal diffusion of plasma from the ionosphere along magnetic field lines, re-filling the plasmasphere within several hours. The strong magnetic field of Jupiter means that the inner Jovian magnetosphere is more rigid (i.e. sheltered from gusts in the solar wind). Moreover, the main plasma source is restricted to the L-shell of Io ($L \approx 5.9$). Therefore the re-distribution of material in the inner magnetosphere must occur on much longer timescales by diffusion across L-shells. The diffusive process is thought to be via a random, small scale interchange (or 'mixing') of fluxtubes. Gold(1959) pointed out that in a low-β plasma whole fluxtubes can be interchanged without a change in magnetic energy. If the energy of the plasma in two fluxtubes differs, the interchange motion may release or absorb energy. In the former case the system is unstable (Sonnerup and Laird,1963). At Jupiter fluxtube interchange is under the influence of the centrifugal force which aids outward but inhibits inward transport (Siscoe and Summers,1981). Therefore plasma preferentially diffuses radially outwards from Io's L-shell while the inward diffusion is very slow.

Since fluxtube interchange conserves magnetic flux the diffusing quantity is the plasma density per unit magnetic flux which, for a dipole field, is directly proportional to NL^2 where N is the density per unit L-shell. The steady state diffusion equation is then

$$L^2 \frac{d}{dL}\left[\frac{D}{L^2}\frac{d(NL^2)}{dL}\right] + \text{Sources} + \text{Losses} = 0\ , \tag{9.13}$$

where D is the diffusion coefficient(Dungey,1965; Siscoe,1979). The radial profile of NL^2 derived from the Voyager plasma data (Bagenal and Sullivan,1981) has a maximum at 5.7 R_J which supports the idea of diffusion of plasma from a single source near Io. The asymmetry of the transport process, imposed by the centrifugal potential, is evident in the slope of the NL^2 profile. The steeper slope inside $5.7R_J$ implies a larger diffusion coefficient (Richardson and Siscoe,1981).

Although the basic ideas of diffusive transport in the inner magnetosphere have been confirmed we are a long way from a full description. It is clear that slow inward diffusion allows time for the plasma to radiatively cool but it is not possible to match both the low temperatures and the observed ionic composition in the inner torus with the simple model described above; the data suggest there must be an additional source of plasma and a considerable energy sink well inside Io's L-shell (Richardson and Siscoe,1983; Bagenal,1985). Radially outward from Io, where diffusion is quite rapid, there is insufficient time for the plasma to either cool by radiation or come to thermodynamic equilibrium as it diffuses outwards. With comparable timescales (tens of days) for plasma transport, production, energy losses and thermal equilibrium it is a formidable task to derive a full description of the multi-component plasma as a function of radial distance in the warm torus. Furthermore, the erratic nature of Io's volcanos, the ultimate source of the torus material, would suggest a non-steady state. Indeed, the emissions from the torus plasma show a considerable degree of temporal variability (Brown et al.,1983).

Still farther out in the Jovian magnetosphere, the simple picture of radial transport by fluxtube interchange is complicated by a substantial pressure of energetic particles and a significant departure of the magnetic field from that of a dipole.

(iv) Plasmasheet

As the plasma diffuses outwards from the Io plasma torus it remains confined to the centrifugal equator forming a thin, disc-shaped, plasmasheet, tilted by $\sim 7°$ from the rotation equator. Thus, when Voyager traversed the magnetosphere, the plasma detector measured enhanced densities and lower temperatures (Figure 14c & d) every time the disc rotated past the spacecraft. Although the bulk motion of the plasma remained azimuthal, the velocity ($\underline{V}_\phi = \underline{\Omega}' \times \underline{r}$) lagged behind strict rotation ($\underline{\Omega} \times \underline{r}$), indicating that the conductivity of the ionosphere is insufficient to carry the currents required to enforce rigid corotation (Figure 14a). At the same time, the magnetic field lines are stretched out by the mechanical stresses on the plasma. Alternatively,one can think of the mechanical stresses on the plasma giving rise to charge-dependent drifts which constitute an azimuthal or "ring" current. For a thin plasmasheet the steady-state magnetic field configuration and plasma distribution are given by considering the balance of stresses. Locally the force density, viewed either as arising from the Maxwell stresses of the stretched-out field lines or equivalently as the $\underline{J} \times \underline{B}$ force of the magnetic field acting on the current sheet,must balance the mechanical stresses on the plasma (Figure 15) (McNutt, 1983). Globally, the large-scale pattern of the currents and the source of the mechanical stresses need to be considered. The present task of finding a self-consistent quantitative model, compatible with the Voyager and Pioneer observations, is reviewed by Vasyliunas(1983b).

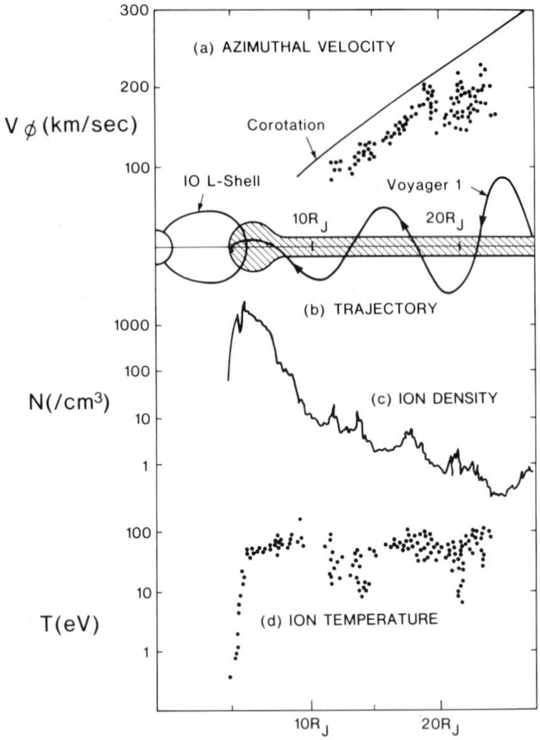

Figure 9.14 Voyager 1 ion measurements in the Jovian plasmasheet. (a) Azimuthal velocity; (b) the spacecraft trajectory in magnetic co-ordinates; (c) ion density; (d) ion temperature.(After Belcher,1983).

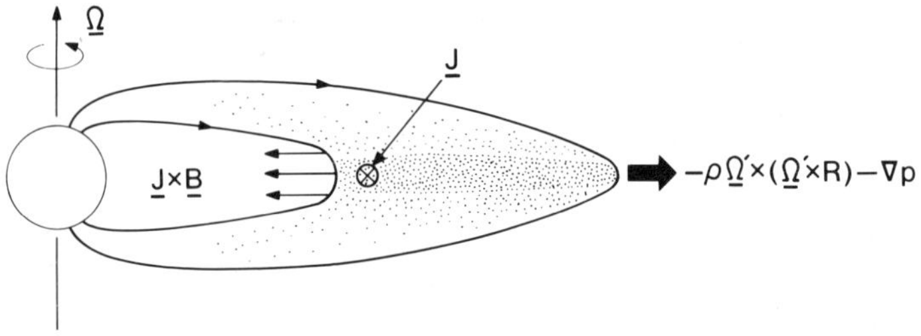

Figure 9.15 Local radial force balance in the plasma sheet (after McNutt,1983).

(v) Magnetotail

There is a limit to the distance at which a fluxtube of given plasma content can be maintained at corotation by an inward magnetic stress. This distance is thought to be where the corotation speed is of the order of the local Alfven speed. On the dayside of the magnetosphere the solar wind may provide sufficient inward stress to keep the fluxtube inside the magnetosphere. On the nightside the fluxtube would be free to move outward down the tail in a planetary wind (Hill, Dessler & Michel,1974). Vasyliunas(1983) suggests that the planetary wind will extend the magnetic field lines until they reconnect, breaking off plasmoids which then travel down the magnetotail (Figure 16).

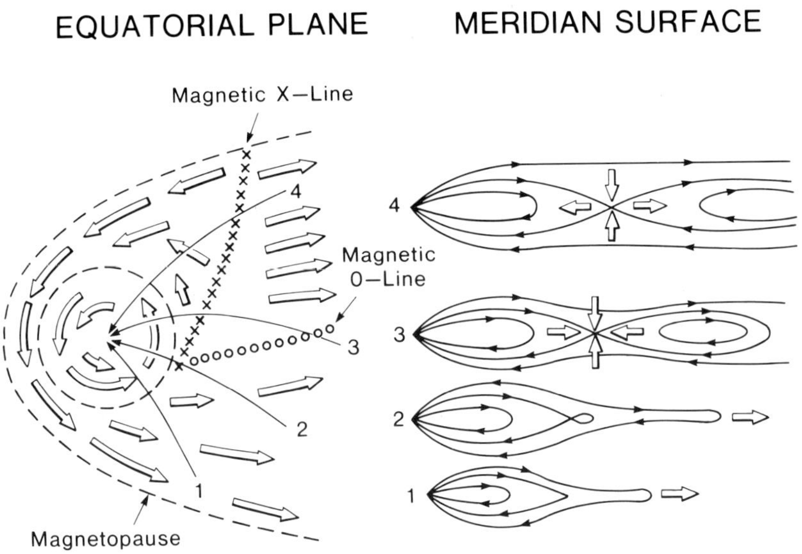

Figure 9.16 A sketch of plasma flow (large arrows) in the equatorial plane (left) and a sequence of meridian surfaces (right) expected from the planetary wind model (from Vasyliunas,1983b).

In conclusion, it is clear that the magnetosphere of Jupiter is radically different from the magnetosphere of the Earth; a large fraction of the Jovian plasma population comes from the satellite Io and rotation dominates the plasma dynamics. Recent reviews of various aspects of the physics of the Jovian magneosphere can be found in Dessler(1983).

9.2.6 Saturn

In the two years between September 1979 and August 1981 the magnetosphere of Saturn was explored by three spacecraft, Poineer 11 and Voyagers 1 & 2. Pioneer 11 detected a planetary field ($M \simeq 560\ M_0$) with a symmetry axis very closely aligned (within 1°) to the planet's rotation axis. Overall, the magnetosphere of Saturn was found to be similar to the Jovian magnetosphere: satellites are the major source of magnetospheric plasma and the plasma dynamics are dominated by the planet's rotation. Nevertheless the magnetosphere of Saturn is considerably smaller and the plasma sources much weaker than at Jupiter. Moreover, the dayside magnetopause at Saturn was found close to the Chapman-Ferraro distance (equation 9.2) at 20 R_S which indicates that it is the planetary field which stands off the solar wind, with a little contribution from the internal plasma pressure (Figure17).

(i) Sources of plasma

The detection of substantial densities of oxygen ions as well as protons

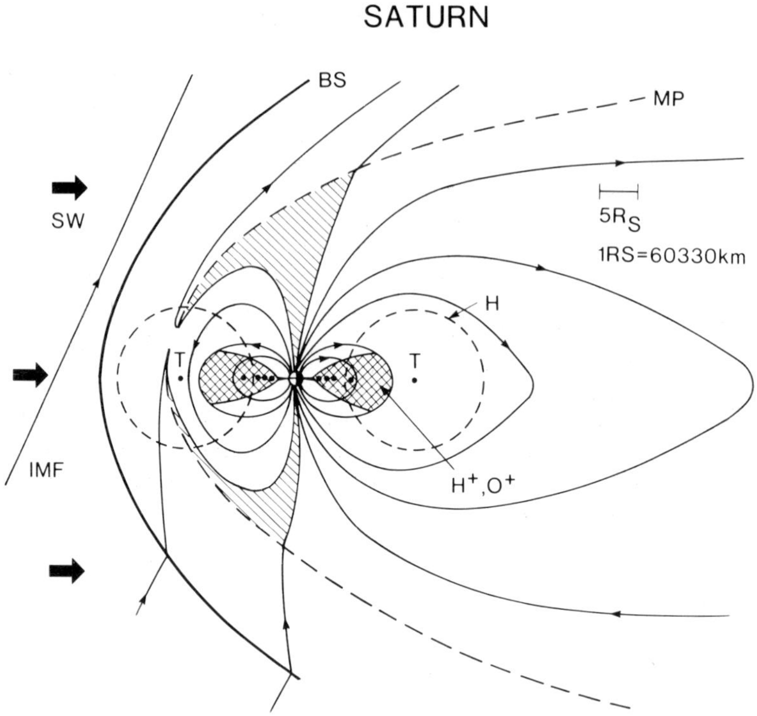

Figure 9.17 The magnetosphere of Saturn. The single hatching shows the polar cusp. The double hatching shows the toroidal region of satellite ions. The dashed circle indicates the cross-section of the neutral hydrogen torus around the orbit of Titan (T).

in the region of the orbits of Saturn's icy satellites (Rhea, Dione and Tethys at 8.8, 6.3 and 4.9 R_S respectively) indicates that the satellites must be major sources of plasma. Recent studies of the Voyager plasma data (Richardson,1985) have revealed the amount (NL^2) of O^+ ions has a maximum at $L \approx 10$ which suggests tha ion production is not a direct result of plasma interaction with the satellites but probably involves a cloud of neutral atoms that extends beyond the satellite orbits. These fairly small (radii $\leqslant$ 800 km) icy satellites have negligible atmospheres and are probably poor electrical conductors so that the corotating plasma and energetic particles impinge on the surface, sputtering off the neutral atoms. In contrast, in the outer magnetosphere, the much larger satellite Titan (radius $\approx$ 2575 km) has a substantial nitrogen atmosphere and is a direct source of heavy ions N^+ and possibly N_2^+ or H_2CN^+ (Hartle et al.,1982). The Voyager ultraviolet spectrometer detected strong emission from a large toroidal cloud of neutral hydrogen between 8 and 25 R_S (Broadfoot et al.,1981). This neutral H cloud must be a source of magnetospheric protons in addition to some, as yet undetermined, combination of icy satellite, solar wind, and ionospheric sources.

(ii) Interaction of magnetospheric plasma with Titan

The passage of Voyager 1 close to Titan provided a unique opportunity to study the interaction between a plasma flow and a body with an atmosphere. Since the incident magnetospheric plasma flow was transalfvenic ($M_A \approx 1.9$) and subsonic ($M_s \approx 0.57$) no upstream bowshock was observed and the plasma flowed quite smoothly round the satellite. However it is clear that the upper atmosphere of Titan is not sufficiently ionised (or, more specifically, the electrical conductivity is not sufficient) to allow the plasma just to slip around the satellite. The impinging plasma interacts with the uppermost atmosphere, picking up N^+, H^+, $N_2{}^+$,etc; from Titan's exosphere. The mass loading of the magnetospheric fluxtubes slows them down and causes them to drape over Titan and form a wake downstream. The wake then resembles an induced magnetotail with the northern and southern lobes containing oppositely-directed field lines. The mass loading appeared to be more effective on the sunlit side of Titan, resulting in an asymmetric wake (Figure 18).

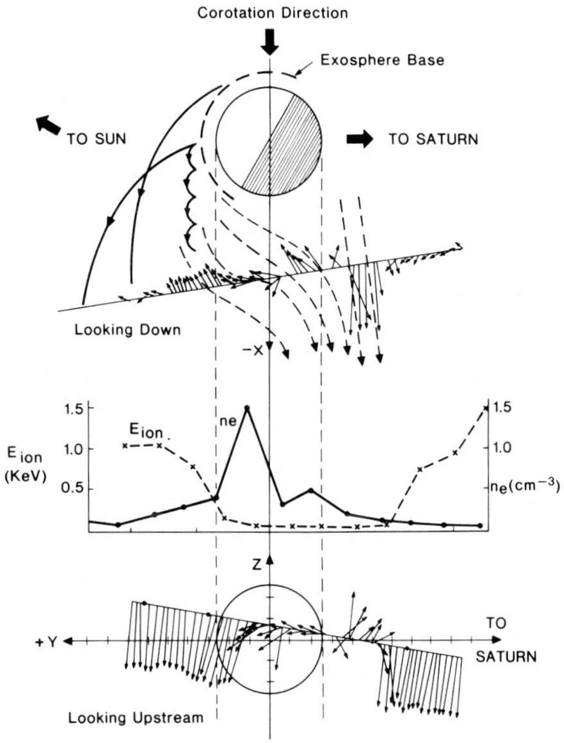

Figure 9.18 The interaction of the magnetospheric plasma with Titan.(Top) The Voyager trajectory with the observed X-Y components of magnetic field (arrows) and inferred plasma flow (dashed lines). The cycloidal motions of hot(cold) ions with large (small) gyro-radii are shown on the sunward side of Titan. (Middle) A plot of the average ion energy (E_{ion}) and electron density (n_e) measured by Voyager as it passed through Titan's wake. (Bottom) The perturbations of the magnetic field (Y-Z components) show how the planetary field is draped over the satellite. After Neubauer et al.(1984) and Bridge et al.(1981).

(iii) Morphology

Inside L≈15 the plasma is concentrated near the equator forming on extensive plamsa disc. With the plasma temperature decreasing radially inwards the thickness of the plasma sheet (characterised by the scale height H in equation 9.12) decreases from $4R_S$ at L≈10 to $0.2R_S$ at L≈3 where the plasma temperature is less than 1 eV (Lazarus and McNutt, 1983; Sittler et al,1983). This picture of a tapered plasma sheet is consistent with the idea of the plasma cooling as it diffuses in from a source in the outer part of the sheet, and as the plasma cools it must collapse towards the equator. Indeed, the equatorial plasma density was observed to increase from less than 1 cm^{-3} at L≈10 to over 100 cm^{-3} at L≈3. In this inner region the plasma flow is largely azimuthal with a lag behind strict corotation, presumably due to mass loading, in the source region. Between 8 and 16 R_S the plasma has a substantial component at higher energies (10's of keV). The ions and electrons at these energies have a differential azimuthal drift motion producing a 10^7 Amp ring current which distorts the magnetic field from a dipole by effectively stretching out the field lines in the equatorial region. Thus at Saturn it is the pressure gradient of the hot component rather than the mechanical stresses on the cold plasma (as for example in the Jovian magnetosphere) that balences the JxB force under steady-state conditions (see equation 9.8) (McNutt,1983).

Outside 15 R_S the plasma properties (density, temperature and bulk motion) are highly variable. Embedded in a background of tenuous, hot plasma there appear to be dense, cold blobs or streams. In this outer region the flow is rather erratic with considerable radial and vertical flows (Richardson,1985). The cause of the variability of the plasma in this outer region is not known in any detail but it is clear that the solar wind could be affecting a thick boundary layer in the dayside magnetosphere. On the nightside the solar wind drags the magnetic field out into an extensive magnetotail, similar to the Jovian magnetotail.

In summary, it appears that the magnetosphere of Saturn is similar to the Jovian magnetosphere in that it is large, the plasma motions are dominated by the planet's rotation and satellites are the main source of plasma. However the fact that saturn's satellites are much weaker plasma sources than Io (by about 10^{-2}) and are distributed throughout the magnetosphere makes a quantitative description less straightforward than at Jupiter. Reviews of various aspects of Saturn's magnetosphere can be found in Gehrels and Mathews(1984).

Finally, it is interesting to note that,although the total energy contents of the magnetospheres of Earth, Jupiter and Saturn vary by orders of magnitude, the ratio of the energy in the trapped particle population to the energy in the magnetic field is approximately the same for the three magnetospheres. It appears that the energy in the plasma builds up to only 1/1000 of the magnetic field energy at which point some sort of plasma instability is presumably triggered, preventing further build up (Schardt,1984).

9.2.7 Uranus

Until Voyager 2 flies past Uranus in January 1986 any discussion of a Uranian magnetosphere is largely speculative. The only indication to date that Uranus has a magnetic field comes from recent International Ultraviolet Explorer observations of Lyman-α emission (Clarke,1982; Durrance and Moos,1982). It is not possible to account for the emission by resonant scattering of solar Lyman-α emission, which makes charged-particle excitation the most likely cause of the emission. The flux of solar wind particles over the disc of Uranus only provides a few percent of the energy of the emission. The observers raised the appealing explanation that the solar wind particles from a much larger area are focussed by a planetary magnetic field. If the emission is confined to an auroral region between 80° and 90° latitude then the observations imply an average surface intensity similar to the measured surface intensities of the Saturnian and Jovian aurorae. Assuming that the energy of the precipitating particles is converted to auroral emission at an efficiency of 5% then the observed auroral intensities require a total precipitated power of about 3×10^{11} watts. To provide this power to Uranus' ionosphere the planet's magnetic field would effectively have to funnel the solar wind impinging on a disc of radius 25 R_U, which implies a very large magnetosphere. Thus the range of observed emission intensities suggest the magnetic field at the surface of Uranus is between 1 and 10 Gauss. The value predicted by the rather dubious empirical relationship, magnetic Bode's law is 1.7 Gauss (Hill and Michel, 1975).

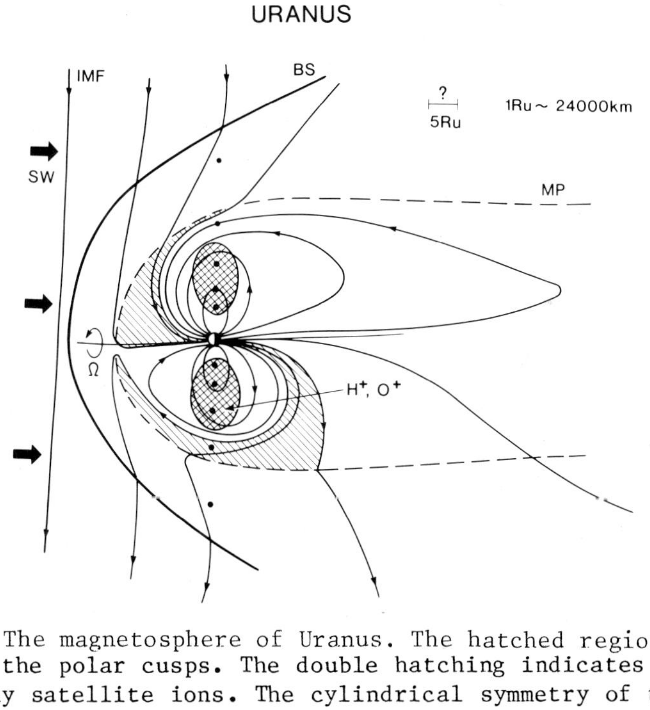

Figure 9.19 The magnetosphere of Uranus. The hatched region shows the position of the polar cusps. The double hatching indicates the expected region of any satellite ions. The cylindrical symmetry of the magnetotail is disturbed by reconnection of the planetary magnetic field with the IMF.

Taking the emissions to imply the existence of a magnetosphere we next consider the possible characteristics of the Uranian magnetosphere. Unlike the other planets, Uranus' axis of rotation lies close to the ecliptic plane. Therefore, for a significant fraction of Uranus' 84-year orbit (including the present epoch) the planet has a pole pointing towards the sun (Figure 19). Siscoe(1971,1975) pointed out that such a 'pole-on' configuration would produce a magnetosphere with several novel features. Firstly, the sunward-pointing pole would form a cusp funnelling solar wind onto the dayside ionosphere. Secondly, the alignment of the dipole axis with the solar wind would produce an unusual magnetic field topology in the magnetotail. Instead of the usual situation with two adjacent lobes of oppositely-directed magnetic field separated by a planar current sheet, with the 'pole-on' configuration at Uranus the two lobes would form concentric cylinders separated by a solenoidal current sheet. Thirdly, with a polar cusp near the sub-solar point, there is always a region on the magnetopause where the planetary magnetic field is anti-parellel to the IMF. Thus, unlike the Earth's magnetosphere where the occurrence of magnetic reconnection at the magnetopause depends strongly on the direction of the IMF, reconnection could occur at Uranus' magnetopause for all IMF orientations.

In spite of the planet's smaller radius (26,000 km) and slower rotation rate the planet's rotation probably controls the internal plasma flows (i.e. $R_c/R_m \geqslant 1$). Thus, despite the possibility of extensive reconnection, it is unlikely that the solar wind drives a large-scale convection system. Two very different models have been proposed for the internal dynamics of Uranus' magnetopshere. Hill et al.(1983) assumed the absence of a strong internal source of plasma and suggested that the rotating ionosphere is electrodynamically coupled to the solar wind by field-aligned (Birkeland) currents in a manner analogous to a Faraday disc dynamo. With the ionosphere acting as the disc dynamo the currents transfer planetary rotational energy and angular momentum to the magnetosheath. Thus the magnetosheath plasma flow would be helical under the combined influences of the solar wind and the planetary rotation. From their analytical model of the disc dynamo Hill et al.(1983) calculated that the particles carrying the Birkeland currents into the ionosphere could readily provide enough power to account for the observed auroral emission.

An alternative model has been proposed by Cheng(1984) who predicts that the Uranian magnetosphere will be dominated by satellite plasma sources and internally-driven convection, on a scale similar to that of Saturn's magnetosphere. Cheng(1984) points out that there are four icy satellites within 22.6 R_U that are comparable in size with Rhea, Dione and Tethys, the main sources of plasma in Saturn's magnetosphere. Cheng(1984) argues that charged-particle sputtering of these icy satellites could provide a comparable plasma population in Uranus' magnetosphere. The crucial issue is whether the charged-particle mechanism can be self-generating or whether a seed population of energetic particles, either from another internal source (such as the dense atmosphere of Titan in Saturn's case) or from the solar wind, is necessary.

9.2.8 Neptune and Pluto

Delving further into the realms of speculation we turn to the two outermost planets of the solar system. Neptune is very similar to Uranus though its rotation axis has the more usual orientation, perpendicular to the ecliptic. If Neptune has a substantial magnetic field then it is possible that Neptune's large satellite Triton ($R \geqslant 1500$ km) is embedded in the planet's magnetosphere,and with an atmosphere it could play a similar role to Titan in Saturn's magnetosphere. If all goes to plan, Voyager 2 will visit Neptune in 1989, first passing very close (0.4 R_N) over the north pole of the planet and then passing close to Triton.

No bigger than the Moon, Pluto ($1300 \leqslant R_p \leqslant 2000$ km) and its satellite Charon may have escaped from orbit around Neptune and barely qualify as a separate planetary system. Although it seems unlikely that Pluto would have an intrinsic magnetic field, even a weak field would carve a relatively large magnetopsheric cavity in the very tenuous ($n_{sw} \approx 5 \times 10^{-3}$ particles cm^{-3}) solar wind at 39 A.U.

9.3 CONCLUSIONS

In this chapter I have attempted to outline our understanding of the plasma environments around various planetary bodies. Considering planetary magnetospheres as a family of related objects one looks for underlying physical principles which allow comparison between them, ranging from the small cavity around Mercury to the giant magnetosphere of Jupiter. For example, a major factor governing magnetospheric structure and plasma dynamics is the relative importance of (internal) forces derived from planetary rotation compared with (external) forces driven by the solar wind. It is also important to consider the various particle populations in different regions of the magnetospheres, their origins (i.e. the solar wind, planetary ionospheres or satellites embedded in the magnetospheric plasma) and how they produce phenomena such as plasma waves, aurorae, radio emissions,etc., which are common to planetary magnetospheres. These studies can be extended to include the interaction regions surrounding unmagnetised bodies immersed in a plasma flow such as Venus, a planet with a deep atmosphere sitting in the supersonic solar wind, or Io, a satellite orbitting in the transsonic flow of magnetospheric plasma. To date, the magnetospheres of six planets have been explored directly with spacecraft carrying different selections of instruments to measure electric and magnetic fields plus particle detectors covering a wide range of energies. Some of the planetary magnetospheres have also been observed remotely from the Earth via their electromagnetic emissions ranging from X-ray to radio wavelengths. It is clear that comparative studies are in their infancy and we are likely to see them develop considerably over the next decade as spacecraft further explore the solar system.

Acknowledgements. Much of my understanding of the subject has come from lengthy discussions with colleagues in the space plasma groups at M.I.T. and Imperial College. In particular I am indebted to Ralph McNutt and David Southwood. I would like to thank John Richardson and Mark Saunders for giving material before publication. I am also grateful to Hilary Todd for her assistance with the manuscript.

REFERENCES

Bagenal F & J D Sullivan (1981) *J. Geophys. Res.* 86, 8447.
Bagenal F (1983) *J. Geophys. Res.* 88, 3013.
Bagenal F (1985) *J. Geophys. Res.* 90, 311.
Bagenal F, R L McNutt, J W Belcher, H S Bridge & J D Sullivan (1985) *J. Geophys. Res.* 90, 1755.
Belcher J W, C K Geortz, J D Sullivan & M H Acuna (1981) *Geophys. Res. Lett.* 86, 8508.
Belcher J W (1983) in Dessler(1983).
Biermann L (1957) *Observatory* 77, 109.
Bigg E K (1964) *Nature* 203, 1008.
Brice N M & G A Ionnidis (1970) *Icarus* 13, 173.
Bridge H S, J W Belcher, A J Lazarus, S Olbert, J D Sullivan, F Bagenal, P Garzis, R H Hartle, K W Ogilvie, J D Scudder, E C Sittler, A Eviatar, G L Siscoe, C K Goertz, V M Vasyliunas (1981) *Science* 212, 217.
Broadfoot A L, B R Sandel, D E Shemansky, J B Holberg, G R Smith, D F Strobel, J C McConnell, S Kumar, D M Hunten, S K Atreya, T M Donahue, H W Moos, J L Bertaux, J E Blamont, R B Pomphrey, S Linick (1981) *Science* 212, 206.
Brown R A (1974) *Exploration of the planetary system,* ed. A Woszczyk & C Iwaniszewska, D.Reidel.
Brown R A (1976) *Astrophys. J.* 206, L179.
Brown R A (1981) *Astrophys. J.* 244, 1072.
Brown R A, C B Pilcher & D F Strobel (1983) in Dessler(1983).
Burke B F & K L Franklin (1955) *J. Geophys. Res.* 60, 213.
Chapman S & V C A Ferraro (1931) *Terr. Magn. Atmos. Elect.* 36, 77.
Cheng A F (1984) *Workshop on Uranus and Neptune* ed. J T Bergstrahl, NASA conference publication 2330.
Clarke J T (1982) *Astrophys. J.* 263, L105.
Dessler A J (1983) *Physics of the Jovian magnetosphere,* Cambridge University Press.
Dolginov Sh Sh (1978) *Geophys. Res. Lett.* 5, 93.
Dungey J W (1965) *Space Sci. Rev.* 4, 199.
Durrance S T & H W Moos (1982) *Nature* 299, 428.
Durrance S T, P D Feldman & H A Weaver (1983) *Astrophys. J.* 267, L125
Gehrels T & M S Mathews (1984) *Saturn,* U Arizona Press, Tucson, Arizona.
Gledhill J A (1967) *Nature* 214, 155.
Gold T (1959) *J. Geophys. Res.* 64, 1219.
Goldreich P & D Lynden-Bell (1969) *Astrophys. J.* 156, 59
Goertz C K & P A Deift (1973) *Planet. Space Sci.* 21, 1399.

Goertz C K (1980) *J. Geophys. Res.* 85, 2949.
Gurnett D A (1972) *Astrophys. J.* 175, 525.
Gurnett D A & C K Goertz (1981) *J. Geophys. Res.* 86, 717.
Hill T W, A J Dessler & F C Michel (1974) *Geophys. Res. Lett.* 1, 3.
Hill T W & F C Michel (1975) *Rev. Geophys. & Space Phys.* 13, 967.
Hill T W, A J Dessler & M E Rassbach (1983) *Planet. Space Sci.* 31, 1187.
Kupo I , Y Mekler & A Eviatar (1976) *Astrophys. J.* 205, L51.
Lazarus A J & R L McNutt (1983) *J. Geophys. Res.* 88, 8831.
Marshall L & W F Libby (1967) *Nature* 214, 126.
McNutt R L (1983) *Adv. Space Res.* 3, 55.
McNutt R L (1984) *Magnetospheres in the solar system,* European Space Agency report ESA SP-207.
Melrose D B (1967) *Planet. Space Sci.* 15, 381.
Ness N F (1979) *Solar system plasma physics,* III, ed. C F Kennel, L J Lanzarotti & E N Parker, North-Holland.
Neubauer F M (1980) *J. Geophys. Res.* 85, 1171.
Neubauer F M, D A Gurnett, J D Scudder & R E Hartle (1984) in Gehrels and Matthews(1984)
Neugebauer M & C W Snyder (1962) *Science* 138, 1095.
Phillips J L, J G Luhmann & C T Russell (1984) *J. Geophys. Res.* 89,10676.
Piddington J H & J F Drake (1968) *Nature* 217, 925.
Richardson J D & G L Siscoe (1981) *J. Geophys. Res.* 86, 8485.
Richardson J D & G L Siscoe (1983) *J. Geophys. Res.* 88, 2001.
Richardson J D (1985) *J. Geophys. Res.* in press.
Russell C T (1978) *Geophys. Res. Lett.* 5, 85.
Russell C T & R C Elphic (1979) *Nature* 279, 616.
Russell C T, R C Elphic & J A Slavin (1980) *J. Geophys. Res.* 85, 8319.
Russell C T, J G Luhmann, R C Elphic & M Naugebauer (1982) *Comets* ed. L L Wilkening, U Arizona Press.
Russell C T, J G Luhmann, J R Spreiter & S S Stahara (1984) *J. Geophys. Res.* 89, 2997.
Saunders M A & C T Russell (1985) *J. Geophys. Res.* in press.
Schardt A W, K W Behannon, R P Lepping, J F Carbary, A Eviatar, G L Siscoe (1984) in Gehrels and Matthews(1984).
Sittler E C, K W Ogilvie & J D Scudder (1983) *J. Geophys. Res.* 88, 8847.
Siscoe G L (1971) *Planet. Space Sci.* 19, 483.
Siscoe G L (1975) *Icarus* 24, 311.
Siscoe G L (1979) *Solar system plasma physics,* III ed. C F Kennel, L J Lanzarotti & E N Parker, North-Holland.
Siscoe G L & L Christopher (1975) *Geophys. Res. Lett.* 2, 158.
Siscoe G L, N F Ness & C M Yeates (1975) *J. Geophys. Res.* 80, 4359.
Siscoe G L & D Summers (1981) *J. Geophys. Res.* 86, 8471.
Slavin J A & R E Holzer (1982) *J. Geophys. Res.* 87, 10285.
Sonnerup B U O & M J Laird (1963) *J. Geophys. Res.* 68, 131.
Southwood D J, M G Kivelson, R J Walker & J A Slavin (1980) *J. Geophys. Res.* 85, 5959.
Spreiter J R, M C Marsh & A L Summers(1970) *Cosmic Electrodynamics* 1, 5.

Stern D P & N F Ness (1982) *Ann. Rev. Astron. Astrophys.* 20, 139.
Taylor H E, K W Behannon & N F Ness (1968) *J. Geophys. Res.* 73, 6763.
Vasyliunas V M (1983a) *Solar-terrestrial Physics* ed. R L Carovillano & J M Forbes, D Reidel.
Vasyliunas V M (1983b) in Dessler(1983).

CHAPTER 10

COMETS

A D Johnstone
Mullard Space Science Laboratory
University College London
Holmbury St Mary
Dorking, Surrey

10.1 INTRODUCTION TO COMET STRUCTURE

Comets are the only objects in the solar system, outside the Earth, which the unaided eye perceives to have a non-spherical structure (figure 1). They are sometimes large enough to be resolved as more than a point of light and then can be seen to have a characteristic structure with a head and a long, slender tail. The head is approximately spherical and is up to 10^5 km in diameter and the tail, which is directed away from the Sun, has a visible length of up to 10^8 km. The tail is not a single structure but consists of two tails with entirely different physical causes. One appears white from the sunlight scattered off the surface of the grains of dust which form it. The spectral distribution of the scattered light contains information about the size distribution of the grains but not about the chemical composition of the dust. The second tail is coloured because it emits light at the wavelengths characteristic of the molecules of gas of which it is composed. The molecules are found to be ionised, consisting mainly of CO^+, H_2O^+, OH^+. The dominant ion CO^+ gives the tail its characteristic blue appearance in photographs. Since the particles are charged the second tail is known as the plasma tail.

The two tails are not parallel because the forces which shape them are different but the separation is rarely seen as clearly as in figure 1. In that picture the dust tail is the broad, curved, featureless one. The two main forces acting on the billions of dust grains are solar gravity and solar radiation pressure. The former keeps them in the heliocentric orbit, like the comet from which they came, while the sunlight pushes the particles outwards as it scatters off them.

The plasma tail is longer, and filamentary with many wave-like features and stretches nearly, but not quite, radially away from the Sun. The forces acting on the plasma tail are electromagnetic and form the main subject of this chapter.

The spherical head section, or coma, like the ion tail, appears in the light of its characteristic molecules. Spectral measurements reveal neutral molecules such as H, OH, O, C, CH, CN as well as the ionised molecules already mentioned.

Figure 10.1 Comet Mrkos photographed in 1957 by the 48 inch Schmidt telescope of the Hale Observatories. The broad featureless dust tail is clearly distinguished from the filamentary plasma tail.

The most successful model of cometary structure is known as the icy conglomerate model or, more colloquially, the dirty snowball model (Whipple, 1950). In this model it is postulated that the comet has, at its centre, a solid nucleus which is a conglomerate of various ices of the lighter elements in compounds such as H_2O, NH_3, CO_2 and other ices that normally have a relatively high vapour pressure at room temperature. Frozen in the icy matrix are the tiny dust particles. As the nucleus approaches the Sun the ice sublimates and the escaping gas which goes to form the visible coma blows the released dust particles out of the weak gravitational grasp of the nucleus. The nucleus itself has never been resolved optically. When close enough to the Earth to be seen it is cloaked by the coma produced by the sublimating gas. When far enough from the Sun for there to be no coma the nucleus appears as a single point of light. Observations under the latter conditions give the product of the albedo and the surface area. By making assumptions about the value of the albedo, the surface area and hence the size of the nucleus can be estimated. The recent observations of the nucleus of Halley's comet, on the inbound leg of its orbit towards the Sun give the radius, if the nucleus is spherical, to be 3.1 kms (Sicardy et al., 1983). There is no way to estimate the mass but if the density is $\rho = 1300 kg/m^3$ and the nucleus is spherical with a radius of 3.1 kms the mass would be $M = 1.6 \times 10^{14}$ kg.

During each orbit the nucleus loses mass as it passes perihelion so that its active life must be finite. The mass loss rate of the short period (3.3 yrs) comet Encke has been estimated to be 10^{10} kg per orbit (Wyckoff, 1982) of which 10% is in the form of dust. The total mass is of the order of 10^{13} kg so that the layer removed each orbit is approximately 0.6m thick. The lifetime should be thousands of orbits.

10.2 INTERACTION BETWEEN THE SOLAR WIND AND THE COMET

Two observations of comet tails convinced Biermann (1951) that there was a stream of particles flowing continuously outward from the Sun. First, the tail was not directed radially away from the Sun but at a small angle to the radius vector. The angle of aberration was proportional to the comet's orbital velocity around the Sun (Hoffmeister, 1943). The tail was found to be parallel to the relative velocity vector between the comet and an outwardly flowing stream moving at 1000 km/s. Secondly, some observations of the formation of envelopes on the sunward side of the nucleus were interpreted as requiring repulsive forces, directed away from the Sun, of more than 1000 times the strength of solar gravity. Biermann suggested that the force was produced by collisions between the stream of solar particles and the cometary particles. The force on the cometary particles is

$$M_c \frac{dV}{dt} = e^2 n_s V_s / \sigma \approx n_s V_s \times 10^{-15} \ m/s^2 , \qquad (10.1)$$

where n_s, v_s are the number density and velocity of the solar stream and σ is its' electrical conductivity. He knew, from the work of Chapman and others, that intermittent streams of particles from the Sun caused geomagnetic storms and from their estimates of the flux (Unsold and Chapman, 1949), $10^9 < n_s < 10^{11}$ m^{-3} and $v_s \sim 10^6$ m/s, obtained accelerations of 1-100 m/s in accordance with the observations. Since the tails of comets are always directed away from the Sun, Biermann argued that the solar streams must flow continuously and not intermittently as previously supposed. This prompted Parker (1958) to make a theoretical study of the solar corona to see if it could support this continuous outflow. Parker's theoretical conclusion that it could and his estimates of the velocity and temperature were confirmed a few years afterwards by measurements from spacecraft. The velocity averaged 450 kms/s, a little smaller than the figure used by Biermann, but the density was much smaller, usually less than 10^7 particles/m^3. This meant that the solar repulsive force produced by Biermann's method was several orders of magnitude too weak to have the required effect. It is therefore necessary to return to the original observations which prompted the calculation of the solar repulsive force for a new interpretation.

Eddington (1910) observed the formation of paraboloidal envelopes on the sunward side of Comet Morehouse (1908). He found (figure 2) that a succession of envelopes formed, more or less at the same height on a given day, and then immediately contracted towards the nucleus. Each envelope appeared first at the vertex, on the sun-comet line, and then spread outwards in both directions. The envelopes were found to develop continuously into tail rays. In Eddington's day, the only model available for the interpretation of these results was the fountain theory which proposed that the envelopes were the result of the superposition of many particle trajectories. The particles were emitted at some speed from the nucleus and were then reflected by an unknown solar repulsive force. From the observations the initial velocity and the strength of the force could be calculated. The theory had great problems matching the shape of the envelopes but contained enough variable parameters that it could be made to fit, albeit somewhat uncomfortably. Biermann used the collisional effects to produce the force but as one author (Fokker, 1953) wrote "one gets the impression that the explanation, at least of the regular streamer structure, has to come from an entirely fresh idea".

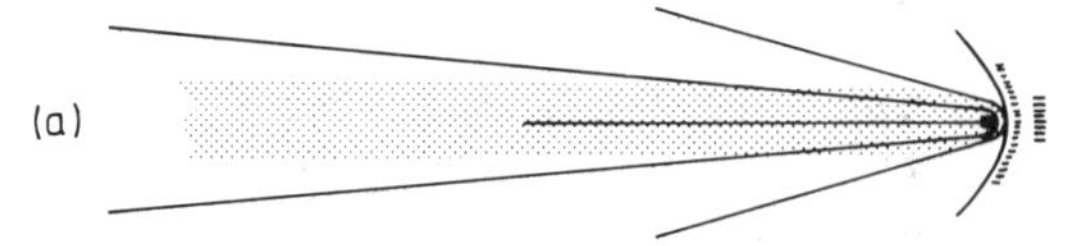

Figure 10.2 A time sequence showing the evolution of tail rays. They first appear upstream near the nose. They contract towards the head as they extend sideways before being pulled back by the solar wind to form the tail (Ip & Axford, 1982).

The fresh idea came from the realisation of the important role played by the magnetic field (Alfven, 1957). If the ion gyroradius in the magnetic field is small compared with the size of an object in the flow then the magnetic field is strong enough to provide the necessary long-range forces between the particles to give fluid-like behaviour around the object.

In the solar wind the ion gyroradius is less than 100 km while the collision mean free path, for the mechanism envisaged by Biermann (1951), is of the order of 1 astronomical unit. In other words, magnetic forces are much more effective than collisions.

Alfven (1957) showed that as the solar wind is decelerated near the comet, the magnetic field lines frozen into the plasma become wrapped around the head and dragged downstream to form the tail (figure 3). The envelopes and tail streamers are to be identified with magnetic flux tubes. The mechanism of formation is totally different from the fountain theory so the calculation of the repulsive forces is no longer appropriate. The magnetic formation of the streamers is discussed in more detail in section 10.8.1.

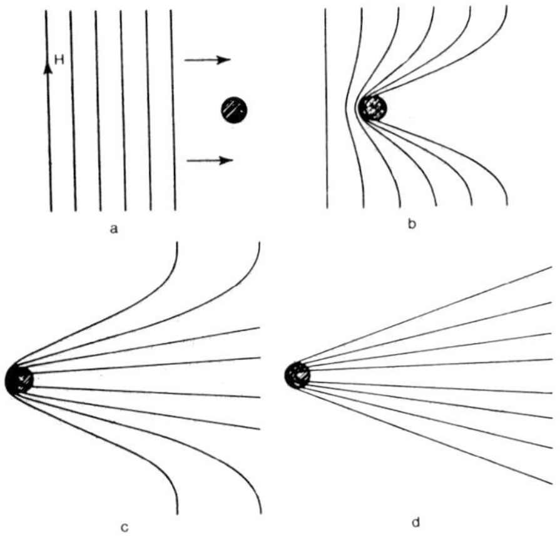

Figure 10.3 The four successive diagrams a, b, c, d show magnetic field lines frozen into the solar wind being caught up in the coma while the ends are pulled downstream to form the tail (Alfven, 1957).

The magnetic theory also provides a more accurate explanation for the tail direction. In the collision-driven theory, the force is a viscous-like one, dependent on the relative velocity between the two populations of particles. Initially, the cometary particles would follow their own velocity of ejection from the nucleus until dragged up to speed by the solar wind. In the magnetic theory, the tail behaves like a windsock, attached to the head by magnetic field lines, and flowing behind along the direction of the relative velocity vector. Since the orbital velocity is known, measurements of the tail aberration can be used to deduce the solar wind velocity (Brandt et al., 1973). The average value obtained (402 ± 12 km/s) is close to the average of direct in-situ measurements.

The coma and tail of a comet is a place where two very different plasmas meet and interact. One is a supersonic, magnetised, tenuous hydrogen plasma from the Sun; the other is the cool, unmagnetised, denser plasma of more massive particles formed by the ionisation of cometary gas. The characteristics of the first of these plasmas, the solar wind, is well-known (Chapter 8) having been monitored almost continuously for more than 20 years. The characteristics of the second plasma, the cometary plasma can, as yet, only be determined by inference from indirect observations and theoretical modelling. The situation will change after the in-situ measurements by the series of spacecraft due to fly through the coma of Halley's Comet in 1986. Until then knowledge of the cometary plasma is limited. In the next two sections what we know about the formation of this plasma by evaporation of the nucleus is discussed. First we derive the production rate of neutral gas in the coma and then describe the various ways it may be ionised. We then obtain the production rate of ionisation throughout the coma.

10.3 PRODUCTION OF NEUTRAL GAS

10.3.1 Vaporisation

The gas is produced by the vaporisation, or sublimation, of the ices in the cometary nucleus by radiant heat from the Sun. The equation of thermal balance governing this process has the form

$$F(1-A)\cos\theta/r^2 = (1-A')\sigma T^4 + \sum_i Z_i(T)L_i(T)\ . \tag{10.2}$$

The term on the left-hand side gives the heat absorbed from solar radiation,

where F is the solar heat flux,
A is the cometary albedo at visible wavelengths,
r is the heliocentric distance of the comet.

The terms on the right-hand side give respectively the radiation of heat by the comet and the heat taken up by the latent heat of vaporisation of molecules, where

σ is the Stefan Boltzmann constant,
A' is the cometary albedo at infrared wavelengths,
T is the surface temperature of the comet,
Z_i is the production rate of species i,
L_i is the latent heat of vaporisation of species i.

The equation applies to unit area of the nucleus surface.

The vaporisation rate Z_i is given by the net outflow at the surface under conditions of molecular effusion

$$Z_i = \frac{1}{4} n_{ei} \langle V_{ti} \rangle = \frac{1}{4} (P_e/kT_i)(8kT_i/\pi M_i)^{\frac{1}{2}} = P_e/(2\pi M_i kT_i)^{\frac{1}{2}}\ , \tag{10.3}$$

where n_{ei} is the density of species i at the equilibrium vapour pressure P_e and $\langle V_{ti} \rangle$ is the thermal velocity

The vapour pressure for ice varies strongly with temperature. Near 200K, the likely surface temperature of an active nucleus near 1 a.u. (Delsemme, 1982), a 20K variation in temperature increases P_e by a factor of 20.

Close enough to the Sun, the vaporisation term in eqn(10.2) will dominate, and when averaged over the surface of a rotating nucleus the equation becomes

$$F(1-A)S/r^2 = QL , \qquad (10.4)$$

where Q is the total production rate of all species molecules per second, S is the surface area of the nucleus, and L the average latent heat. Figure 4 shows the calculated vaporisation rate for a number of possible cometary molecules. Each curve shows the inverse square law behaviour predicted by eqn(10.4) at small heliocentric distances.

The actual gas production rate of comets can be estimated from the variation in brightness of the comet with heliocentric distance. Figure 5 gives the light curve for Comet Encke which follows the inverse square law close to the Sun and exhibits a steep increase in brightness at the distance expected for a nucleus composed of water ice.

The vaporisation rate of water ice at 1 a.u. depends on the unknown albedo of the surface.

From figure 4, using reasonable estimates of the albedo, it is approximately 3×10^{17} molecules cm^{-2} s^{-1}. The production rate of H_2O molecules at 1 a.u. will then be, for a nucleus of 2 km radius

$$Q = 1.5 \times 10^{29} \text{ molecules/sec} . \qquad (10.5)$$

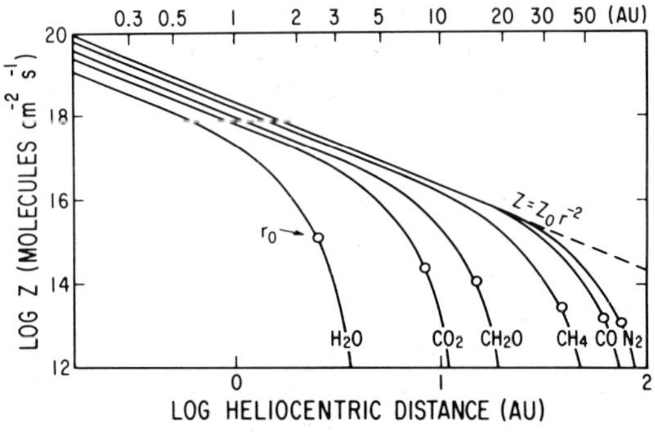

Figure 10.4 The vaporisation rate Z, for various snows as a function of heliocentric distance assuming a rotating nucleus. The H_2O vaporisation first becomes significant at a distance of 2 a.u. All rates become proportional to r^{-2} close enough to the Sun (Delsemme, 1982).

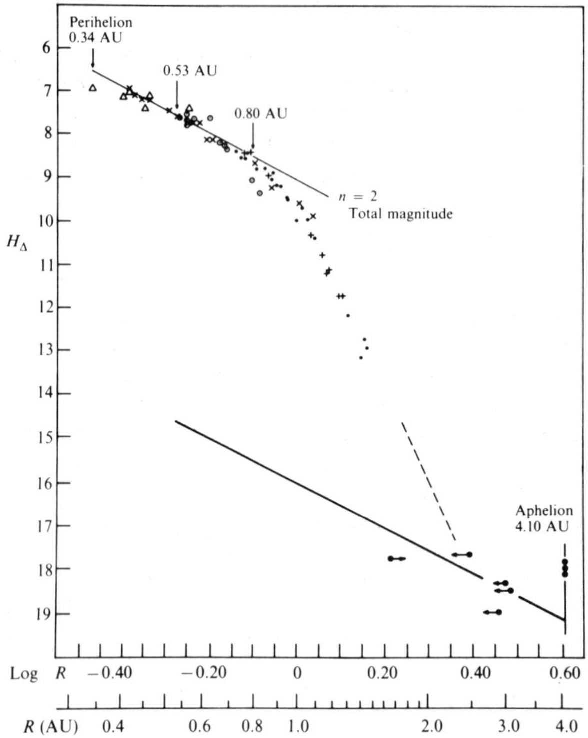

Figure 10.5 The variation in the brightness of comet Encke with heliocentric distance. The brightness increases sharply inside 2 au as would be expected for a water-ice comet. Near perihelion the curve follows the expected inverse square law (Brandt and Chapman, 1981)

10.3.2 Neutral gas density

In the calculations of the gas density which follows, the expressions refer to the density of a single species, though for convenience this will not be stated explicitly.

The density at a cometocentric distance ℓ is, ignoring losses by ionisation

$$n(\ell) = Q/4\pi V_e \ell^2 . \qquad (10.6)$$

where we have assumed that the gas expands radially outwards at a velocity V_e following evaporation. The expansion velocity is difficult to calculate, but since the gas is expanding into a vacuum it will be approximately equal to the thermal velocity $\langle V_t \rangle$ of the molecules. For CO at 200 K, $\langle V_e \rangle = 387$ m/s. The mean free path is equal to the distance from the nucleus when

$$\ell = Q\sigma/4\pi V_e \sim 3000 \text{ kms} , \qquad (10.7)$$

where σ is the collisional cross-section which for a typical molecule is of the order of 10^{-19} m^2.

Closer to the nucleus the gas is collision-dominated and many chemical and photochemical reactions taken place. They may be grouped into four major types of phenomenon (Delsemme, 1982).

1. Photoexcitation of molecules by sunlight
2. Photodissociation of the parent molecules emitted from the surface of the nucleus
3. Photoionisation of molecules, or the dissociated molecular fragments
4. Fast ion-molecular reactions which shuffle the fragments toward molecular states of lower internal energies.

Item 3 photoionisation is discussed in more detail in the section 10.4.1. The other processes, though of great importance to the structure of the cometary ionosphere (Giguere et al., 1982) do not significantly affect the ionisation rate and will not be discussed here where we concentrate on the solar wind interaction. However some aspects do affect the gross plasma structure. The temperature of the gas varies as a result of the exothermic reactions taking place and radiative cooling. It is impossible to make accurate estimates of the magnitude of the effect of the resultant temperature on the basis of current knowledge. The outward flow velocity of the gas is even more uncertain because it depends on the complex fluid dynamics of the flow as well as the temperature. Using an approach based on fluid dynamics Delsemme (1982) showed that the terminal velocity of the gas should be $1.77 \langle V_t \rangle$ which for CO is

$$V_e = 1.77 \langle V_t \rangle = 685 \text{ m/s} \qquad (10.8)$$

A curious feature of the reactions is that the parent molecules, which evaporate from the nucleus, cannot be identified in the optical spectra. None of the molecules or radicals which can be seen in those spectra exist in the solid state so they could not have come from the nucleus. Since the probable parent molecules, H_2O and CO_2 do not have emission lines in the visible range (Fernandez and Jockers, 1983), their role as the parents can only be inferred from the products of the chemical reactions.

10.4 IONISATION

10.4.1 Ionisation processes

The neutral gas released from the nucleus must be ionised before it can be influenced by the electric and magnetic fields in interplanetary space. The relative importance of the various ionisation processes in the cometary ionosphere is not well-known. Some of the processes can be clearly identified and the ionisation rates established. Other rates can only be guessed.

The process which can be most reliably identified is photoionisation. The rate depends on the particular molecule or radical but is usually close to 3×10^{-7} s^{-1} at 1 a.u. (table 10.1). Put another way, inverting the ionisation rates, the average lifetime of a neutral molecule before ionisation is more than 10^6 secs. The rate is proportional to the solar ultraviolet flux and so varies as (r^{-2}) where r is the heliocentric distance.

Table 10.1

Molecule	Photoionisation (s^{-1})	Charge-exchange ($N_s V_s = 3 \times 10^8$)
CO	3.1×10^{-7}	8.4×10^{-7}
H_2O	3.3×10^{-7}	6.3×10^{-7}
H	0.7×10^{-7}	6.0×10^{-7}

The second process is charge-exchange in which cometary neutrals exchange an electron with a solar wind proton. This does not increase the total amount of ionisation but it ionises a neutral cometary particle, and creates a fast neutral hydrogen atom. The rate depends on the solar wind proton flux as well as the type of molecule. When the solar wind flux is high the charge-exchange rate is higher than the photoionisation rate, although the particle's lifetime as a neutral is still more than 10^6 secs. There is evidence from cometary observations that the ionisation rate is much faster at times. The brightness of CO^+ features in the coma increases on a time scale of the order of 1000 secs. Therefore (Wurm, 1961) there must be other, faster, mechanisms. Among those that have been suggested are:

(a) electron collisional ionisation
(b) critical velocity ionisation

The former requires a population of energetic electrons which could be produced by the thermalisation of solar wind electrons in the comet's bow shock (Axford, 1964) or by auroral-like acceleration in the tail of the comet (Ip and Mendis, 1976). Since the fluxes of such electrons cannot be quantified with present knowledge the effect on the ionisation rate can only be guessed.

The second mechanism above, critical velocity ionisation, was proposed by Alfven (1960). When a neutral gas passes through a magnetised plasma, with a relative velocity greater than a critical value V_c,

$$V_c = (2eV_i/M)^{\frac{1}{2}} , \qquad (10.9)$$

there is a rapid ionisation of the neutrals. The critical velocity is the velocity at which the neutral's kinetic energy is equivalent to the ionisation potential. This happens despite the large collision mean free paths which should make the interaction collisionless (Mendis and Houpis, 1982).

10.4.2 Size of the coma

The ionisation rates are important not because they determine the amount of ionisation but because they control its distribution. Ultimately all the neutral atoms will become ionised.

The radial size of the neutral coma, the spherical visible region around the nucleus is the outflow velocity × neutral particle lifetime. The neutral particle density $n(\ell)$, including ionisation unlike eqn (10.6), is

$$n(\ell) = Q \exp(-\ell\nu/V_e)/4\pi V_e \ell^2 , \qquad (10.10)$$

where ν is the ionisation rate. Outside the collision-dominated region, the equation must be applied to each species independently, because of differences in the outflow velocity V_e and the ionisation rate ν. For example, for CO, with M = 28 amu, T = 200K, V_e = 688 m/s, $\nu = 11.5 \times 10^{-7}$ s^{-1} (table 10.1) the radius of the cloud to the (1/e) level is

$$\ell_{CO} = V_e/\nu \approx 6 \times 10^5 \text{kms} . \qquad (10.11)$$

The hydrogen coma is much larger than this. Atomic hydrogen is produced by the dissociation of H_2O and OH in the collisional region. The reaction is exothermic so the hydrogen atoms have an outflow velocity of approximately 8 km/s, greater than would be expected for a temperature of 200 K. The corresponding size of the coma is, with $\nu = 6.7 \times 10^{-7}$ (table 10.1)

$$\ell_H = 1.2 \times 10^7 \text{ kms} . \qquad (10.12)$$

This, much larger, hydrogen coma can only be observed in ultraviolet light, which in turn can only be observed from above the Earth's atmosphere. Spaceborne detectors have detected this enormous hydrogen coma (Keller, 1976), many times the size of the visible coma (figure 6).

The importance of these dimensions is that it shows how far from the nucleus that ionisation can be created. The ion production rate, as a function of distance from the nucleus, is

$$A = \nu n(\ell) = Q\nu \exp(-\ell\nu/V_e)/4\pi V_e \ell^2 . \qquad (10.13)$$

Thus CO^+ ions are created in significant numbers more than 6×10^5 kms, and H^+ more than 1.2×10^7 kms from the nucleus.

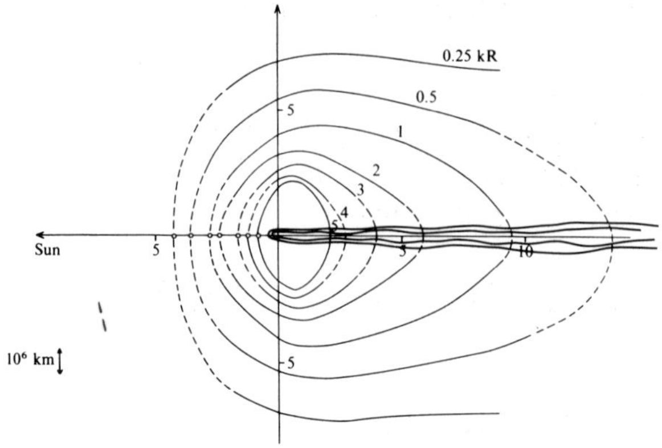

Figure 10.6 The hydrogen coma as shown by the Lyman-alpha isophotes of Comet Bennett (Keller, 1976) is compared with the approximate size of the coma and tail in CO^+ light

10.5 ION PICKUP

10.5.1 Ion pickup trajectories

Immediately the neutral particle becomes ionised it is subjected to the electric and magnetic fields in the solar wind. To derive the ion trajectory we use the coordinate system of figure 7. The y axis is parallel to the magnetic field B, and the yz plane contains the solar wind velocity vector V_s. The frame is moving at a velocity $V_s \sin \phi$ in the z direction. This is the velocity of the field lines sometimes called the Hoffman-Teller frame.

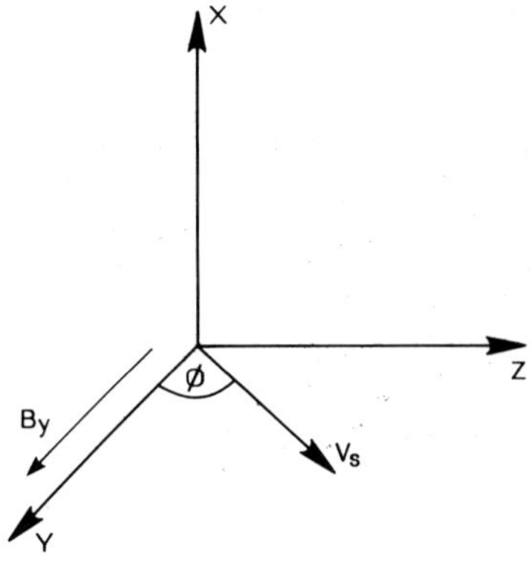

Figure 10.7 The frame of reference used to calculate ion pickup trajectories.

In this frame the electric field, E, given by

$$\underline{E} = -\underline{V}_s \times \underline{B} = 0 \, . \tag{10.14}$$

The equations of motion are then

$$M\ddot{x} = -q\, V_z\, B_y \; ; \quad M\ddot{y} = 0 \; ; \quad M\ddot{z} = q\, V_x\, B_y \, , \tag{10.15}$$

with the solution

$$V_z = A \cos \omega_c t + B \sin \omega_c t \, , \tag{10.16}$$

where $\omega_c = q\, B/M$.

At the moment of ionisation the ion has a velocity V_e in the rest frame. As we shall see this velocity is negligible in comparison with the velocity it receives from the acceleration it undergoes in the solar wind. In the H-T frame, V_e may be ignored and the initial velocity in the z direction is $-V_s \sin \phi$. Substituting in the general solution and then transforming back to the rest frame gives the following components of the ion velocity

$$\left. \begin{aligned} V_x &= V_s \sin \phi \sin \omega_c t \, , \\ V_y &= 0 \, , \\ V_z &= V_s \sin \phi \,(1 - \cos \omega_c t) . \end{aligned} \right\} \tag{10.17}$$

The motion consists of (a) a gyration about B with a velocity $V_s \sin \phi$ and (b) a drift perpendicular to B with a velocity $V_s \sin \phi$. The motion is cycloidal in a plane perpendicular to B.

The gyroperiod, and the distance travelled in a complete cycle of the motion are

$$2\pi/\omega_c = 183 \text{ secs, } 6.5\text{s} \, , \tag{10.18}$$

$$2\pi V_s \sin\phi/\omega_c = 50{,}000 \text{ kms, } 1850 \text{ kms} \, , \tag{10.19}$$

when $B = 10$nT, $V_s \sin \phi = 283$ km/s and the two sets of figures are given for CO^+ and H^+, respectively.

For CO^+ the distance given by eqn(10.19) is one tenth the size of the CO^+ coma eqn(10.11) and comparable with the spatial scale of variation in the magnetic field. The actual motion will therefore depart significantly from this cycloid in a uniform field. The H^+, which is produced further away from the nucleus, has a much smaller gyroradius, and therefore room to develop cycloidal trajectories.

One feature of the trajectory is that the maximum velocity of the particle is $2 V_s \sin \phi$. If $V_s = 400$ km/s, $\phi = 90^o$, and the ion is CO_2^+ with a mass of 44 the maximum energy is 140 keV, very much more than the solar wind energy of 1 keV.

The momentum, and kinetic energy of the motion, averaged over a complete cycle are easily calculated;

$$\langle p \rangle = MV_s \sin \phi \, \hat{k} \;\; ; \;\; \langle W \rangle = MV_s^2 \sin^2\phi \; . \qquad (10.20)$$

The momentum is perpendicular to the magnetic field and is equivalent to the particle moving with the field line velocity. The kinetic energy is twice the kinetic energy associated with this average momentum because the particle also has kinetic energy of $\frac{1}{2}M V_s^2 \sin^2 \phi$ associated with its gyration.

The shape of the velocity distribution function of the implanted cometary ions is a ring in velocity space. The ring is in the plane perpendicular to the magnetic field and is centred on the component of the solar wind velocity perpendicular to the field and with a radius of the same velocity.

10.5.2 Stability of the distribution

Such a distribution of ions is highly unstable. All the ions have a pitch angle of 90° creating a distribution with a lot of free energy. Wu and co-workers (Wu and Davidson 1972, Hartle and Wu, 1973) calculated the growth rate of both electromagnetic and electrostatic waves for frequencies below the electron gyrofrequency. The linear growth rate for transverse electromagnetic waves is

$$\gamma \sim (\omega_{ip}/2)(V_s \sin\phi/c) \; , \qquad (10.21)$$

where ω_{ip} is the ion plasma frequency.

For CO^+ densities of 10^{-2} cm^{-2} the exponentiation time $1/\gamma$ is approximately 300 secs; for H^+ the corresponding figure is 70 secs. Both figures are a few times the gyroperiod (eqn (10.18)).

Their analysis did not take into account a finite spread of velocities parallel to the magnetic field which is found to exert a stabilising effect, or at least to slow down the growth of the instability. The parallel spread arises from two sources; the thermal velocities of the neutral particles before ionisation (i.e. the small component V_e we ignored in the analysis of sec 10.5.1) and the non-uniformity of the magnetic field. A variation of the field strength parallel to the field line (i.e. diverging magnetic field lines) accelerates the particles along the field line by the magnetic mirror force,

$$M \frac{dV_{\parallel}}{dt} = -\mu \frac{\partial B}{\partial s} , \qquad (10.22)$$

where μ is the magnetic moment of the particle which is conserved in the motion. Thus at any point along the field line there is a mixture of particles, ionised at different points along the field line, with a range of pitch angles.

Galeev (1982) argues that this spread of velocities stabilises the electrostatic and electromagnetic instabilities and the net result of the possible effects is a small amount of energy diffusion.

If the instabilities were effective then the result would be that the population of implanted ions would rapidly become coupled to the solar wind flow, both parallel and perpendicular to the field, with average momentum and kinetic energy given respectively by

$$\langle p \rangle = MV_s \quad , \quad \langle W \rangle = \tfrac{1}{2}MV_s^2 . \qquad (10.23)$$

At the present time we cannot determine whether eqn(10.20) or eqn(10.23) more accurately describe the average motion of the cometary ions after injection, or whether the truth lies somewhere between these two extremes (Ip and Axford, 1982).

10.6 PRINCIPAL PLASMA REGIMES

10.6.1 Main regions

The basic source plasmas in the cometary plasma environment, the solar wind and cometary gas, have now been established.

The interaction region can be divided into three distinct regions each with different properties (figure 8). In the immediate neighbourhood of the nucleus is the inner coma where the plasma of cometary origin dominates. It is bounded by a current layer across which a pressure balance between the internal ionisation and the solar wind is maintained. Inside, the plasma is dense and the magnetic field is very small since this contact surface deflects the magnetic flux frozen into the solar wind around it and there is not likely to be an intrinsic magnetic field in the comet's nucleus. The structure of the contact surface is completely unknown at the present time. It could be a thin structure of the order of an ion gyroradius like the ionopause at Venus (Russell et al, 1982) or could be much thicker with a more gradual transition. There will be a velocity shear across the surface with the decelerated solar wind flowing past on the outside ~ 50 Km/s and drift velocities comparable with the outflow velocity of 1 Km/s on the inside.

The second regime is in the region far from the comet which is dominated by the solar wind. Some of the neutral gas from the comet, which has travelled unperturbed through magnetic and electric fields of the inner cometary regions, becomes ionised far upstream in the freely-flowing solar wind.

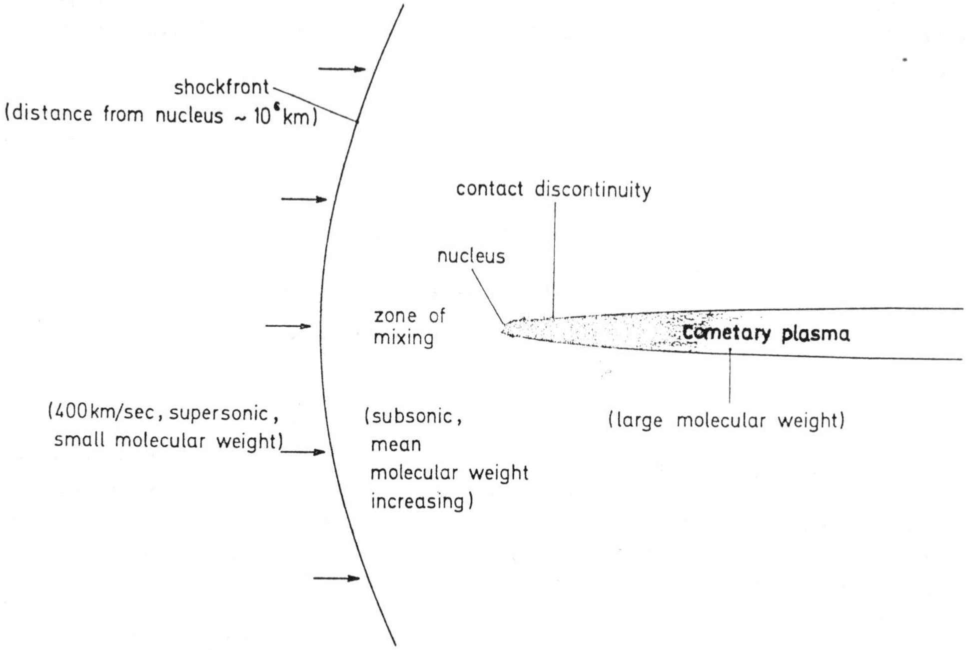

Figure 10.8 The principal plasma regions near a comet (Schmidt & Wegmann, 1982)

The addition of a slow moving mass to the flow, decelerates the solar wind because momentum is conserved. The distance from the comet at which this occurs depends on the outflow velocity and the lifetime of the neutral particle (section 10.4). From the figures given in eqns.(10.11) and (10.12) the hydrogen gets furthest upstream, but its effect on the solar wind is relatively weak because of its small mass. The more massive CO^+ ions do not go as far, but a small fraction by number, say 1%, adds 28% to the mass density in the flow. This fraction penetrates more than 2.7×10^6 kms away from the nucleus. As the flow approaches the comet it is decelerated by increasing amounts.

In between these two regions there is a region of mixed plasma bounded, it is almost certain, on the upstream side by a bow shock wave, similar to those known to exist upstream from the major planets, and on the downstream side by the contact surface.

On crossing the bow shock the plasma will be heated, compressed, and decelerated.

10.6.2 The contact surface

To locate the contact surface the various contributions to the pressure balance on either side of the surface must be evaluated as a function of the radial position. The factors discussed briefly here are described in more detail by Schmidt & Wegmann (1982).

Outside the contact surface the main term is the stagnation pressure P_s of the hypersonic solar wind which can be calculated given that $\gamma=2$,

$$P_s = 0.84\, \rho_s^2\, V_s^2 . \tag{10.24}$$

The addition of mass to the flow by ionisation reduces this pressure but not by much.

Numerical results described in Section 10.7 put it at

$$P_s = 0.6\, \rho_s^2\, V_s^2 . \tag{10.25}$$

The other main external pressure term is provided by the magnetic field. This is effected in two ways. First, as the flow velocity decreases and the density increases, the magnetic field becomes compressed. The gas-dynamic calculations (Sect. 10.7) show that the magnetic pressure $B^2/8\pi$ increases towards the nucleus along the axis of symmetry as $(1/R)$ until it becomes comparable with the gas pressure. Then the field influences the flow by the second effect. The magnetic field lines are curved around the head (figure 3) and exert a volume force towards the nucleus. This is due to the transfer of momentum along magnetic field lines by the so-called Maxwell stresses from the more distant solar wind at the side of the comet. The numerical results show that the total magnetic pressure now increases towards the nucleus as $(1/R^{1/2})$.

Inside the contact surface the main pressure term is the supersonic outflow of ionisation.

There is no magnetic field so the ions continue to flow outwards with the same velocity as the neutral particles. The total out-flux, over the sphere, of ions and neutrals is Q, the gas production rate. The proportion converted to ions since leaving the nucleus is $Q\nu\ell/V_e$ so the density of the ions which can exert the pressure is

$$n = (Q\nu\ell/V_e)/4\pi\ell^2 V_e . \tag{10.26}$$

The stagnation pressure from the outward flow is, in analogy with the solar wind flow, therefore

$$P_c = 0.84\, M_c Q\nu/4\pi\ell . \tag{10.27}$$

The supersonic outflow from the nucleus must pass through an inner shock before encountering the contact surface but the effect of the layer between the main shock and the contact surface has not been taken into account here. The expression eqn(10.27) is only strictly valid at the inner shock itself, but this approximation only affects the value slightly.

The pressure of the cometary ionisation will be affected by the heating effects of the ion-molecule reactions mentioned in Section 10.3.2. and by radiative cooling. The effect is difficult to quantify but has been included in a notional way in figure 9.

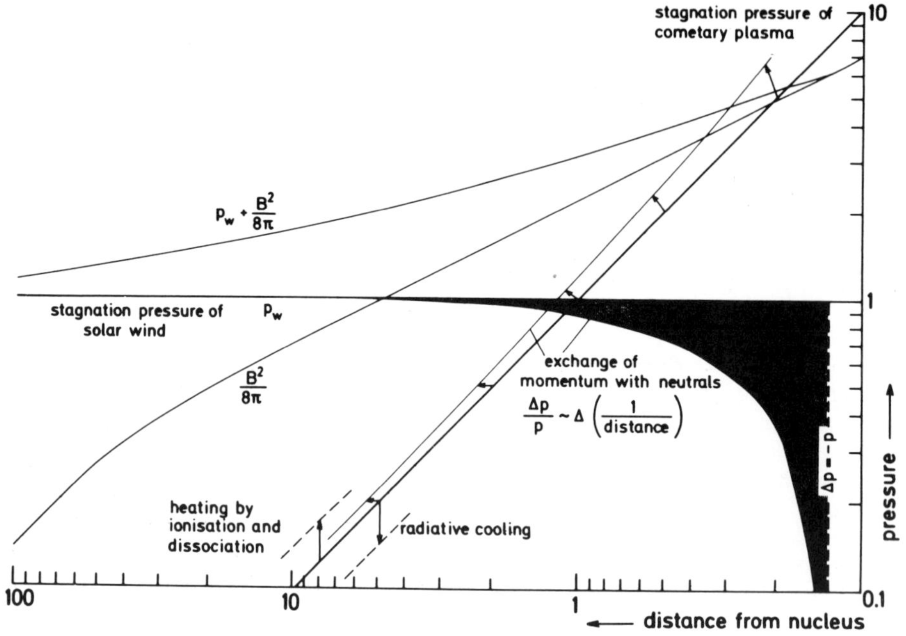

Figure 10.9 A comparable assessment of the various pressure terms affecting the pressure balance at the contact discontinuity (Schmidt & Wegmann, 1982)

Finally there is an exchange of momentum between the outflowing neutrals and ions. This affects the plasma on both sides of the surface. An evaluation of the magnitude is given in Schmidt & Wegmann (1982). Their results on the overall pressure balance are summarised in figure 9. Taking just the major components, the internal and external stagnation pressures the stand-off distance of the contact surface from the nucleus ℓ_c is given by

$$\ell_c = M_c Q \nu / 4\pi \rho_s V_s^2 . \tag{10.28}$$

Using the values $Q = 5 \times 10^{29}\ s^{-1}$, $\nu = 3 \times 10^{-7}\ s^{-1}$, $M_c = 28$ (CO^+), $\rho_s = 5\ cm^{-3}$, $V_s = 450$ Km/s gives $\ell_c = 300$ kms.

In figure 10 (Schmidt & Wegmann 1982) the stand-off distance has been calculated by making the arbitrary assumption that heating in the inner coma increases the internal pressure by an order of magnitude.

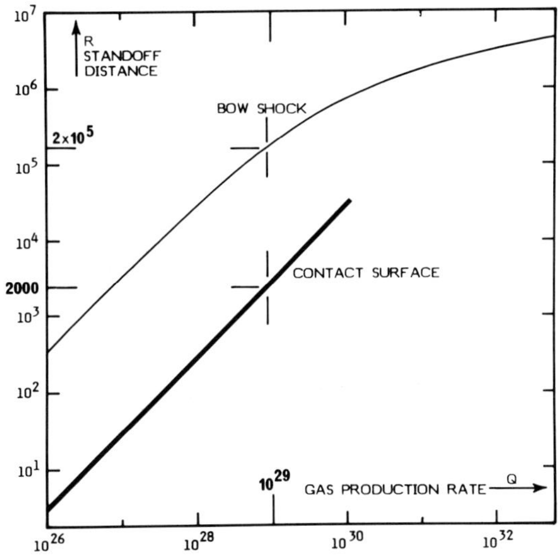

Figure 10.10 Calculated stand-off distance for contact surface and bow shock along the stagnation streamline (Schmidt & Wegmann, 1982). The contact surface is based on eqn 10.28 assuming that internal heating increases the pressure by an order of magnitude. The bow shock position is based on eqn 10.32.

10.6.3 Bow Shock

The bow shock upstream from Earth and the other planets occurs because the solar wind is deflected around an obstacle. At Earth that obstacle is the magnetopause formed by the geomagnetic dipole field; at Venus it is the ionopause formed by currents in the ionosphere. In such cases the stand-off distance of the bow shock is proportional to the size of the obstacle and is approximately 1.4 times its radius. It is expected that a bow shock will form upstream from a comet but for different reasons. It is not the deflection of the flow around the contact surface but the result of the mass-loading of the solar wind. If the deceleration due to mass-loading is too fast a shock is generated in order to allow transverse expansion of the gas. Wallis (1973) has shown that the critical point in the flow occurs when the mass flux $\rho'v'$ is given by

$$(\rho' V' / \rho_s V_s) = \gamma^2 / (\gamma^2 - 1) \ , \qquad (10.29)$$

where $\rho_s V_s$ is the upstream, undisturbed mass flux in the solar wind.

Eqn.(10.29) may also be expressed in the form that

$$\text{flux added} = \rho_s V_s/(\gamma^2-1)\ . \qquad (10.30)$$

The appropriate value of γ is not known but should be between 5/3 and 2, so an added flux of between 0.56 and 0.33 of the solar wind flux takes the flow to the critical point. If the ions are CO^+ with a mass of 28 their number density needs only to be between 2% and 1.2% of the solar wind density.

The mass added is the integral of the rate of ionisation from the critical point upstream to infinity, i.e (Schmidt & Wegmann, 1982).

$$\int_{\ell_s}^{\infty} QM_c \nu \exp(-\nu\ell/V_e)/4\pi\ell^2 V_e\ .\ d\ell$$

$$= (QM_c\nu^2/4\pi V_e^{\,2}) \int_{\ell_s\nu/V_e}^{\infty} \exp(-y)/y^2\ .\ dy\ . \qquad (10.31)$$

If $\ell_s \ll V_e/\nu$ the integral is approximately $V_e/\ell_s\nu$ giving the stand-off distance to the bow shock ℓ_s

$$\ell_s = QM_c\nu(\gamma^2-1)/4\pi V_e \rho_s V_s\ . \qquad (10.32)$$

This value is plotted in figure 10 and compared with the results of a number of mhd numerical models. Using the values given in eqn(10.33) and $\gamma = 2$, $V_e = 685$ m/s gives

$$\ell_s = 5.9 \times 10^5 \text{ kms}\ , \qquad (10.33)$$

which is much greater than the stand-off distance of a deflection bow shock. The deflection by the contact surface will therefore have no effect on the shape of the bow shock.

10.7 MAGNETOHYDRODYNAMIC FLOW AT A COMET

10.7.1 Numerical solution of the mhd equations

The flow of the solar wind past a comet can be described in the mhd limit by a set of equations which express the conservation of mass, momentum, energy flux and magnetic flux and include terms describing a source of particles.

They are (Biermann et al., 1967):

$$\frac{\partial \rho}{\partial t} + \mathrm{div}\rho\underline{v} = AM_c \ ,$$

$$\frac{\partial}{\partial t} \rho\underline{v}+(\underline{v}.\mathrm{grad})\rho\underline{v}+\rho\underline{v}\ \mathrm{div}\underline{v}+\mathrm{grad}P - (\mathrm{curl}\underline{B})\mathrm{x}\underline{B}/4\pi = AM_c\underline{V}_e \ ,$$

$$(10.34)$$

$$\frac{\partial}{\partial t}(\rho v^2 + \frac{P}{\gamma-1} + \frac{B^2}{8\pi})+\mathrm{div}[\underline{v}(\frac{\rho v^2}{2} + \frac{\gamma P}{\gamma-1} + \frac{B^2}{4\pi}) - (\underline{v}.\underline{B})\underline{B}/4\pi] = AM_c V_e^2 /2 \ ,$$

$$\frac{\partial}{\partial t}\underline{B} - \mathrm{curl}(\underline{v} \times \underline{B}) = 0 \ .$$

The source terms describe the addition of cometary ions at the rate given by eqn(10.13),with momentum and kinetic energy appropriate to an ion flowing outwards with a velocity V_e. It is assumed that the new ions immediately take up the same velocity as the solar wind ions, i.e. the momentum and energy are given by eqn 10.23 The values are an average over the various species involved. The accuracy of the numerical methods employed to solve the equations does not warrant a more precise formulation of a summation over the species.

The numerical integration of these equations is difficult and involves making a number of assumptions which will not be discussed here. The reader is referred to papers by Biermann et al (1967), Schmidt & Wegmann (1980) for the details.

The results are illustrated in figure 11. They show a flow around the comet which is retarded most near the comet as expected and which drapes the magnetic field lines round the head of the comet as envisaged by Alfven (1957), figure 3. Figure 12 shows the profile of the plasma parameters which the Giotto spacecraft might encounter along its trajectory if the model is correct.

10.7.2 Validity of the mhd approach

The mhd formulation given in the previous section assumes that the plasma can be treated as a single fluid with bulk parameters ρ, v and P. While this is reasonable for the solar wind alone, the addition of the cometary ions with a totally different velocity distribution (Section 10.5.1.) makes this assumption questionable. If the instabilities described in Section 10.5.2. operate quickly enough the ring distribution of the cometary ions become thermalised and share their energy with the solar wind. In that case the mhd approach is accurate. However even if the instabilities act on time scales of the order of a gyroperiod, as when eqn(10.21) applied, it may not be fast enough. The gyroperiod of a CO^+ ion is 183 secs and its gyroradius is 50,000 kms comparable with the temporal and spatial scales of the flow pattern. If the effects analysed by Wu and Davidson (1972) are stabilised by the finite velocity spread as argued by Galeev (1982) then the mhd approach can only be considered as a rough first approximation. The actual trajectories of the ions could then have a gross influence on the results not allowed for by the mhd solutions.

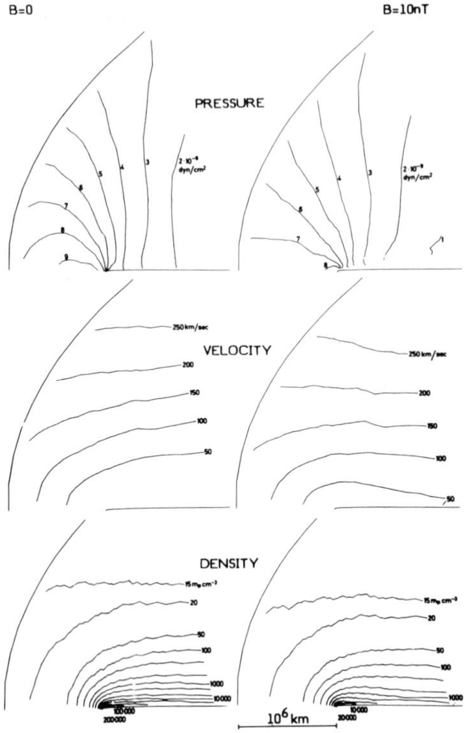

Figure 10.11 The flow profiles obtained from a numerical solution of the mhd equations, with and without all interplanetary magnetic field included (Schmidt & Wegmann, 1982).

The appropriate approximation will only become apparent when in-situ observations of ion velocity distributions in the cometary environment are available.

10.8 SPECIAL FEATURES OF THE MORPHOLOGY

10.8.1 Rays, tail streamers

Some of the early observations and theories of rays, envelopes or tail streamers were discussed in Section 10.2. It is now obvious, following the suggestion of Alfven (1957) that the rays are aligned with the magnetic field and are essentially magnetic structures. They usually occur in symmetric pairs, on opposite sides of the head. The rays are nearly straight, not paraboloid as required by the fountain model, and rotate towards the axis of the tail (Wurm, 1968). This can be explained by the end of the field lines being taken downstream by the solar wind while the nose of the envelope remains held in the coma.

The formation of the rays can be explained, qualitatively at least, in the following way. The ray, or envelope, starts to form near the nose of the contact surface where the flow is at its slowest. The plasma trapped in the flux tube spreads out sideways because the field lines diverge and so the magnetic mirror force (eqn 10.22) accelerates the particles sideways and tailwards. The whole flux tube is gradually

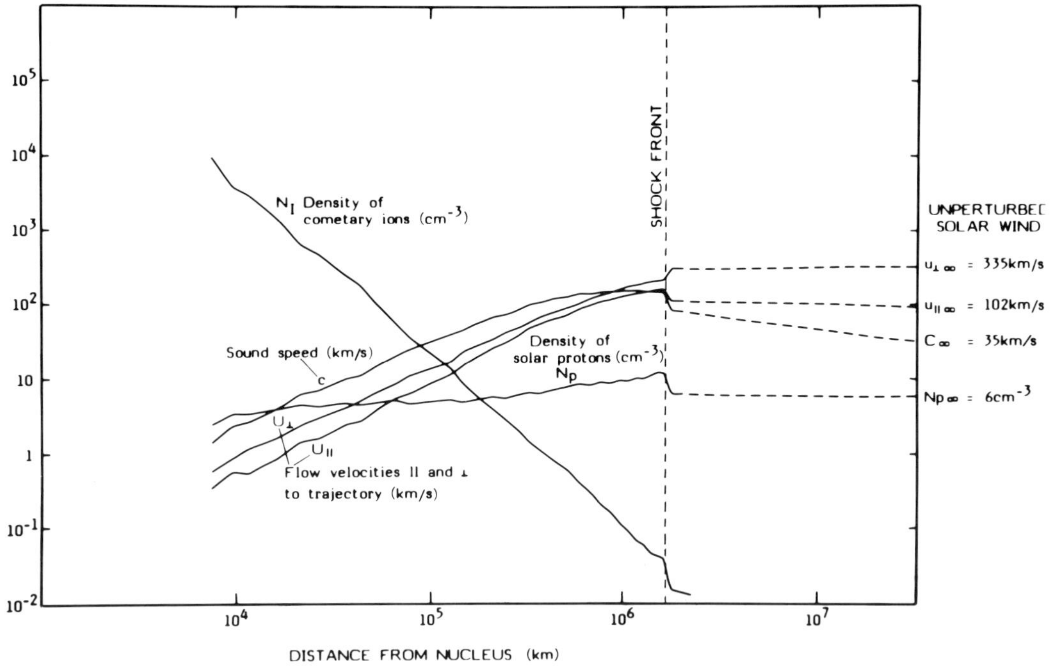

Figure 10.12 The variation of density, flow velocity and sound speed that might be measured along the trajectory of Giotto, as it approaches at an angle of 107° to the Sun-comet line, according to the mhd model of Schmidt & Wegmann (Johnstone et al., 1982).

pulled downstream, around the contact surfaces so that it appears to be contracting inwards. One question that remains is - why are they visible as discrete structures? Why does the process described above not occur continuously? This cannot be answered with any certainty but it has been suggested (Schmidt & Wegmann, 1982) that conditions near the nose of the rays are appropriate for the occurrence of the interchange instability. The result of this would be to break up the sheets enveloping the contact surface into separate filaments. Other possibilities are that the formation of the structures could be associated with fluctuations in the gas production rate, the solar wind flux (Jockers 1982), the operation of a non-steady ionisation process such as those mentioned in Section 10.4.1., or in the strength or direction of the interplanetary magnetic field (Ness & Donn, 1965). In fact the formation of filaments is a very common phenomenon in space plasmas and is seen most obviously in auroral structures. An analogy has also been drawn between cometary tail rays and magnetic flux ropes near the Venusian ionopause (Russell et al 1982).

10.8.2 Disconnection events

Figure 13 shows a disconnection event in the tail of Comet Morehouse. Part of the tail becomes detached from the head of the comet and then drifts away downstream. A new tail is then formed to replace it. Two explanations have been put forward for this phenomenon. The first is that oppositely directed magnetic field lines in tail rays which are adjacent to one another reconnect with one another. The downstream portion is then detached from the head and moves away from the comet. Similar events have been detected at large distances down the geomagnetic tail by ISEE-3 (Hones et al., 1984). The second explanation (Neidner & Brandt, 1978) is that it is associated with the passage of a sector boundary in the interplanetary magnetic field. At the boundary the interplanetary magnetic field reverses. As this reversal reaches the stagnation region, reconnection of the field lines occurs and the two magnetic field loops are pulled tailwards on either side of the head.

10.8.3 Dusty plasmas

Individual dust particles in the cometary ionosphere can become charged to a few volts as a result of bombardment by plasma particle. The equilibrium voltage is set where the net current to the dust particle is zero. The important current components are likely to include bombardment by the different populations of ions, by electrons, and the emission of photoelectrons. The charged dust particle will be influenced by the magnetic and electric fields, and may through its charge affect the behaviour of the plasma. Little attention has so far been paid to this phenomenon (Wallis & Hassan, 1983).

10.9 CONCLUSION

In preparing this chapter it is impossible not to be aware that it is likely to be the most short-lived in this book. Within 18 months of the time of writing, five spacecraft carrying plasma instrumentation will have passed through the Coma and tail of a comet while they make in-situ measurements of the conditions. Every previous occasion on which exploratory in-situ measurements have been made in solar system plasmas has revealed the inadequacy of our attempts to predict the conditions. The measurements about to be made will almost certainly do the same. In this chapter we have attempted to outline the important types of process, and therefore provide a crude framework within which the observations can be considered, but we may expect the picture to be changed substantially by the end of 1986.

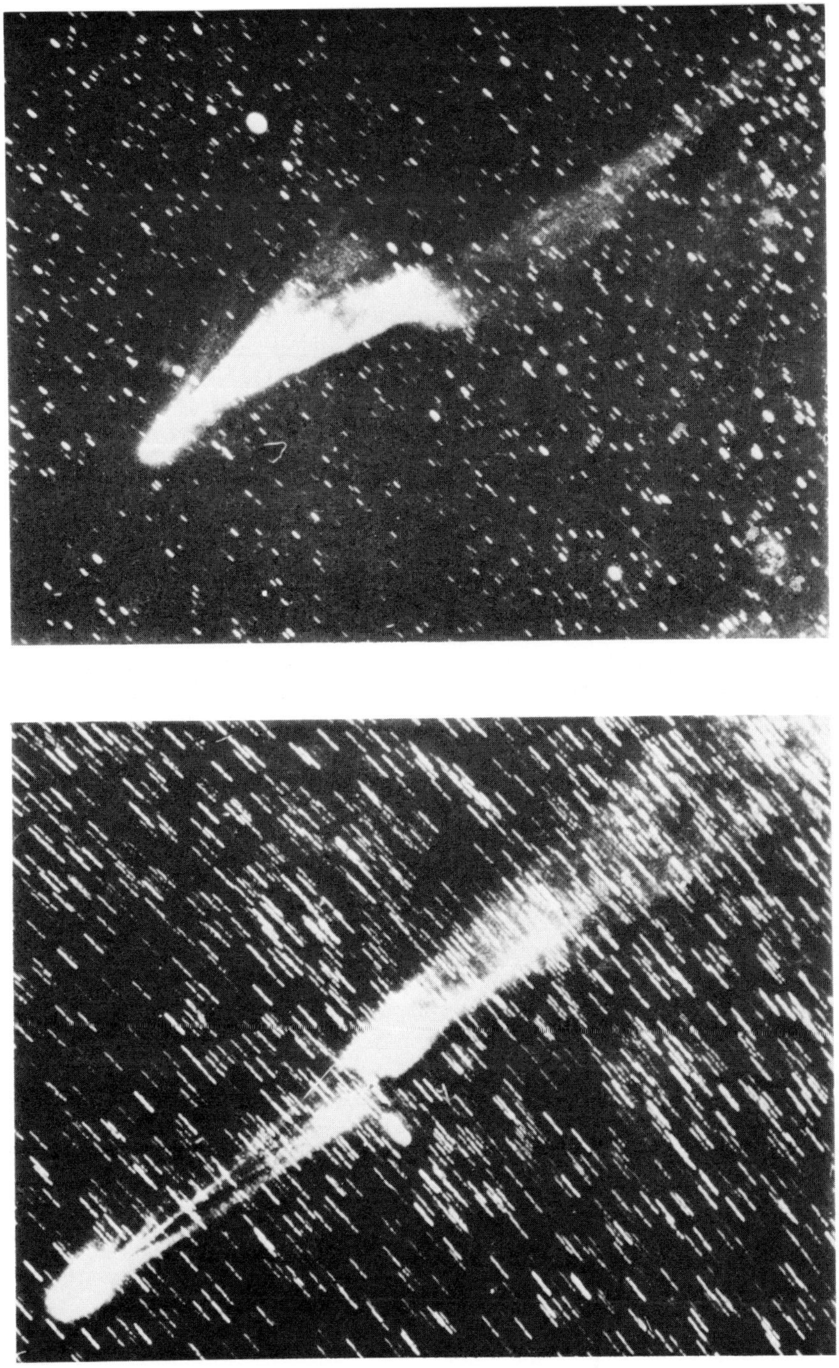

Figure 10.13 A disconnection event seen in the tail of Comet Morehouse in 1908. The photograph at the top was taken the day before the bottom one.

REFERENCES

Alfven H (1957) Tellus 9, 92.
Alfven H (1960) Rev. Mod. Phys. 32, 710.
Axford I (1964) Planet. Sp. Sci. 12, 719.
Biermann L (1951) Zs fur Astrophysik 29, 274.
Biermann L, Brosowski B, & Schmidt H U (1967) Solar Phys. 1, 254.
Brandt J C, Harrington R S, Roosen R G (1973) Astrophys. J. 184, 27.
Brandt J C & Chapman R (1981) Introduction to Comets, Cambridge University Press
Brandt J C & Mendis D A (1979) Solar System Plasma Physics Volume II, Eds. C F Kennel, L J Lanzerotti & E N Parker, North Holland Publishing Co.
Delsemme A H (1982) Comets (ed. L. Wilkening), p.85. Univ. Arizona Press, Tucson.
Eddington A S (1910) Mon. Not. R. Astron. Soc. 70, 442.
Fernandez J A & Jockers K (1983) Rep. Prog. Phys.
Fokker A D (1953) Mem. Soc. R. Sci. Liege XIII, 241.
Galeev A (1982) Cometary Exploration Proc. Int. Conf. Cometary Exploration Ed. T. Gombosi, Budapest, p.243.
Hartle R E & Wu C S (1973) J. Geophys. Res. 78, 5802.
Hoffmeister C (1943) Zs. fur Astrophysik 22, 265.
Hones E W, Birn J, Baker D N, Bame S J, Feldman W C, McComas D J, Zwickle R D, Slavin V A, Smith E J, Tsurtani B T (1984) Geophys. Res. Lett., 11, 1046.
Huebner W F, Giguere P T & Slattery W L (1982) Comets (op cit) p.496.
Ip W-H & Axford W I (1982) Comets (op cit) p.588.
Ip W-H & Mendis D A (1976) Icarus 29, 147.
Jockers K (1982) ESO Workshop on Comet Halley, p.193
Johnstone A D et al., (1982) Cometary Exploration (op cit) p.301
Keller H U (1976) Sp. Sci. Rev. 18, 641.
Mendis D A & Houpis H L F (1982) Rev. Geophys. Sp. Phys. 20, 885.
Mendis D A & Ip W-H (1977) Sp. Sci. Rev. 20 145
Neidner M B & Brandt J C (1978) Astrophys. J. 223, 655.
Ness N F & Donn B D (1965) Mem. in 8°, Soc. R. Sci. Liege XII, 343.
Parker E N (1958) Astrophys. J. 128 664.
Russell C T, Luhmann J G, Elphic R C, & Neugebauer M (1982) Comets (op cit) p.561.
Schmidt H U & Wegmann R (1980) Comp. Phys. Comm. 19 309.
Schmidt H U & Wegmann R (1982) Comets (op cit) p.538.
Sicardy B, Guerin J, Lecacheux J, Baudrand J, Combes M, Picat J P, Lelievre G, Lemannier J P (1983) Astron. Astrophys., 121, L4.
Unsold A & Chapman S (1949) The Observatory 69, 219.
Wallis M K (1973) Planet Sp. Sci. 21, 1647.
Wallis M K & Hassan M H A (1983) Astron. Astrophys. 121, 10.
Whipple F (1950) Astrophys. J. 111, 375.
Wu C S & Davidson (1972) J. Geophys. Res. 77, 5399.
Wurm K (1961) Astron. J. 66, 362.
Wurm K (1968) Icarus 8, 287.
Wyckhoff S (1982) Comets (op cit) p.3.

General References

Mendis D A & Houpis H L F (1982) Rev. Geophys. Sp. Phys., 20, 885.

Brandt J C & Mendis D A (1979) Solar System Plasma Physics, Vol. 2, Ed. by Kennel, Lanzerotti & Parker, North Holland Publishing Co.

Mendis D A & Ip W J (1977) Space Sci. Rev. 20, 145.

INDEX

GEOPHYSICS AND ASTROPHYSICS MONOGRAPHS

Editor:

BILLY M. McCORMAC (Lockheed Palo Alto Research Laboratory)

Editorial Board:

R. GRANT ATHAY (High Altitude Observatory, Boulder)
W. S. BROECKER (Lamont-Doherty Geological Observatory, New York)
P. J. COLEMAN, Jr. (University of California, Los Angeles)
G. T. CSANADY (Woods Hole Oceanographic Institution, Mass.)
D. M. HUNTEN (University of Arizona, Tucson)
C. DE JAGER (The Astronomical Institute at Utrecht, Utrecht)
J. KLECZEK (Czechoslovak Academy of Sciences, Ondřejov)
R. LÜST (President Max-Planck-Gesellschaft zur Förderung der Wissenschaften, München)
R. E. MUNN (University of Toronto, Toronto)
Z. ŠVESTKA (The Astronomical Institute at Utrecht, Utrecht)
G. WEILL (Service d'Aéronomie, Verrieres-le-Buisson)

1. R. Grant Athay, *Radiation Transport in Spectral Lines.*
2. J. Coulomb, *Sea Floor Spreading and Continental Drift.*
3. G. T. Csanady, *Turbulent Diffusion in the Environment.*
4. F. E. Roach and Janet L. Gordon, *The Light of the Night Sky.*
5. H. Alfvén and G. Arrhenius, *Structure and Evolutionary History of the Solar System.*
6. J. Iribarne and W. Godson, *Atmospheric Thermodynamics.*
7. Z. Kopal, *The Moon in the Post-Apollo Era.*
8. Z. Švestka, *Solar Flares.*
9. A. Vallance Jones, *Aurora.*
10. C.-J. Allègre and G. Michard, *Introduction to Geochemistry.*
11. J. Kleczek, *The Universe.*
12. E. Tandberg-Hanssen, *Solar Prominences.*
13. A. Giraud and M. Petit, *Ionospheric Techniques and Phenomena.*
14. E. L. Chupp, *Gamma Ray Astronomy.*
15. W. D. Heintz, *Double Stars.*
16. A. Krüger, *Introduction to Solar Radio Astronomy and Radio Physics.*
17. V. Kourganoff, *Introduction to Advanced Astrophysics.*
18. J. Audouze and S. Vauclaire, *An Introduction to Nuclear Astrophysics.*
19. C. de Jager, *The Brightest Stars.*
20. J. L. Sérsic, *Extragalactic Astronomy.*
21. E. R. Priest, *Solar Magnetohydrodynamics.*

V. A. Bronshten, *Physics of Meteoric Phenomena.*
L. R. Lyons and D. J. Williams, *Quantitative Aspects of Magnetospheric Physics.*